Genetic Algorithms in Electromagnetics

THE WILEY BICENTENNIAL–KNOWLEDGE FOR GENERATIONS

Each generation has its unique needs and aspirations. When Charles Wiley first opened his small printing shop in lower Manhattan in 1807, it was a generation of boundless potential searching for an identity. And we were there, helping to define a new American literary tradition. Over half a century later, in the midst of the Second Industrial Revolution, it was a generation focused on building the future. Once again, we were there, supplying the critical scientific, technical, and engineering knowledge that helped frame the world. Throughout the 20th Century, and into the new millennium, nations began to reach out beyond their own borders and a new international community was born. Wiley was there, expanding its operations around the world to enable a global exchange of ideas, opinions, and know-how.

For 200 years, Wiley has been an integral part of each generation's journey, enabling the flow of information and understanding necessary to meet their needs and fulfill their aspirations. Today, bold new technologies are changing the way we live and learn. Wiley will be there, providing you the must-have knowledge you need to imagine new worlds, new possibilities, and new opportunities.

Generations come and go, but you can always count on Wiley to provide you the knowledge you need, when and where you need it!

WILLIAM J. PESCE
PRESIDENT AND CHIEF EXECUTIVE OFFICER

PETER BOOTH WILEY
CHAIRMAN OF THE BOARD

Genetic Algorithms in Electromagnetics

Randy L. Haupt
Applied Research Laboratory
Pennsylvania State University

Douglas H. Werner
Department of Electrical Engineering
Applied Research Laboratory
Materials Research Institute
Pennsylvania State University

IEEE PRESS

WILEY-INTERSCIENCE
A John Wiley & Sons, Inc., Publication

Published by John Wiley & Sons, Inc., Hoboken, New Jersey.
Published simultaneously in Canada.

Wiley Bicentennial Logo: Richard J. Pacifico

Library of Congress Cataloging-in-Publication Data:

Haupt, Randy L.
Genetic algorithms in electromagnetics / by Randy L. Haupt and Douglas H. Werner.
p. cm.
A Wiley-Interscience publication.
Includes bibliographical references and index.
ISBN: 978-0-471-48889-7
1. Antenna arrays—Design. 2. Electromagnetism—Mathematical models. 3. Genetic algorithms. I. Werner, Douglas H., 1960– II. Title.
TK7871.6.H37 2007
621.30285'631—dc22

2006046738

Printed in the United States of America.

10 9 8 7 6 5 4 3 2 1

We dedicate this book to our wives,
Sue Ellen Haupt and Pingjuan L. Werner.

Contents

Preface

Most books on electromagnetics describe how to solve particular problems using classical analysis techniques and/or numerical methods. These books help in formulating the objective function that is used in this book. The objective function is the computer algorithm, analytical model, or experimental result that describes the performance of an electromagnetic system. This book focuses primarily on the optimization of these objective functions. Genetic algorithms (GAs) have proved to be tenacious in finding optimal results where traditional techniques fail. This book is an introduction to the use of GAs to optimizing electromagnetic systems.

This book begins with an introduction to optimization and some of the commonly used numerical optimization routines. Chapter 1 provides the motivation for the need for a more powerful "global" optimization algorithm in contrast to the many "local" optimizers that are prevalent. The next chapter introduces the GA to the reader in both binary and continuous variable forms. MATLAB® commands are given as examples. Chapter 3 provides two step-by-step examples of optimizing antenna arrays. This chapter serves as an excellent introduction to the following chapter, on optimizing antenna arrays. GAs have been applied to the optimization of antenna arrays more than has any other electromagnetics topic. Chapter 5 somewhat follows Chapter 4, because it reports the use of a GA as an adaptive algorithm. Adaptive and smart arrays are the primary focal points, but adaptive reflectors and crossed dipoles are also presented. Chapter 6 explains the optimization of several different wire antennas, starting with the famous "crooked monopole." Chapter 7 is a review of the results for horn, reflector, and microstip patch antennas. Optimization of these antennas entails computing power significantly greater than that required for wire antennas. Chapter 8 diverges from antennas to present results on GA optimization of scattering. Results include scattering from frequency-selective surfaces and electromagnetic bandgap materials. Finally, chapter 9 presents ideas on operator and parameter selection for a GA. In addition, particle swarm optimization and multiple objective optimization are explained in detail. The Appendix contains some MATLAB® code for those who want to try it, followed by a chronological list of publications grouped by topic.

State College, Pennsylvania

RANDY L. HAUPT
DOUGLAS H. WERNER

Acknowledgments

A special thanks goes to Lana Yezhova for assisting in the preparation and editing of the manuscript for this book, and to Rodney A. Martin for the cover art. The authors would also like to express their sincere appreciation to the following individuals, each of whom provided a valuable contribution to the material presented in this book:

Zikri Bayraktar
Daniel W. Boeringer
Jeremy A. Bossard
Matthew G. Bray
Jennifer Carvajal
James W. Culver
Steven D. Eason
K. Prakash Hari
Douglas J. Kern
Ling Li
Derek S. Linden
Kenneth A. O'Connor
Joshua S. Petko
Shawn Rogers
Thomas G. Spence
Pingjuan L. Werner
Michael J. Wilhelm

1

Introduction to Optimization in Electromagnetics

As in other areas of engineering and science, research efforts in electromagnetics have concentrated on finding a solution to an integral or differential equation with boundary conditions. An example is calculating the radar cross section (RCS) of an airplane. First, the problem is formulated for the size, shape, and material properties associated with the airplane. Next, an appropriate mathematical description that exactly or approximately models the airplane and electromagnetic waves is applied. Finally, numerical methods are used for the solution. One problem has one solution. Finding such a solution has proved quite difficult, even with powerful computers.

Designing the aircraft with the lowest RCS over a given frequency range is an example of an optimization problem. Rather than finding a single solution, optimization implies finding many solutions then selecting the best one. Optimization is an inherently slow, difficult procedure, but it is extremely useful when well done. The difficult problem of optimizing an electromagnetics design has only recently received extensive attention.

This book concentrates on the genetic algorithm (GA) approach to optimization that has proved very successful in applications in electromagnetics. We do not think that the GA is the best optimization algorithm for all problems. It has proved quite successful, though, when many other algorithms have failed. In order to appreciate the power of the GA, a background on the most common numerical optimization algorithms is given in this chapter to familiarize the reader with several optimization algorithms that can be applied to

Genetic Algorithms in Electromagnetics, by Randy L. Haupt and Douglas H. Werner

electromagnetics problems. The antenna array has historically been one of the most popular optimization targets in electromagnetics, so we continue that tradition as well.

The first optimum antenna array distribution is the binomial distribution proposed by Stone [1]. As is now well known, the amplitude weights of the elements in the array correspond to the binomial coefficients, and the resulting array factor has no sidelobes. In a later paper, Dolph mapped the Chebyshev polynomial onto the array factor polynomial to get all the sidelobes at an equal level [2]. The resulting array factor polynomial coefficients represent the Dolph–Chebyshev amplitude distribution. This amplitude taper is optimum in that specifying the maximum sidelobe level results in the smallest beamwidth, or specifying the beamwidth results in the lowest possible maximum sidelobe level. Taylor developed a method to optimize the sidelobe levels and beamwidth of a line source [3]. Elliot extended Taylor's work to new horizons, including Taylor-based tapers with asymmetric sidelobe levels, arbitrary sidelobe level designs, and null-free patterns [4]. It should be noted that Elliot's methods result in complex array weights, requiring both an amplitude and phase variation across the array aperture. Since the Taylor taper optimized continuous line sources, Villeneuve extended the technique to discrete arrays [5]. Bayliss used a method similar to Taylor's amplitude taper but applied to a monopulse difference pattern [6]. The first optimized phase taper was developed for the endfire array. Hansen and Woodyard showed that the array directivity is increased through a simple formula for phase shifts [7].

Iterative numerical methods became popular for finding optimum array tapers beginning in the 1970s. Analytical methods for linear array synthesis were well developed. Numerical methods were used to iteratively shape the mainbeam while constraining sidelobe levels for planar arrays [8–10]. The Fletcher–Powell method [11] was applied to optimizing the footprint pattern of a satellite planar array antenna. An iterative method has been proposed to optimize the directivity of an array via phase tapering [12] and a steepest-descent algorithm used to optimize array sidelobe levels [13]. Considerable interest in the design of nonuniformly spaced arrays began in the late 1950s and early 1960s. Numerical optimization attracted attention because analytical synthesis methods could not be found. A spotty sampling of some of the techniques employed include linear programming [14], dynamic programming [15], and steepest descent [16]. Many statistical methods have been used as well [17].

1.1 OPTIMIZING A FUNCTION OF ONE VARIABLE

Most practical optimization problems have many variables. It's usually best to learn to walk before learning to run, so this section starts with optimizing one variable; then the next section covers multiple variable optimization. After describing a couple of single-variable functions to be optimized, several single variable optimization routines are introduced. Many of the multidimensional optimization routines rely on some version of the one-dimensional optimization algorithms described here.

Optimization implies finding either the minimum or maximum of an objective function, the mathematical function that is to be optimized. A variable is passed to the objective function and a value returned. The goal of optimization is to find the combination of variables that causes the objective function to return the highest or lowest possible value.

Consider the example of minimizing the output of a four-element array when the signal is incident at an angle ϕ. The array has equally spaced elements ($d = \lambda/2$) along the x axis (Fig. 1.1). If the end elements have the same variable amplitude (a), then the objective function is written as

$$\mathrm{AF}_1(a) = 0.25\left|a + e^{j\Psi} + e^{j2\Psi} + ae^{j3\Psi}\right| \tag{1.1}$$

where $\Psi = kdu$
$k = 2\pi/\lambda$
λ = wavelength
$u = \cos\phi$

A graph of AF_1 for all values of u when $a = 1$ is shown in Figure 1.2. If $u = 0.8$ is the point to be minimized, then the plot of the objective function as a function of a is shown in Figure 1.3. There is only one minimum at $a = 0.382$.

Another objective function is a similar four-element array with uniform amplitude but conjugate phases at the end elements

$$\mathrm{AF}_2(\delta) = 0.25\left|e^{j\delta} + e^{j\Psi} + e^{j2\Psi} + e^{-j\delta}e^{j3\Psi}\right| \tag{1.2}$$

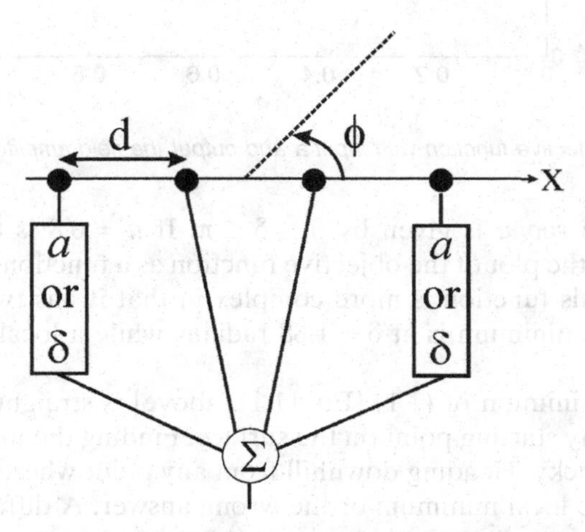

Figure 1.1. *Four-element array with two weights.*

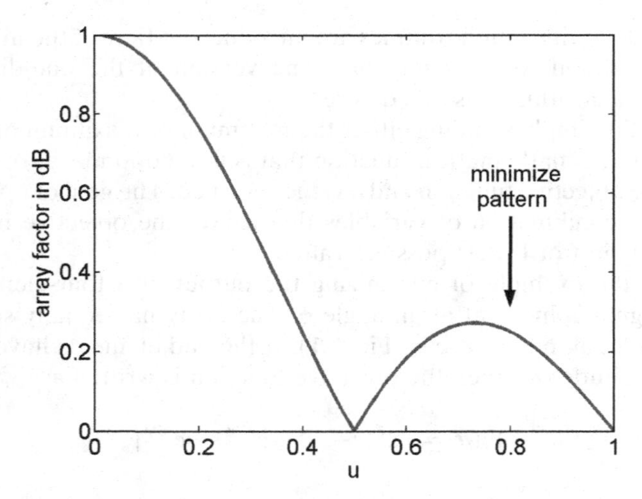

Figure 1.2. *Array factor of a four-element array.*

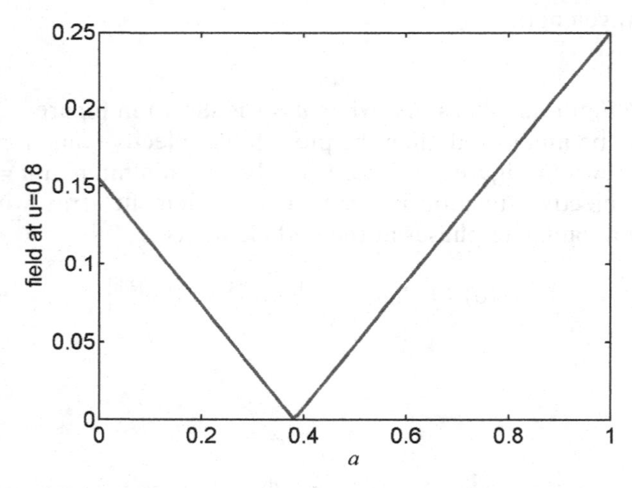

Figure 1.3. *Objective function with input* a *and output the field amplitude at* u = *0.8.*

where the phase range is given by $0 \le \delta \le \pi$. If $u = 0.8$ is the point to be minimized, then the plot of the objective function as a function of δ is as shown in Figure 1.4. This function is more complex in that it has two minima. The global or lowest minimum is at $\delta = 1.88$ radians while a local minimum is at $\delta = 0$.

Finding the minimum of (1.1) [Eq. (1.1), above] is straightforward—head downhill from any starting point on the surface. Finding the minimum of (1.2) is a little more tricky. Heading downhill from any point where $\delta < 0.63$ radian (rad) leads to the local minimum or the wrong answer. A different strategy is needed for the successful minimization of (1.2).

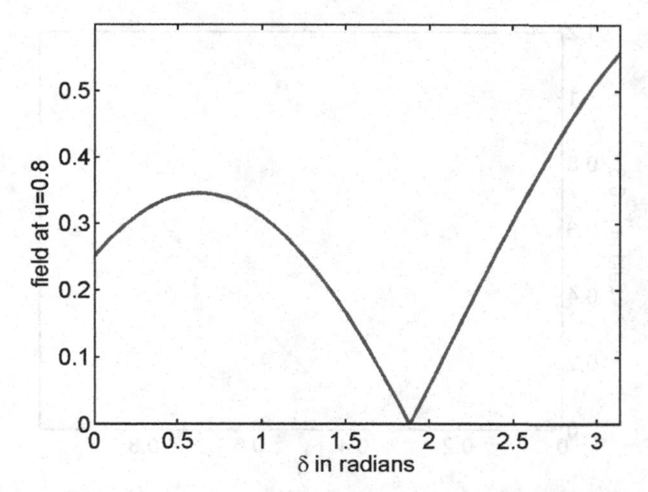

***Figure 1.4.** Objective function with input δ and output the field amplitude at u = 0.8.*

1.1.1 Exhaustive Search

One way to feel comfortable about finding a minimum is to check all possible combinations of input variables. This approach is possible for a small finite number of points. Probably the best example of an exhaustive search is graphing a function and finding the minimum on the graph. When the graph is smooth enough and contains all the important features of the function in sufficient detail, then the exhaustive search is done. Figures 1.3 and 1.4 are good examples of exhaustive search.

1.1.2 Random Search

Checking every possible point for a minimum is time-consuming. Randomly picking points over the interval of interest may find the minimum or at least come reasonably close. Figure 1.5 is a plot of AF_1 with 10 randomly selected points. Two of the points ended up close to the minimum. Figure 1.6 is a plot of AF_2 with 10 randomly selected points. In this case, six of the points have lower values than the local minimum at $\delta = 0$. The random search process can be refined by narrowing the region of guessing around the best few function evaluations found so far and guessing again in the new region. The odds of all 10 points appearing at $\delta < 0.63$ for AF_2 is $(0.63/\pi)^{10} = 1.02 \times 10^{-7}$, so it is unlikely that the random search would get stuck in this local minimum with 10 guesses. A quick random search could also prove worthwhile before starting a downhill search algorithm.

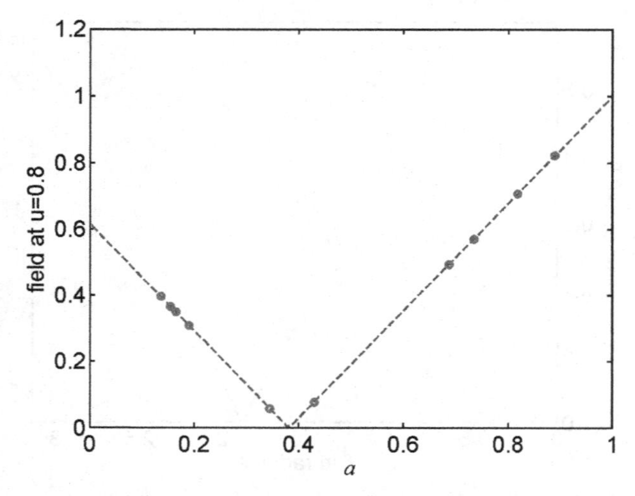

Figure 1.5. *Ten random guesses (circles) superimposed on a plot of AF_1.*

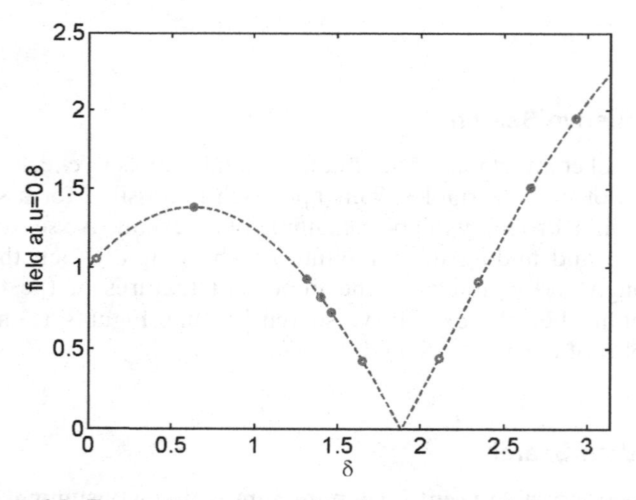

Figure 1.6. *Ten random guesses (circles) superimposed on a plot of AF_2.*

1.1.3 Golden Search

Assume that a minimum lies between two points a and b. Three points are needed to detect a minimum in the interval: two to bound the interval and one in between that is lower than the bounds. The goal is to shrink the interval by picking a point (c) in between the two endpoints (a and b) at a distance Δ_1 from a (see Fig. 1.7). Now, the interval is divided into one large interval and one small interval. Next, another point (d) is selected in the larger of the two subintervals. This new point is placed at a distance Δ_2 from c. If the new point

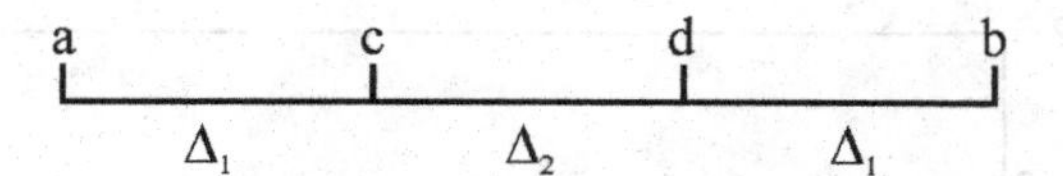

Figure 1.7. *Golden search interval.*

on the reduced interval ($\Delta_1 + \Delta_2$) is always placed at the same proportional distance from the left endpoint, then

$$\frac{\Delta_1}{\Delta_1 + \Delta_2 + \Delta_1} = \frac{\Delta_2}{\Delta_1 + \Delta_2} \tag{1.3}$$

If the interval is normalized, the length of the interval is

$$\Delta_1 + \Delta_2 + \Delta_1 = 1 \tag{1.4}$$

Combining (1.3) and (1.4) yields the equation

$$\Delta_1^2 - 3\Delta_1 + 1 = 0 \tag{1.5}$$

which has the root

$$\Delta_1 = \frac{\sqrt{5} - 1}{2} = 0.38197\ldots \tag{1.6}$$

This value is known as the "golden mean" [18].

The procedure above described is easy to put into an algorithm to find the minimum of AF_2. As stated, the algorithm begins with four points (labeled 1–4 in Fig. 1.8). Each iteration adds another point. After six iterations, point 8 is reached, which is getting very close to the minimum. In this case the golden search did not get stuck in the local minimum. If the algorithm started with points 1 and 4 as the bound, then the algorithm would have converged on the local minimum rather than the global minimum.

1.1.4 Newton's Method

Newton's method is a downhill sliding technique that is derived from the Taylor's series expansion for the derivative of a function of one variable. The derivative of a function evaluated at a point x_{n+1} can be written in terms of the function derivatives at a point x_n

$$f'(x_{n+1}) = f'(x_n) + f''(x_n)(x_{n+1} - x_n) + \frac{f'''(x_n)}{2!}(x_{n+1} - x_n)^2 + \cdots \tag{1.7}$$

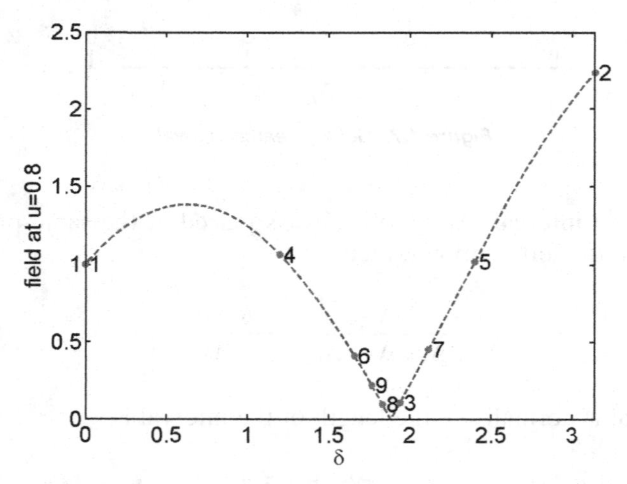

Figure 1.8. *The first eight function evaluations (circles) of the golden search algorithm when minimizing AF_2.*

Keeping only the first and second derivatives and assuming that the next step reaches the minimum or maximum, then (1.7) equals zero, so

$$f'(x_n) + f''(x_n)(x_{n+1} - x_n) = 0 \tag{1.8}$$

Solving for x_{n+1} yields

$$x_{n+1} = x_n - \frac{f'(x_n)}{f''(x_n)} \tag{1.9}$$

If no analytical derivatives are available, then the derivatives in (1.9) are approximated by a finite-difference formula

$$x_{n+1} = x_n - \frac{\Delta[f(x_{n+1}) - f(x_{n-1})]}{2[f(x_{n+1}) - 2f(x_n) + f(x_{n-1})]} \tag{1.10}$$

where

$$\Delta = |x_{n+1} - x_n| = |x_n - x_{n-1}| \tag{1.11}$$

This approximation slows the method down but is often the only practical implementation.

Let's try finding the minimum of the two test functions. Since it's not easy to take the derivatives of AF_1 and AF_2, finite-difference approximations will be used instead. Newton's method converges on the minimum of AF_1 for every starting point in the interval. The second function is more interesting,

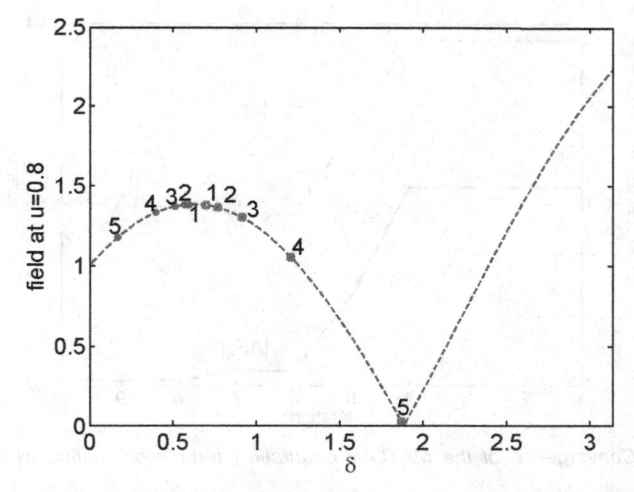

Figure 1.9. *The convergence of Newton's algorithm when starting at two different points.*

though. Figure 1.9 shows the first five points calculated by the algorithm from two different starting points. A starting point at $\delta = 0.6$ radians results in the series of points that heads toward the local minimum on the left. When the starting point is $\delta = 0.7$ rad, then the algorithm converges toward the global minimum. Thus, Newton's method is known as a *local search algorithm*, because it heads toward the bottom of the closest minimum. It is also a non-linear algorithm, because the outcome can be very sensitive to the initial starting point.

1.1.5 Quadratic Interpolation

The techniques derived from Taylor's series assumed that the function is quadratic near the minimum. If this assumption is valid, then we should be able to approximate the function by a quadratic polynomial near the minimum and find the minimum of that quadratic polynomial interpolation [19]. Given three points on an interval (x_0, x_1, x_2), the extremum of the quadratic interpolating polynomial appears at

$$x_3 = \frac{f(x_0)\left(x_1^2 - x_2^2\right) + f(x_1)\left(x_2^2 - x_0^2\right) + f(x_2)\left(x_0^2 - x_1^2\right)}{2f(x_0)(x_1 - x_2) + 2f(x_1)(x_2 - x_0) + 2f(x_2)(x_0 - x_1)} \tag{1.12}$$

When the three points are along the same line, the denominator is zero and the interpolation fails. Also, this formula can't differentiate between a minimum and a maximum, so some caution is necessary to insure that it pursues a minimum.

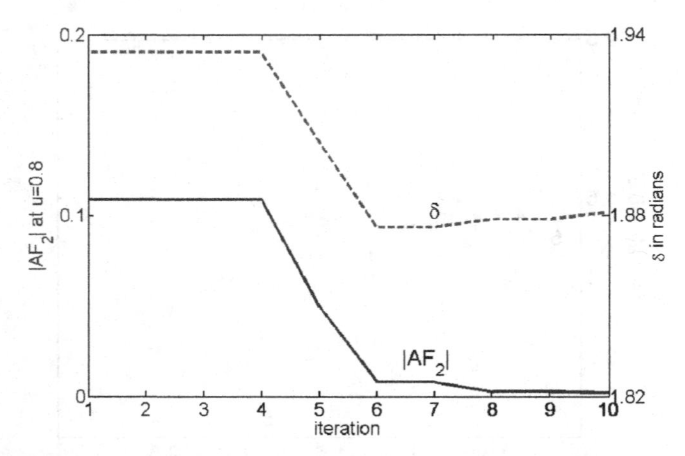

Figure 1.10. *Convergence of the MATLAB quadratic interpolation routine when minimizing AF_2.*

MATLAB uses a combination of golden search and quadratic interpolation in its function *fminbnd.m*. Figure 1.10 shows the convergence curves for the field value on the left-hand vertical axis and the phase in radians on the right-hand vertical axis. This approach converged in just 10 iterations.

1.2 OPTIMIZING A FUNCTION OF MULTIPLE VARIABLES

Usually, arrays have many elements; hence many variables need to be adjusted in order to optimize some aspect of the antenna pattern. To demonstrate the complexity of dealing with multiple dimensions, the objective functions in (1.1) and (1.2) are extended to two variables and three angle evaluations of the array factor.

$$\mathrm{AF}_3(a_1, a_2) = \frac{1}{6}\sum_{m=1}^{3}\left|a_2 + a_1 e^{j\Psi_m} + e^{j2\Psi_m} + e^{j3\Psi_m} + a_1 e^{j4\Psi_m} + a_2 e^{j5\Psi_m}\right| \tag{1.13}$$

$$\mathrm{AF}_4(\delta_1, \delta_2) = \frac{1}{6}\sum_{m=1}^{3}\left|e^{j\delta_2} + e^{j\delta_1}e^{j\Psi_m} + e^{j2\Psi_m} + e^{j3\Psi_m} + e^{j\delta_1}e^{j4\Psi_m} + e^{j\delta_2}e^{j5\Psi_m}\right| \tag{1.14}$$

Figure 1.11 is a diagram of the six-element array with two independent adjustable weights. The objective function returns the sum of the magnitude of the array factor at three angles: ϕ_m = 120°, 69.5°, and 31.8°. The array factor for a uniform six-element array is shown in Figure 1.12. Plots of the objective function for all possible combinations of the amplitude and phase weights appear in Figures 1.13 and 1.14. The amplitude weight objective func-

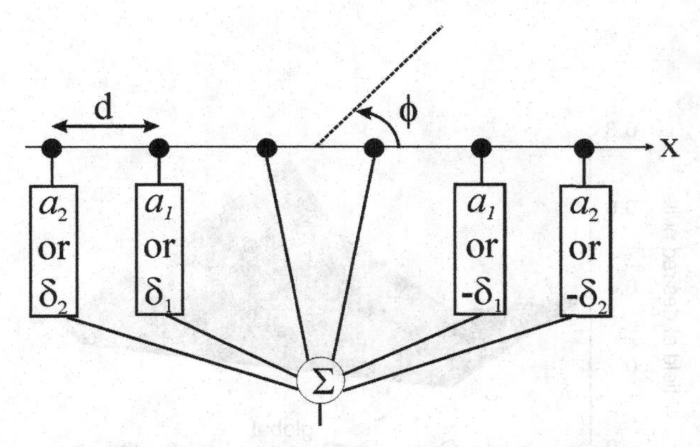

Figure 1.11. *A six-element array with two independent, adjustable weights.*

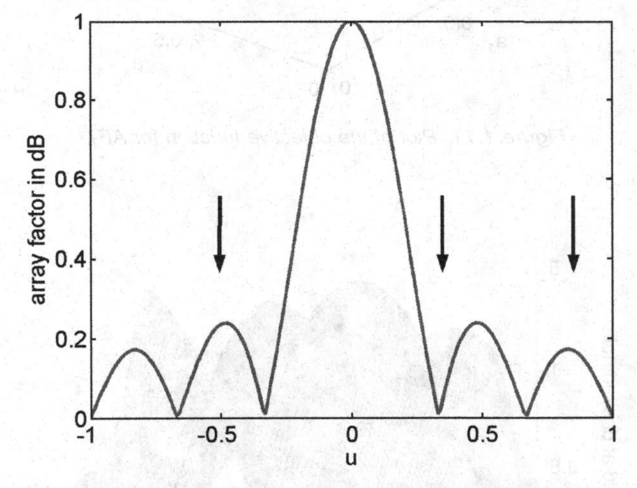

Figure 1.12. *The array factor for a six-element uniform array.*

tion has a single minimum, while the phase weight objective function has several minima.

1.2.1 Random Search

Humans are intrigued by guessing. Most people love to gamble, at least occasionally. Beating the odds is fun. Guessing at the location of the minimum sometimes works. It's at least a very easy-to-understand method for minimization—no Hessians, gradients, simplexes, and so on. It takes only a couple of lines of MATLAB code to get a working program. It's not very elegant, though, and many people have ignored the power of random search in the

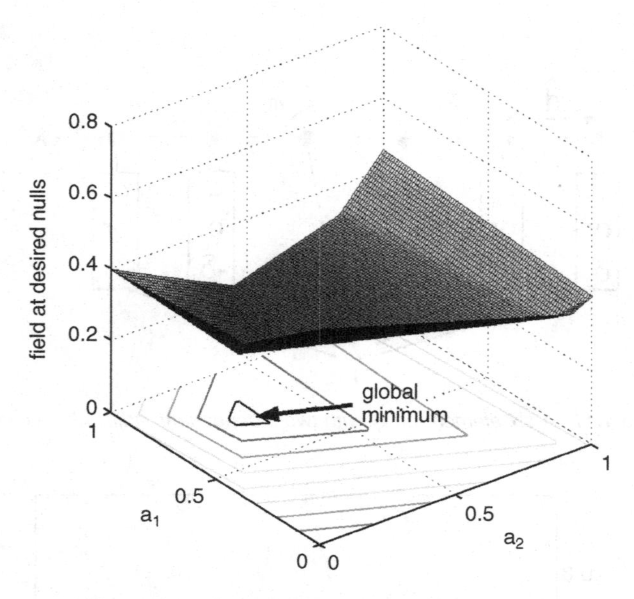

Figure 1.13. *Plot of the objective function for AF_3.*

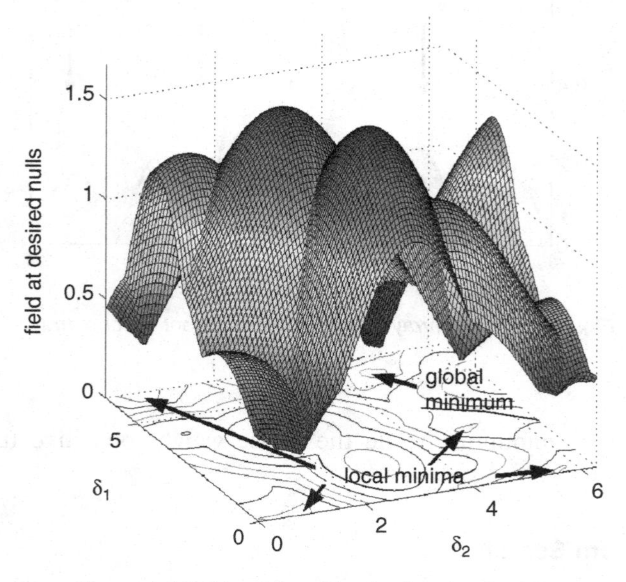

Figure 1.14. *Plot of the objective function for AF_4.*

development of sophisticated minimization algorithms. We often model processes in nature as random events, because we don't understand all the complexities involved. A complex cost function more closely approximates nature's ways, so the more complex the cost function, the more likely that random

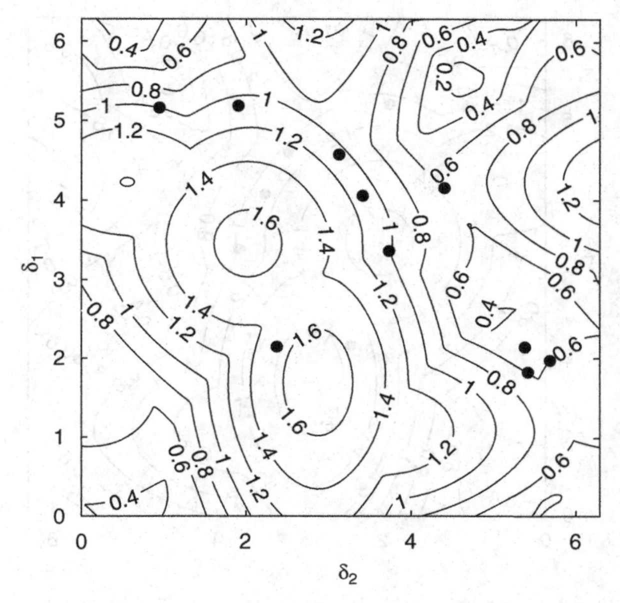

Figure 1.15. Ten random guesses for AF_4.

guessing plays an important part in finding the minimum. Even a local optimizer makes use of random starting points. Local optimizers are made more "global" by making repeated starts from several different, usually random, points on the cost surface.

Figure 1.15 shows 10 random guesses on a contour plot of AF_4. This first attempt clearly misses some of the regions with minima. The plot in Figure 1.16 results from adding 20 more guesses. Even after 30 guesses, the lowest value found is not in the basin of the global minimum. Granted, a new set of random guesses could easily land a value near the global minimum. The problem, though, is that the odds decrease as the number of variables increases.

1.2.2 Line Search

A line search begins with an arbitrary starting point on the cost surface. A vector or line is chosen that cuts across the cost surface. Steps are taken along this line until a minimum is reached. Next, another vector is found and the process repeated. A flowchart of the algorithm appears in Figure 1.17. Selecting the vector and the step size has been an area of avid research in numerical optimization. Line search methods work well for finding a minimum of a quadratic function. They tend to fail miserably when searching a cost surface with many minima, because the vectors can totally miss the area where the global minimum exists.

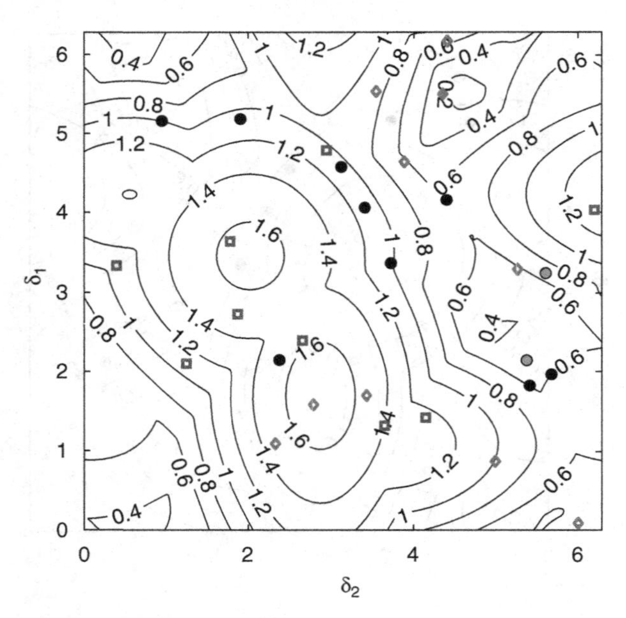

Figure 1.16. *Thirty random guesses for AF_4.*

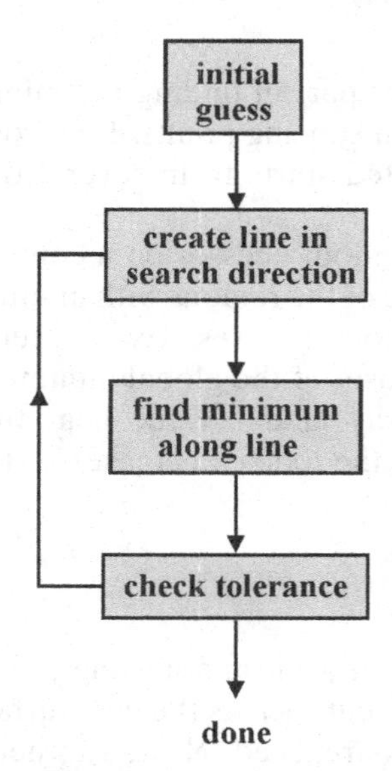

Figure 1.17. *Flowchart of a line search minimization algorithm.*

The easiest line search imaginable is the coordinate search method. If the function has two variables, then the algorithm begins at a random point, holds one variable constant, and searches along the other variable. Once it reaches a minimum, it holds the second variable constant and searches along the first variable. This process repeats until an acceptable minimum is found. Mathematically, a two-dimensional cost function follows the path given by

$$f(v_1^0, v_2^0) \rightarrow f(v_1^1, v_2^0) \rightarrow f(v_1^1, v_2^1) \rightarrow f(v_1^2, v_2^1) \rightarrow \cdots \tag{1.15}$$

where $v_n^{m+1} = v_n^m + \ell_n^{m+1}$ and ℓ_m^{n+1} is the step length calculated using a formula. This approach doesn't work well, because it does not exploit any information about the cost function. Most of the time, the coordinate axes are not the best search directions [20]. Figure 1.18 shows the paths taken by a coordinate search algorithm from three different starting points on AF_4. A different minimum was found from each starting point.

The coordinate search does a lot of unnecessary wiggling to get to a minimum. Following the gradient seems to be a more intuitive natural choice for the direction of search. When water flows down a hillside, it follows the gradient of the surface. Since the gradient points in the direction of maximum increase, the negative of the gradient must be followed to find the minimum. This observation leads to the method of steepest descent given by [19]

$$v^{m+1} = v^m - \alpha^m \nabla f(v^m) \tag{1.16}$$

where α^m is the step size. This formula requires only first-derivative information. Steepest descent is very popular because of its simple form and often excellent results. Problems with slow convergence arise when the cost function

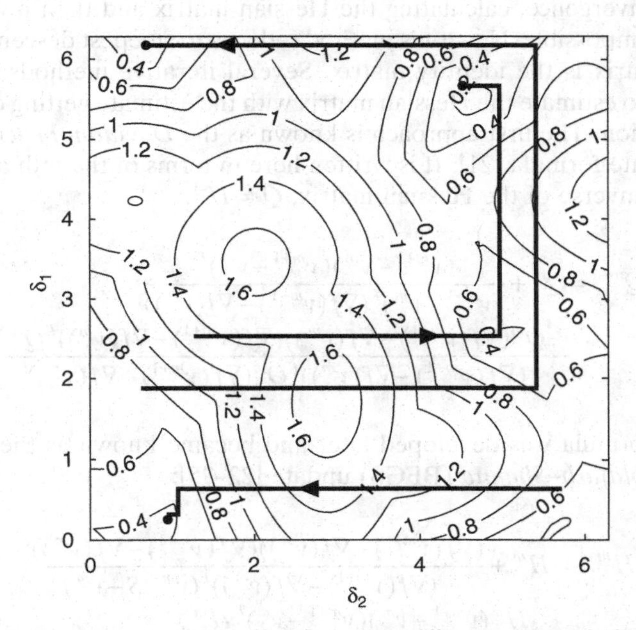

Figure 1.18. *A coordinate search algorithm finds three different minima of AF_4 when starting at three different points.*

has narrow valleys, since the new gradient is always perpendicular to the old gradient at the minimum point on the old gradient.

Even more powerful methods are possible if second-derivative information is used. Starting with the Taylor series expansion of the function

$$f(v^{m+1}) = f(v^m) + \nabla f(v^m)(v^{m+1} - v^m)^T + 0.5(v^{m+1} - v^m)\mathrm{H}(v^{m+1} - v^m)^T + \cdots \tag{1.17}$$

where v^m = point about which Taylor series is expanded
v^{m+1} = point near v^m
^{T} = transpose of vector (in this case row vector becomes column vector)
H = Hessian matrix with elements given by $h_{ij} = \partial^2 f/\partial v_i \partial v_j$

Taking the gradient of (1.17) and setting it equal to zero yields

$$\nabla f(v^{m+1}) = \nabla f(v^m) + (v^{m+1} - v^m)\mathrm{H} = 0 \tag{1.18}$$

which leads to

$$v^{m+1} = v^m - \alpha^m \mathrm{H}^{-1} \nabla f(v^m) \tag{1.19}$$

This formula is known as *Newton's method*. Although Newton's method promises fast convergence, calculating the Hessian matrix and then inverting it is difficult or impossible. Newton's method reduces to steepest descent when the Hessian matrix is the identity matrix. Several iterative methods have been developed to estimate the Hessian matrix with the estimate getting closer after every iteration. The first approach is known as the *Davidon–Fletcher–Powell* (DFP) update formula [21]. It is written here in terms of the mth approximation to the inverse of the Hessian matrix, $Q = H^{-1}$:

$$Q^{m+1} = Q^m + \frac{(v^{m+1} - v^m)(v^{m+1} - v^m)^T}{(v^{m+1} - v^m)^T(\nabla f(v^{m+1}) - \nabla f(v^m))} - \frac{Q^m(\nabla f(v^{m+1}) - \nabla f(v^m))(\nabla f(v^{m+1}) - \nabla f(v^m))^T Q^m}{(\nabla f(v^{m+1}) - \nabla f(v^m))^T Q^m (\nabla f(v^{m+1}) - \nabla f(v^m))} \tag{1.20}$$

A similar formula was developed later and became known as the *Broyden–Fletcher–Goldfarb–Shanno* (BFGS) update [22–25]:

$$H^{m+1} = H^m + \frac{(\nabla f(v^{m+1}) - \nabla f(v^m))(\nabla f(v^{m+1}) - \nabla f(v^m))^T}{(\nabla f(v^{m+1}) - \nabla f(v^m))^T(v^{m+1}S - v^m)} - \frac{H^m(v^{m+1} - v^m)(v^{m+1} - v^m)^T H^m}{(v^{m+1} - v^m)^T H^m (v^{m+1} - v^m)} \tag{1.21}$$

As with the DFP update, the BFGS update can be written for the inverse of the Hessian matrix.

A totally different approach to the line search is possible. If a problem has N dimensions, then it might be possible to pick N orthogonal search directions that could result in finding the minimum in N iterations. Two consecutive search directions, u and v, are orthogonal if their dot product is zero or

$$u \cdot v = uv^T = 0 \tag{1.22}$$

This result is equivalent to

$$uHv^T = 0 \tag{1.23}$$

where H is the identity matrix. If (1.23) is true and H is not the identity matrix, then u and v are known as *conjugate vectors* or *H-orthogonal vectors*. A set of N vectors that have this property is known as a *conjugate set*. It is these vectors that will lead to the minimum in N steps.

Powell developed a method of following these conjugate directions to the minimum of a quadratic function. Start at an arbitrary point and pick a search direction. Next, Gram–Schmidt orthogonalization is used to find the remaining search directions. This process is not very efficient and can result in some search directions that are nearly linearly dependent. Some modifications to Powell's method make it more attractive.

The best implementation of the conjugate directions algorithm is the conjugate gradient algorithm [26]. This approach uses the steepest descent as its first step. At each additional step, the new gradient vector is calculated and added to a combination of previous search vectors to find a new conjugate direction vector

$$v^{m+1} = v^m + \alpha^m \ell^m \tag{1.24}$$

where the step size is given by

$$\alpha^m = -\frac{\ell^{mT} \nabla f(v^m)}{\ell^{mT} H \ell^m} \tag{1.25}$$

Since (1.25) requires calculation of the Hessian, α^m is usually found by minimizing $f(v^m + \alpha^m \ell^m)$. The new search direction is found using

$$\ell^{m+1} = -\nabla f^{m+1} + \beta^{m+1} \ell^m \tag{1.26}$$

The Fletcher–Reeves version of β^m is used for linear problems [18]:

$$\beta^m = \frac{\nabla f^T(v^{m+1}) \nabla f(v^{m+1})}{\nabla f^T(v^m) \nabla f(v^m)} \tag{1.27}$$

This formulation converges when the starting point is sufficiently close to the minimum. A nonlinear conjugate gradient algorithm that uses the Polak–Ribiere version of β^m also exists [18]:

$$\beta^m = \max\left\{\frac{[\nabla f(v^{m+1}) - \nabla f(v^m)]^T \nabla f(v^{m+1})}{\nabla f^T(v^m)\nabla f(v^m)}, 0\right\} \tag{1.28}$$

The nonlinear conjugate gradient algorithm is guaranteed to converge for linear functions but not for nonlinear functions.

The problem with conjugate gradient is that it must be "restarted" every N iterations. Thus, for a nonquadratic problem (most problems of interest), conjugate gradient starts over after N iterations without finding the minimum. Since the BFGS algorithm does not need to be restarted and approaches superlinear convergence close to the solution, it is usually preferred over conjugate gradient. If the Hessian matrix gets too large to be conveniently stored, however, conjugate gradient shines with its minimal storage requirements.

1.2.3 Nelder–Mead Downhill Simplex Algorithm

Derivatives and guessing are not the only way to do a downhill search. The Nelder–Mead downhill simplex algorithm moves a simplex down the slope until it surrounds the minimum [27]. A simplex is the most basic geometric object that can be formed in an N-dimensional space. The simplex has $N + 1$ sides, such as a triangle in two-dimensional space. The downhill simplex method is given a single starting point (v^0). It generates an additional N points to form the initial simplex using the formula

$$v_n^0 = v_1^0 + \mu \ell_n \tag{1.29}$$

where the ℓ_n are unit vectors and μ is a constant. If the simplex surrounds the minimum, then the simplex shrinks in all directions. Otherwise, the point corresponding to the highest objective function is replaced with a new point that has a lower objective function value. The diameter of the simplex eventually gets small enough that it is less than the specified tolerance and the solution is the vertex with the lowest objective function value. A flowchart outlining the steps to this algorithm is shown in Figure 1.19.

Figure 1.20 shows the path taken by the Nelder–Mead algorithm starting with the first triangle and working its way down to the minimum of AF_3. Sometimes the algorithm flips the triangle and at other times it shrinks or expands the triangle in an effort to surround the minimum. Although it can successfully find the minimum of AF_3, it has great difficulty finding the global

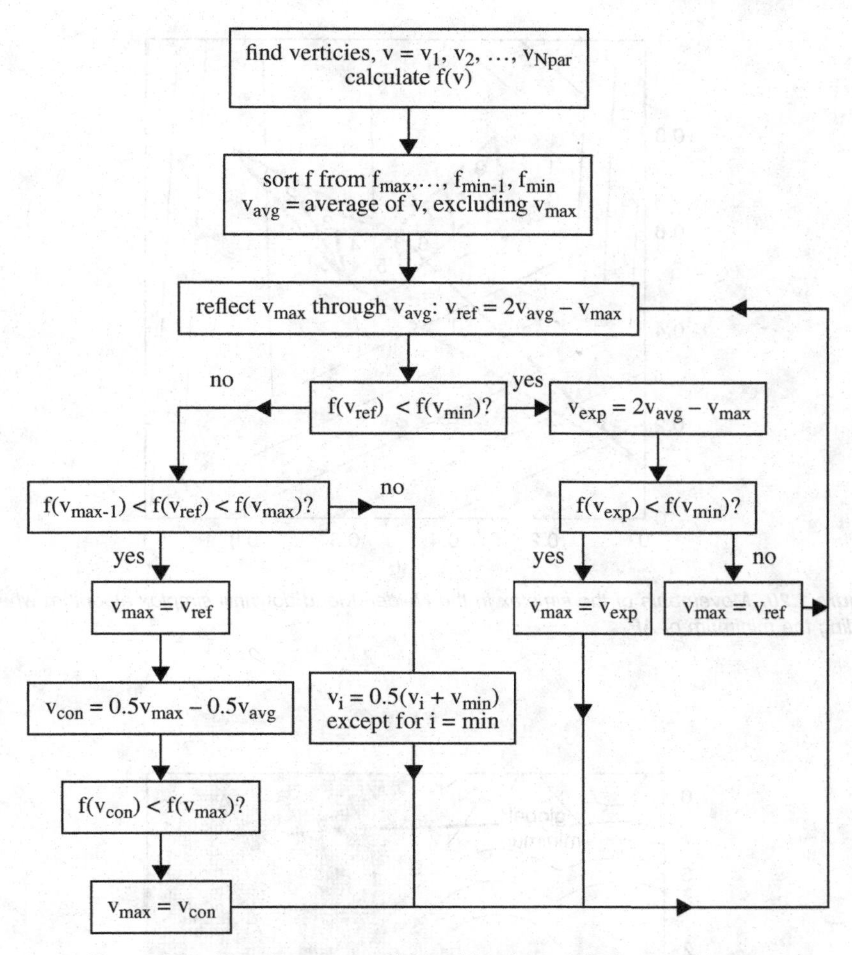

Figure 1.19. *Flowchart for the Nelder–Mead downhill simplex algorithm.*

minimum of AF_4. Its success with AF_4 depends on the starting point. Figure 1.21 shows a plot of the starting points for the Nelder–Mead algorithm that converge to the global minimum. Any other point on the plot converges to one of the local minima. There were 10,201 starting points tried and 2290 or 22.45% converged to the global minimum. That's just slightly better than a one-in-five chance of finding the true minimum. Not very encouraging, especially when the number of dimensions increases. The line search algorithms exhibit the same behavior as the Nelder–Mead algorithm in Figure 1.21.

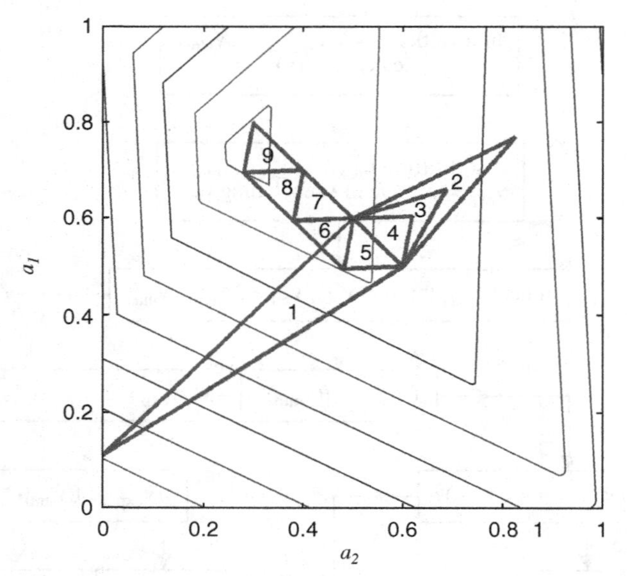

Figure 1.20. *Movements of the simplex in the Nelder–Mead downhill simplex algorithm when finding the minimum of* AF_3.

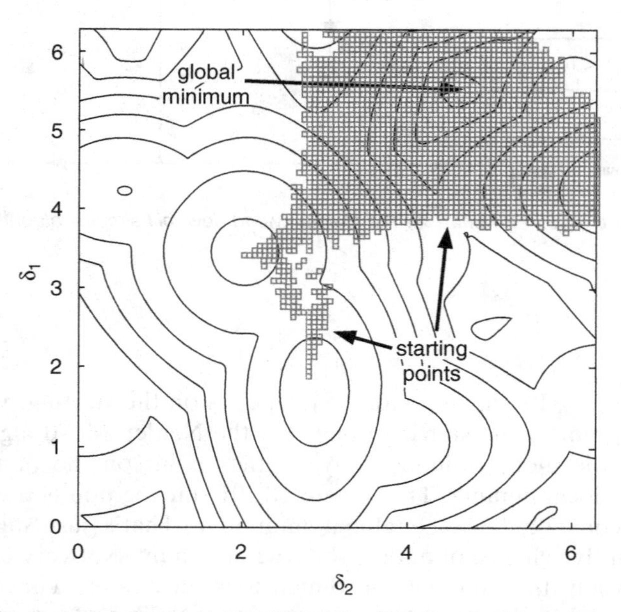

Figure 1.21. *Starting points for the Nelder–Mead algorithm denoted by small squares converge to the global minimum. All other starting points converge to a local minimum.*

1.3 COMPARING LOCAL NUMERICAL OPTIMIZATION ALGORITHMS

This section compares four local optimization approaches against a more formidable problem. Consider a linear array along the x axis with an array factor given by

$$\mathrm{AF}(u, \lambda, w, x, N_{\mathrm{el}}) = \sum_{n=1}^{N_{el}} w_n e^{jkx_n u} \tag{1.30}$$

where N_{el} = number of elements in the array
$w_n = a_n \exp(j\delta_n)$ = complex element weight
x_n = distance of element n from the origin

Some of these variables may be constants. For instance, if all except u are constant, then AF returns an antenna pattern that is a function of angle.

The optimization example here minimizes the maximum relative sidelobe level of a uniformly spaced array using amplitude weighting. Assume that the array weights are symmetric about the center of the array; thus the exponential terms of the symmetric element locations can be combined using Euler's identity. Also assume that the array has an even number of elements. With these assumptions, the objective function is written as a function of the amplitude weights

$$\mathrm{AF}_5(a) = 2\max\left|\sum_{n=1}^{N} a_n \cos[(n-0.5)\Psi_m]\right|, \quad u > u_b \tag{1.31}$$

where u_b defines the extent of the main beam. This function is relatively simple, except for finding the appropriate value for u_b. For a uniform aperture, the first null next to the mainlobe occurs at an angle of about λ/(size of the aperture) $\cong \lambda/[d(N+1)]$. Amplitude tapers decrease the efficiency of the aperture, thus increasing the width of the mainbeam. As a result, u_b depends on the amplitude, making the function difficult to characterize for a given set of input values. In addition, shoulders on the mainbeam may be overlooked by the function that finds the maximum sidelobe level.

The four methods used to optimize AF_5 are

1. Broyden–Fletcher–Goldfarb–Shanno (BFGS)
2. Davidon–Fletcher–Powell (DFP)
3. Nelder–Mead downhill simplex (NMDS)
4. Steepest descent

All of these are available using the MATLAB functions *fminsearch.m* and *fminunc.m*. The analytical solution is simple: $-\infty$ dB. Let's see how the different methods fare.

The four algorithms are quite sensitive to the starting values of the amplitude weights. These algorithms quickly fall into a local minimum, because their theoretical development is based on finding the minimum of a bowl-shaped objective function. The first optimization run randomly generated a starting vector of amplitude weights between 0 and 1. Each algorithm was given 25 different starting values and the results were averaged. Table 1.1 shows the results for a linear array of isotropic point sources spaced 0.5λ apart. A –∞ sidelobe level would ruin a calculation of the mean of the best sidelobe level found over the 25 runs, so the median is as reported in Table 1.1. The mean and median are very close except when a –∞ sidelobe level occurs.

These results are somewhat disappointing, since the known answer is –∞ dB. The local nature of the algorithms limits their ability to find the best or the global solution. In general, a starting point is selected, then the algorithm proceeds downhill from there. When the algorithm encounters too many hills and valleys in the output of the objective function, it can't find the global optimum. Even selecting 25 random starting points for four different algorithms didn't result in finding an output with less than –50 dB sidelobe levels.

In general, as the number of variables increases, the number of function calls needed to find the minimum also increases. Thus, Table 1.1 has a larger median number of function calls for larger arrays. Table 1.2 shows how increas-

TABLE 1.1. Comparison of Optimized Median Sidelobes for Three Different Array Sizes[a]

	22 Elements		42 Elements		62 Elements	
	Median Sidelobe Level (dB)	Median Function Calls	Median Sidelobe Level (dB)	Median Function Calls	Median Sidelobe Level (dB)	Median Function Calls
BFGS	–30.3	1007	–25.3	2008	–26.6	3016
DFP	–27.9	1006	–25.2	2011	–26.6	3015
Nelder Mead	–18.7	956	–17.3	2575	–17.2	3551
Steepest descent	–24.6	1005	–21.6	2009	–21.8	3013

[a]Performance characteristics of four algorithms are averaged over 25 runs with random starting values for the amplitude weights. The isotropic elements were spaced 0.5λ apart.

TABLE 1.2. Algorithm Performance in Terms of Median Maximum Sidelobe Level versus Maximum Number of Function Calls[a]

Algorithm	1000 Function Calls (dB)	3000 Function Calls (dB)	10000 Function Calls (dB)
BFGS	–24.3	–26.6	–28.2
DFP	–24.0	–26.6	–28.3
Nelder–Mead	–17.6	–17.2	–17.5
Steepest descent	–23.3	–21.8	–23.4

[a]Performance characteristics of four algorithms are averaged over 25 runs with random starting values for the amplitude weights. The 42 isotropic elements were spaced 0.5λ apart.

TABLE 1.3. Algorithm Performance in Terms of Median Maximum Sidelobe When the Algorithm Is Restarted Every 2000 Function Calls (5 Times)[a]

Algorithm	10,000 Function Calls (dB)
BFGS	–34.9
DFP	–36.9
Nelder–Mead	–29.1
Steepest descent	–26.1

[a] Performance characteristics of four algorithms are averaged over 25 runs with random starting values for the amplitude weights. The 42 isotropic elements were spaced 0.5λ apart.

ing the maximum number of function calls improves the results found using BFGS and DFP whereas the Nelder–Mead and steepest-descent algorithms show no improvement.

Another idea is warranted. Perhaps taking the set of parameters that produces the lowest objective function output and using them as the initial starting point for the algorithm will produce better results. The step size gets smaller as the algorithm progresses. Starting over may allow the algorithm to take large enough steps to get out of the valley of the local minimum. Thus, the algorithm begins with a random set of amplitude weights, the algorithm optimizes with these weights to produce an "optimal" set of parameters, these new "optimal" set of parameters are used as a new initial starting point for the algorithm, and the process repeats several times. Table 1.3 displays some interesting results when the cycle is repeated 5 times. Again, the algorithms were averaged over 25 different runs. In all cases, the results improved by running the optimization algorithm for 2000 function calls on five separate starts rather than running the optimization algorithm for a total of 10,000 function calls with one start (Table 1.2). The lesson learned here is to use this iterative procedure when attempting an optimization with multiple local minima. The size of the search space collapses as the algorithm converges on the minimum. Thus, restarting the algorithm at the local minimum just expands the search space about the minimum.

An alternative approach known as "seeding" starts the algorithm with a good first guess based on experience, a hunch, or other similar solutions. In general, we know that low-sidelobe amplitude tapers have a maximum amplitude at the center of the array, while decreasing in amplitude toward the edges. The initial first guess is a uniform amplitude taper with a maximum sidelobe level of –13dB. Table 1.4 shows the results of using this good first guess after 2000 function calls. The Nelder–Mead algorithm capitalized on this good first guess, while the others didn't. Trying a triangle amplitude taper, however, significantly improved the performance of all the algorithms. In fact, the Nelder–Mead and steepest-descent algorithms did better than the BFGS and DFP algorithms. Combining the good first guess with the restarting idea in Figure 1.11 may produce the best results of all.

TABLE 1.4. Algorithm Performance in Terms of Median Maximum Sidelobe after 2000 Function Calls When the Algorithm Seeded with a Uniform or Triangular Amplitude Taper[a]

Algorithm	Uniform Taper Seed (dB)	Triangular Taper Seed (dB)
BFGS	−23.6	−35.9
DFP	−26.0	−35.7
Nelder–Mead	−23.9	−39.1
Steepest descent	−21.2	−39.3

[a] The 42 isotropic elements were spaced 0.5λ apart.

1.4 SIMULATED ANNEALING

Annealing heats a substance above its melting temperature, then gradually cools it to produce a crystalline lattice that has a minimum energy probability distribution. The resulting crystal is an example of nature finding an optimal solution. If the liquid cools too rapidly, then the crystals do not form and the substance becomes an amorphous mass with an energy state above optimal. Nature is seldom in a hurry to find the optimal state.

A numerical optimization algorithm that models the annealing process is known as *simulated annealing* [28,29]. The initial state of the algorithm is a single random guess of the objective function input variables. In order to model the heating process, the values of the variables are randomly modified. Higher heat creates greater random fluctuations. The objective function returns a measure of the energy state or value of the present minimum. The new variable set replaces the old variable set if the output decreases. Otherwise the output is still accepted if

$$r \le e^{[f(p_{\text{old}})-f(p_{\text{new}})]/T} \tag{1.32}$$

where r is a uniform random number and T is a temperature value. If r is too large, then the variable values are rejected. A new variable set to replace a rejected variable set is found by adding a random step to the old variable set

$$p_{\text{new}} = dp_{\text{old}} \tag{1.33}$$

where d is a random number with either a uniform or normal distribution with a mean of p_{old}. When the minimization process stalls, the value of T and the range of d decrease by a certain percent and the algorithm starts over. The algorithm is finished when T gets close to zero. Some common cooling schedules include (1) linearly decreasing, $T_n = T_0 - n(T_0 - T_n)/N$; (2) geometrically decreasing, $T_n = 0.99T_{n-1}$; and (3) Hayjek optimal, $T_n = c/\log(1 + n)$, where c = smallest variation required to get out of any local minimum, $0 < n \le N$; T_0 = initial temperature, and T_N = ending temperature.

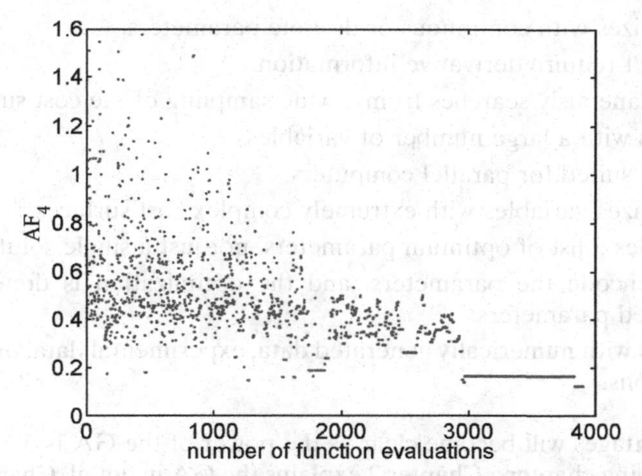

Figure 1.22. Convergence of the simulated annealing algorithm for AF_4.

The temperature is lowered slowly, so that the algorithm does not converge too quickly.

Simulated annealing (SA) begins as a random search and ends with little variations about the minimum. Figure 1.22 shows the convergence of the simulated annealing algorithm when minimizing AF_4. The final value was 0.1228 or −33.8dB. Simulated annealing was first applied to the optimization of antenna arrays in 1988 [30].

SA has proven superior to the local optimizers discussed in this chapter. The random perturbations allow this algorithm to jump out of a local minimum in search of the global minimum. SA is very similar to the GA. It is a random search that has tuning parameters that have tremendous effect on the success and speed of the algorithm. SA starts with a single guess at the solution and works in a serial manner to find the solution. A genetic algorithm starts with many initial guesses and works in a parallel manner to find a list of solutions. The SA algorithm slowly becomes less random as it converges, while the GA may or may not become less random with time. Finally, the GA is more adept at working with continuous, discrete, and integer variables, or a mix of those variables.

1.5 GENETIC ALGORITHM

The rest of this book is devoted to the relatively new optimization technique called the *genetic algorithm* (GA). GAs were introduced by Holland [31] and were applied to many practical problems by Goldberg [32]. A GA has several advantages over the traditional numerical optimization approaches presented in this chapter, including the facts that it

1. Optimizes with continuous or discrete parameters.
2. Doesn't require derivative information.
3. Simultaneously searches from a wide sampling of the cost surface.
4. Works with a large number of variables.
5. Is well suited for parallel computers.
6. Optimizes variables with extremely complex cost surfaces.
7. Provides a list of optimum parameters, not just a single solution.
8. May encode the parameters, and the optimization is done with the encoded parameters.
9. Works with numerically generated data, experimental data, or analytical functions.

These advantages will become clear as the power of the GA is demonstrated in the following chapters. Chapter 2 explains the GA in detail. Chapter 3 gives a step-by-step analysis of finding the minimum of AF_4. Many other more complex examples are presented in Chapters 3–8. For further enlightenment on GAs, please turn to the next chapter.

REFERENCES

1. J. S. Stone, US Patents 1,643,323 and 1,715,433 C.L.
2. C. L. Dolph, A current distribution for broadside arrays which optimizes the relationship between beam width and side-lobe level, *Proc. IRE* **34**:335–348 (June 1946).
3. T. T. Taylor, Design of line source antennas for narrow beamwidth and low side lobes, *IRE AP Trans.* **4**:16–28 (Jan. 1955).
4. R. S. Elliott, Antenna *Theory and Design*, Prentice-Hall, New York, 1981.
5. A. T. Villeneuve, Taylor patterns for discrete arrays, *IEEE AP-S Trans.* **32**(10): 1089–1094 (Oct. 1984).
6. E. T. Bayliss, Design of monopulse antenna difference patterns with low sidelobes, *Bell Syst. Tech. J.* **47**:623–650 (May–June 1968).
7. W. W. Hansen and J. R. Woodyard, A new principle in directional antenna design, *Proc. IRE* **26**:333–345 (March 1938).
8. W. L. Stutzman and E. L. Coffey, Radiation pattern synthesis of planar antennas using the iterative sampling method, *IEEE Trans.* **AP-23**(6):764–769 (Nov. 1975).
9. H. J. Orchard, R. S. Elliot, and G. J. Stern, Optimizing the synthesis of shaped beam antenna patterns, *IEE Proc.* **132**(1):63–68 (Feb. 1985).
10. R. S. Elliot and G. J. Stearn, Shaped patterns from a continuous planar aperture distribution, *IEEE Proc.* **135**(6):366–370 (Dec. 1988).
11. F. Ares, R. S. Elliott, and E. Moreno, Design of planar arrays to obtain efficient footprint patterns with an arbitrary footprint boundary, *IEEE AP-S Trans.* **42**(11):1509–1514 (Nov. 1994).

12. D. K. Cheng, Optimization techniques for antenna arrays, *Proc. IEEE* **59**(12):1664–1674 (Dec. 1971).
13. J. F. DeFord and O. P. Gandhi, Phase-only synthesis of minimum peak sidelobe patterns for linear and planar arrays, *IEEE AP-S Trans.* **36**(2):191–201 (Feb. 1988).
14. J. E. Richie and H. N. Kritikos, Linear program synthesis for direct broadcast satellite phased arrays, *IEEE AP-S Trans.* **36**(3):345–348 (March 1988).
15. M. I. Skolnik, G. Nemhauser, and J. W. Sherman, III, Dynamic programming applied to unequally spaced arrays, *IEEE AP-S Trans.* **12**:35–43 (Jan. 1964).
16. J. Perini, Note on antenna pattern synthesis using numerical iterative methods, *IEEE AP-S Trans.* **12**:791–792 (July 1976).
17. Y. T. Lo, A mathematical theory of antenna arrays with randomly spaced elements, *IEEE AP-S Trans.* **12**(3):257–268 (May 1964).
18. W. H. Press et al., *Numerical Recipes*, Cambridge Univ. Press, New York, 1992.
19. G. Luenberger, *Linear and Nonlinear Programming*, Addison-Wesley, Reading, MA, 1984.
20. H. H. Rosenbrock, An automatic method for finding the greatest or least value of a function, *Comput. J.* **3**:175–184 (1960).
21. M. J. D. Powell, An efficient way for finding the minimum of a function of several variables without calculating derivatives, *Comput. J.* 155–162 (1964).
22. G. C. Broyden, A class of methods for solving nonlinear simultaneous equations, *Math. Comput.* 577–593 (Oct. 1965).
23. R. Fletcher, Generalized inverses for nonlinear equations and optimization, in *Numerical Methods for Non-linear Algebraic Equations*, R. Rabinowitz, ed., Gordon & Breach, London, 1963.
24. D. Goldfarb and B. Lapidus, Conjugate gradient method for nonlinear programming problems with linear constraints, *I&EC Fund.* 142–151 (Feb. 1968).
25. D. F. Shanno, An accelerated gradient projection method for linearly constrained nonlinear estimation, *SIAM J. Appl. Math.* 322–334 (March 1970).
26. J. R. Shewchuk, *An Introduction to the Conjugate Gradient Method without the Agonizing Pain*, Technical Report CMU-CS-94-125, Carnegie Mellon Univ. 1994.
27. J. A. Nelder and R. Mead, *Comput. J.* **7**:308–313 (1965).
28. S. Kirkpatrick, C. D. Gelatt, Jr., and M. P. Vecchi, Optimization by simulated annealing, *Science* **220**:671–680 (May 13, 1983).
29. N. Metropolis et al., Equation of state calculations by fast computing machines, *J. Chem. Phys.* **21**:1087–1092 (1953).
30. T. J. Cornwell, A novel principle for optimization of the instantaneous Fourier plane coverage of correction arrays, *IEEE AP-S Trans.* **36**(8):1165–1167 (Aug. 1988).
31. J. H. Holland, *Adaptation in Natural and Artificial Systems*, Univ. Michigan Press, Ann Arbor, 1975.
32. D. E. Goldberg, *Genetic Algorithms in Search, Optimization, and Machine Learning*, Addison-Wesley, New York, 1989.
33. R. L. Haupt and Sue Ellen Haupt, *Practical Genetic Algorithms*, 2nd ed., Wiley, New York, 2004.

2

Anatomy of a Genetic Algorithm

A genetic algorithm (GA) offers an alternative to traditional local search algorithms. It is an optimization algorithm inspired by the well-known biological processes of genetics and evolution. Genetics is the study of the inheritance and variation of biological traits. Manipulation of the forces behind genetics is found in breeding animals and genetic engineering. Evolution is closely intertwined with genetics. It results in genetic changes through natural selection, genetic drift, mutation, and migration. Genetics and evolution result in a population that is adapted to succeed in its environment. In other words, the population is optimized for its environment.

A combination of genetics and evolution is analogous to numerical optimization in that they both seek to find a good result within constraints on the variables. Input to an objective function is a chromosome. The output of the objective function is known as the *cost* when minimizing. Each chromosome consists of genes or individual variables. The genes take on certain alleles much as the variable have certain values. A group of chromosomes is known as a *population*. For our purposes, the population is a matrix with each row corresponding to a chromosome:

Genetic Algorithms in Electromagnetics, by Randy L. Haupt and Douglas H. Werner

$$\text{population} = \underbrace{\begin{bmatrix} chrom_1 \\ chrom_2 \\ \vdots \\ chrom_N \end{bmatrix}}_{\text{chromosomes}} = \underbrace{\begin{bmatrix} g_{11} & g_{12} & \cdots & g_{1M} \\ g_{21} & g_{22} & & \\ \vdots & & \ddots & \vdots \\ g_{N1} & & \cdots & g_{NM} \end{bmatrix}}_{\text{genes}}$$
$$= \underbrace{\begin{bmatrix} red & 1 & \cdots & 19.132 \\ blue & 4 & & -12.954 \\ \vdots & & \ddots & \vdots \\ green & 2 & \cdots & 0.125 \end{bmatrix}}_{\text{values or alleles}} \tag{2.1}$$

Each chromosome is the input to an objective function f. The cost associated with each chromosome is calculated by the objective function one at a time or in parallel:

$$f\left\{\begin{bmatrix} chrom_1 \\ chrom_2 \\ \vdots \\ chrom_N \end{bmatrix}\right\} = \begin{bmatrix} cost_1 \\ cost_2 \\ \vdots \\ cost_N \end{bmatrix} \tag{2.2}$$

It is the cost that determines the fitness of an individual in the population. A low cost implies a high fitness.

As you will see, GA operations work only with numbers. Thus, nonnumerical values, such as a color or an opinion, must be assigned a number. Often, the numerical values assigned to genes are in binary format. Continuous values have an infinite number of possible combinations of input values, whereas binary values have a very large but finite number of possible combinations of input values. Binary representation is also common when there are a finite number of values for a variable, such as four values of permittivity for a dielectric substrate.

A basic, "no thrills" GA is quite simple and powerful. The algorithm has the following steps:

1. Create an initial population.
2. Evaluate the fitness of each population member.
3. Invoke natural selection.
4. Select population members for mating.
5. Generate offspring.
6. Mutate selected members of the population.
7. Terminate run or go to step 2.

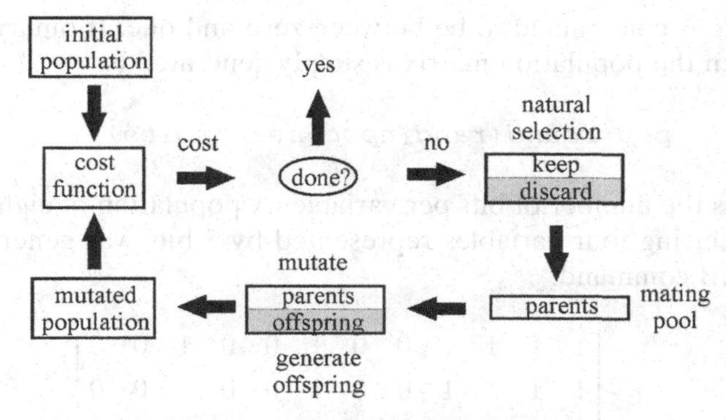

Figure 2.1. Genetic algorithm flowchart.

These steps are shown in the flowchart in Figure 2.1. Each of these steps is discussed in detail in the following sections. MATLAB commands are used to demonstrate the basic concepts.

2.1 CREATING AN INITIAL POPULATION

The initial population is the starting matrix of chromosomes. Each row is a random "guess" at the optimum solution. If *nvar* variables are used to calculate the output of the cost function, then a chromosome in the initial population consists of *nvar* random values assigned to these variables. As an example, a MATLAB command that yields a random population matrix of *npop* chromosomes is given by

```
pop=rand(npop,nvar)
```

A population of eight chromosomes each having four variables was generated from the MATLAB command:

$$pop = \begin{bmatrix} 0.44510 & 0.83812 & 0.30462 & 0.37837 \\ 0.93181 & 0.01964 & 0.18965 & 0.86001 \\ 0.46599 & 0.68128 & 0.19343 & 0.85366 \\ 0.41865 & 0.37948 & 0.68222 & 0.59356 \\ 0.84622 & 0.83180 & 0.30276 & 0.49655 \\ 0.52515 & 0.50281 & 0.54167 & 0.89977 \\ 0.20265 & 0.70947 & 0.15087 & 0.82163 \\ 0.67214 & 0.42889 & 0.69790 & 0.64491 \end{bmatrix} \tag{2.3}$$

Each variable is constrained to be between zero and one. If binary numbers are used, then the population matrix is simply generated by

```
pop=round(rand(npop,nvar*nbits))
```

where *nbits* is the number of bits per variable. A population of eight chromosomes each having four variables represented by 3 bits was generated from the MATLAB command:

$$pop = \begin{bmatrix} 1 & 1 & 1 & 1 & 0 & 0 & 1 & 0 & 0 & 1 & 0 & 0 \\ 1 & 1 & 1 & 1 & 0 & 0 & 1 & 0 & 0 & 0 & 0 & 0 \\ 0 & 0 & 1 & 0 & 0 & 1 & 0 & 1 & 0 & 0 & 1 & 0 \\ 0 & 1 & 1 & 0 & 1 & 1 & 1 & 0 & 1 & 1 & 1 & 0 \\ 0 & 1 & 0 & 1 & 1 & 1 & 1 & 1 & 0 & 1 & 0 & 1 \\ 1 & 0 & 1 & 0 & 0 & 1 & 0 & 1 & 1 & 0 & 1 & 0 \\ 1 & 1 & 0 & 0 & 1 & 0 & 1 & 1 & 1 & 1 & 0 & 0 \\ 0 & 1 & 0 & 1 & 0 & 0 & 0 & 1 & 0 & 1 & 0 & 1 \end{bmatrix} \tag{2.4}$$

2.2 EVALUATING FITNESS

The chromosomes are passed to the cost function *fun* for evaluation. Each chromosome then has an associated cost:

```
cost=fun(pop)
```

An example of a cost function is

$$cost = f(x_1, \ldots, x_N) = \sum_{n=1}^{N} x_n^2 \tag{2.5}$$

Here, we assume that the cost function does the work of translating the variable values into the proper range and/or converting from binary to quantized real numbers. A variable x is bounded by *xhi* and *xlo*, so

```
x=xlo+(xhi-xlo)*pop(1,:)
```

Thus, if the variables for the population in (2.3) are bound by $-5 \le x_n \le 5$, then the first chromosome has the values

$$chrom_1 = [-0.549 \quad 3.3812 \quad -1.9538 \quad -1.2163] \tag{2.6}$$

Using the population in (2.3) as input to (2.5) results in

$$cost = \begin{bmatrix} 17.031 \\ 64.313 \\ 25.308 \\ 6.310 \\ 26.888 \\ 16.219 \\ 35.763 \\ 9.485 \end{bmatrix} \tag{2.7}$$

Binary variables may have to be converted to continuous values by the MATLAB command

```
x=xlo+(xhi-xlo)*([2.^(-[1:nbits])]*reshape(pop(1,:),
  nbits,nvar))
```

If the variables for the population in (2.4) are bound by $-5 \le x_n \le 5$, then the first chromosome has the values

$$chrom_1 = [3.75 \quad 0 \quad 0 \quad 0] \tag{2.8}$$

Sometimes, the binary string does not need conversion, or the string represents a selection rather than a continuous number.

Formulating the cost function is an extremely important step in optimization. Since the function must be called many times to evaluate the cost of the population members, there is usually a tradeoff between calculation accuracy and evaluation time. To reduce convergence time, only relevant variables of the cost function should be included. For instance, when maximizing the gain of a microstrip antenna, the size of the patch is important, while the color of the antenna is not. Some formulations of the cost function are easier to optimize than other formulations. For instance, optimizing the location of the zeros on the unit circle for an array factor works better than optimizing the element weights when minimizing the sidelobe levels of the array factor. Time spent carefully formulating the cost function before optimizing will reap considerable rewards later.

Frequently, the cost function must satisfy more than one goal. As an example, the antenna gain may be maximized while at the same time, the volume of the antenna is minimized. This type of problem is known as *multiple-objective optimization*. A common way of dealing with multiple objectives is to normalize the cost of each objective, weight each cost, then add the weighted costs together to get a single cost for the overall cost function. Thus, the output of the cost function that has N objectives is

$$cost = \sum_{n=1}^{N} w_n c_n \tag{2.9}$$

where $\Sigma_{n=1}^{N} w_n = 1$ and $0 \le c_n \le 1$. Normalizing the cost is important to ensure control over the relative weighting of each cost.

2.3 NATURAL SELECTION

Only the healthiest members of the population are allowed to survive to the next generation. There are two common ways to invoke natural selection. The first is to keep *natsel* healthy chromosomes and discard the rest. First, the cost is sorted in order to determine the relative fitness of the chromosomes:

```
[cost ind]=sort(cost)
```

Sorting the cost in (2.7) leads to

$$[cost\ ind] = \begin{bmatrix} 6.310 & 4 \\ 9.485 & 8 \\ 16.219 & 6 \\ 17.031 & 1 \\ 25.308 & 3 \\ 26.888 & 5 \\ 35.763 & 7 \\ 64.313 & 2 \end{bmatrix} \tag{2.10}$$

The *ind* column vector is the row where the cost resided in (2.7). Next, the population is sorted so that it corresponds to the cost and only *natsel* = 4 chromosomes are retained:

```
pop=pop(ind(1:natsel),:)
cost=cost(1:natsel)
```

The resulting population and cost are given by

$$[pop\ cost] = \begin{bmatrix} 0.41865 & 0.37948 & 0.68222 & 0.59356 & 6.3100 \\ 0.67214 & 0.42889 & 0.69790 & 0.64491 & 9.4852 \\ 0.52515 & 0.50281 & 0.54167 & 0.89977 & 16.2190 \\ 0.44510 & 0.83812 & 0.30462 & 0.37837 & 17.0310 \end{bmatrix} \tag{2.11}$$

A second approach, called *thresholding*, keeps all chromosomes that have a cost below a threshold cost value.

```
pop=pop(find(cost < maxcost));
```

If *maxcost* is the average value of the cost vector, then the result is identical to (2.11). If *maxcost* = 10, then only the first two rows in (2.11) remain. Thresholding avoids the sorting step. Some alternatives for *maxcost* might be the mean of the costs or the median of the costs. The chromosomes that survive form the mating pool, or the group of chromosomes from which parents will be selected to create offspring.

2.4 MATE SELECTION

The most fit members of the population are assigned the highest probability of being selected for mating. The two most common ways of choosing mates are roulette wheel and tournament selection.

2.4.1 Roulette Wheel Selection

The population must first be sorted for roulette wheel selection. Each chromosome is assigned a probability of selection on the basis of either its rank in the sorted population or its cost. Rank order selection is the easiest implementation of roulette wheel selection. The MATLAB code to create the roulette wheel is

```
parents=1:natsel
prob=parents/sum(parents)
odds=[0 cumsum(prob)]
```

When *natsel* = *4*, these MATLAB commands produce

$$\begin{aligned} parents &= [1\ \ 2\ \ 3\ \ 4] \\ prob &= [0.1\ \ 0.2\ \ 0.3\ \ 0.4] \\ odds &= [0\ \ 0.1\ \ 0.3\ \ 0.6\ \ 1] \end{aligned} \tag{2.12}$$

The roulette wheel for a selection pool of four parents is shown in Figure 2.2. Chromosomes with low costs have a higher percent chance of being selected than do chromosomes with higher costs. In this case, the first or best chromosome has a 40% chance of being selected. As more parents are added, the percent chance of a chromosome being selected changes. For instance, Figure 2.3 shows a roulette wheel for eight parents in the mating pool. Now, the best chromosome has a 22% chance of being selected. The roulette wheel needs to be computed only once, since the number of parents in the mating pool remains constant from generation to generation.

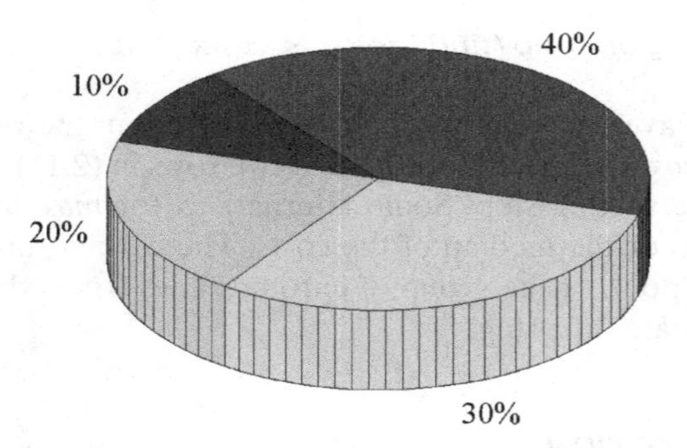

Figure 2.2. Roulette wheel probabilities for four parents in the mating pool.

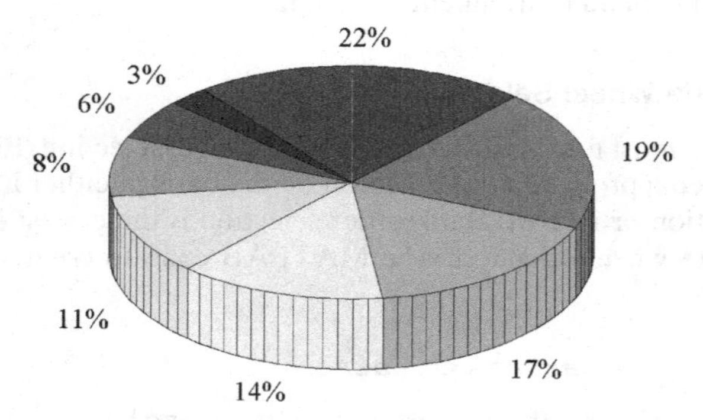

Figure 2.3. Roulette wheel probabilities for eight parents in the mating pool.

It is possible to develop a roulette wheel on the basis of the costs associated with the chromosomes. There are several problems related to this approach:

1. The roulette wheel must be recomputed every generation.
2. If the mutation rate is low, then in later generations all the chromosomes will have approximately the same probability of selection.
3. The costs must be normalized in order to develop the probabilities. The normalization is arbitrary.

As a result, we recommend rank order selection over a cost-based approach.

Once the probability of selection is assigned to each parent, then a uniform random number (r) is generated. For the four parent mating pool, the chromosome selected is based on the value of r:

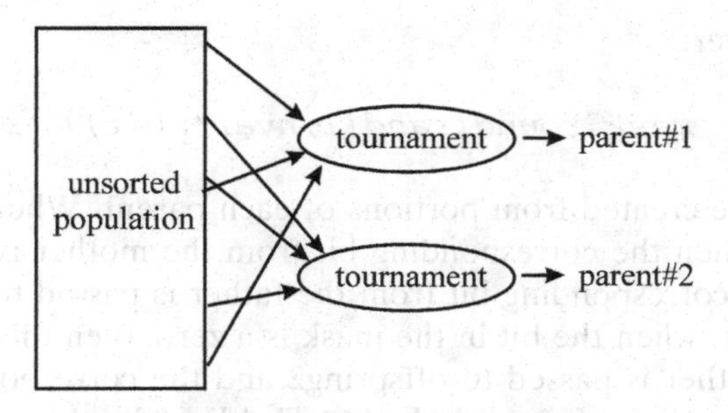

Figure 2.4. Tournament selection.

$$\begin{array}{ll} 0 \le r \le 0.4 & chrom_1 \\ 0.4 < r \le 0.7 & chrom_2 \\ 0.7 < r \le 0.9 & chrom_3 \\ 0.9 < r \le 1.0 & chrom_4 \end{array} \tag{2.13}$$

For example, if $r = 0.5678$, then the second chromosome is selected.

2.4.2 Tournament Selection

A second approach to finding parents randomly selects two small groups of chromosomes from the mating pool (usually two or three per group). The chromosome with the lowest cost in each group becomes a parent. Enough of these tournaments are held to generate the required number of parents. The tournament repeats for every parent needed. Tournament selection works well with thresholding, because the population never needs sorting. Sort speed becomes an issue only with large population sizes. Figure 2.4 diagrams the tournament selection process when three chromosomes are selected for each tournament. Rank order roulette wheel and tournament seletion result in nearly the same probability of selection for the chromosomes [1].

2.5 GENERATING OFFSPRING

Offspring can be generated from selected parents in a number of different ways. For binary chromosomes, uniform crossover is the most general procedure. A mask that consists of ones and zeros is generated for each set of parents. The mask has the same number of bits as the parent chromosomes. Some mask examples include

Single-point crossover:

```
mask=zeros(1,ceil(rand*(nvar*nbit -1))).
  *ones(1,nvar*nbit)
```

Uniform crossover:

```
mask=round(rand(1,nvar*nbit))
```

The offspring are created from portions of each parent. When the bit in the mask is a one, then the corresponding bit from the mother is passed to offspring1 and the corresponding bit from the father is passed to offspring2. In a similar manner, when the bit in the mask is a zero, then the corresponding bit from the mother is passed to offspring2 and the corresponding bit from the father is passed to offspring1. In MATLAB code, this process appears as

```
offspring1=mask.*mother+not(mask).*father
offspring2=not(mask).*mother+mask.*father
```

Many other types of binary crossover are possible.

To demonstrate the concept of binary crossover, consider the following two parents:

```
mother=[1  0  1  0  1  0  1  0  1  0  1  0]
father=[1  1  1  1  1  1  0  0  0  0  0  0]
```

If the mask for single point crossover is given by

```
mask=[1  1  1  0  0  0  0  0  0  0  0  0]
```

then the offspring are

```
offspring1=[1  0  1  1  1  1  0  0  0  0  0  0]
offspring2=[1  1  1  0  1  0  1  0  1  0  1  0]
```

If the mask for double-point crossover is given by

```
mask=[1  1  1  0  0  0  0  1  1  1  1  1]
```

then the offspring are

```
offspring1=[1  0  1  1  1  1  0  0  1  0  1  0]
offspring2=[1  1  1  0  1  0  1  0  0  0  0  0]
```

If the mask for uniform crossover is given by

```
mask=[1  1  0  0  0  1  1  0  1  0  0  1]
```

then the offspring are

$$offspring1 = [1\ \ 0\ \ 1\ \ 1\ \ 1\ \ 0\ \ 1\ \ 0\ \ 1\ \ 0\ \ 0\ \ 0]$$

$$offspring2 = [1\ \ 1\ \ 1\ \ 0\ \ 1\ \ 1\ \ 0\ \ 0\ \ 0\ \ 0\ \ 1\ \ 0]$$

It is possible to use a mask for nonbinary chromosomes as well. This approach shuffles variable values between parent chromosomes to produce the offspring. More common approaches combine the variable values from the parents. One approach is to weight the parents and then add them together to produce offspring [2]

$$\begin{aligned} offspring1 &= \beta\,\text{mother} + (1-\beta)\,\text{father} \\ offspring2 &= (1-\beta)\,\text{mother} + \beta\,\text{father} \end{aligned} \tag{2.14}$$

where $0 \le \beta \le 1$. When $\beta = 0.5$, the result is an average of the variables of the two parents. This linear combination process is done for all variables to the right or to the left of some crossover point, or it can be applied to each variable. The variables can be blended by using the same β or by choosing different values for each variable [3]. These blending methods create values of the variables between the values bracketed by the parents. They do not introduce values beyond the extremes already represented in the population.

An extrapolating method finds offspring that have variable values not bounded by the values of the parents. Linear crossover [4] creates values outside the values of the variables by finding three new values given by

$$offspring1 = 0.5 * mother + 0.5 * father$$

$$offspring2 = 1.5 * mother - 0.5 * father$$

$$offspring3 = -1.5 * mother + 0.5 * father$$

Only two of the three are kept. Any value outside the constraints is discarded. Heuristic crossover [5] is a variation where some random number (β) is chosen on the interval [0,1] and the variables of the offspring are defined by [5]

$$offspring = \beta * (mother - father) + mother$$

Another possibility is to generate different β values for each variable. This method also allows generation of offspring outside the values of the two parent variables. Sometimes values are generated outside the allowed range. If this happens, the offspring is discarded and the algorithm tries another β. Quadratic crossover [6] performs a quadratic interpolation of the cost as a function of each variable. Three parents and their costs are necessary to perform a quadratic fit. The offspring from these three parents is the chromosome that corresponds to the minimum of the quadratic.

Continuous single point crossover closely mimics single-point crossover for binary GAs [1]. It begins by randomly selecting a variable as the crossover point within the parent chromosomes:

```
q=ceil(rand*nvar)
```

Let

$$mother = [m_1 m_2 \cdots m_q \cdots m_{nvar}]$$
$$father = [f_1 f_2 \cdots f_q \cdots f_{nvar}]$$

The selected variables are combined to form new variables that will appear in the offspring

$$b_q = m_q - \beta[m_q - f_q]$$
$$g_q = f_q + \beta[m_q - f_q]$$

where β is also a random value between 0 and 1. The final step is to complete the crossover with the rest of the chromosome as before:

$$offspring1 = [m_1 m_2 \cdots b_q \cdots f_{nvar}]$$
$$offspring2 = [f_1 f_2 \cdots g_q \cdots m_{nvar}]$$

If the first variable of the chromosomes is selected, then only the variables to the right of the selected variable are swapped. If the last variable of the chromosomes is selected, then only the variables to the left of the selected variable are swapped. This method does not allow offspring variables outside the bounds set by the parent unless $\beta > 1$.

To demonstrate the concept of continuous variable crossover, consider the following two parents:

$$mother = [1 \quad 2 \quad 3 \quad 4 \quad 5 \quad 6]$$
$$father = [3 \quad 2 \quad 1 \quad 0 \quad 1 \quad 2]$$

First, perform crossover with one random weighting variable:

```
b=rand=0.78642

chrom3=b*chrom1+(1-b)*chrom2

chrom3=[1.4272  2  2.5728  3.1457  4.1457  5.1457]

chrom4=(1-b)*chrom1+b*chrom2

chrom4=[2.5728  2  1.4272  0.85433  1.8543  2.8543]
```

Next, a random weighting variable is tried for each variable in the chromosome:

```
b=rand(1,6)=[0.30415  0.79177  0.22736  0.24999  0.61258  0.61086]
chrom3=b.*chrom1+(1-b).*chrom2
chrom3=[2.3917  2  1.4547  0.99997  3.4503  4.4434]
chrom4=(1-b).*chrom1+b.*chrom2
chrom4=[1.6083  2  2.5453  3  2.5497  3.5566]
```

Linear crossover results in

```
chrom3=0.5*chrom1+0.5*chrom2
chrom3=[2  2  2  2  3  4]
chrom4=1.5*chrom1-0.5*chrom2
chrom4=[0  2  4  6  7  8]
chrom5=-.5*chrom1+1.5*chrom2
chrom5=[4  2  0  -2  -1  0]
```

Heuristic crossover produces

```
b=0.78642
chrom3=chrom1-b*(chrom1-chrom2)
chrom3=[2.5728  2  1.4272  0.85432  1.8543  2.8543]
chrom4=chrom1+b*(chrom1-chrom2)
chrom4=[-0.57284  2  4.5728  7.1457  8.1457  9.1457]
```

Continuous single-point crossover produces

```
a=round(rand*6)
a=5
chrom3=[chrom1(1:a-1)  chrom1(a)-b*(chrom1(a)-
  chrom2(a))  chrom2(a+1:6)]
chrom3=[1  2  3  4  1.8543  2]
chrom4=[chrom2(1:a-1)  chrom2(a)+b*(chrom1(a)-
  chrom2(a))  chrom1(a+1:6)]
chrom4=[3  2  1  0  4.1457  6]
```

2.6 MUTATION

Mutation induces random variations in the population. The mutation rate is the portion of bits or values within a population that will be changed. A binary mutation changes a one to a zero or a zero to a one. The MATLAB commands to accomplish this are

```
pop(mutindx)=abs(pop(mutindx)-1)
```

Mutation for continuous variables can take many different forms. One way is to totally replace the selected mutated value with a new random value

```
pop(mutindx)=rand(1,nmut);
```

This approach keeps all variable values within acceptable bounds. An alternative is to randomly perturb the chosen variable value. Care must be taken to ensure that the values do not extend outside the limits of the variables.

2.7 TERMINATING THE RUN

This generational process is repeated until a termination condition has been reached. Common terminating conditions are

- Set number of iterations.
- Set time reached.
- A cost that is lower than an acceptable minimum.
- Set number of cost function evaluations.
- A best solution has not changed after a set number of iterations.
- Operator termination.

These processes ultimately result in the next-generation population of chromosomes that is different from the initial generation. Generally the average fitness will have increased by this procedure for the population, since only the best chromosomes from the preceding generation are selected for breeding.

REFERENCES

1. R. L. Haupt and Sue Ellen Haupt, *Practical Genetic Algorithms*, 2nd ed., Wiley, New York, 2004.
2. N. J. Radcliff, Forma analysis and random respectful recombination, *Proc. 4th Int. Conf. Genetic Algorithms*, Morgan Kauffman, San Mateo, CA, 1991.

3. L. Davis, Hybridization and numerical representation, in *The Handbook of Genetic Algorithms*, L. Davis, ed., Van Nostrand Reinhold, New York, 1991, pp. 61–71.
4. A. Wright, Genetic algorithms for real parameter optimization, in *Foundations of Genetic Algorithms*, G. J. E. Rawlins, ed., Morgan Kaufmann, San Mateo, CA, 1991, pp. 205–218.
5. Z. Michalewicz, *Genetic Algorithms + Data Structures = Evolution Programs*, 2nd ed., Springer-Verlag, New York, 1994.
6. A. A. Adewuya, *New Methods in Genetic Search with Real-Valued Chromosomes*, Master's thesis, Massachusetts Institute of Technology, Cambridge, 1996.

3

Step-by-Step Examples

The last chapter provided the details of how a GA works. This chapter presents two detailed array optimization examples. The first example applies a GA to one of the cost functions in Chapter 1. As will be seen, the GA outperforms all the traditional methods. The second example exploits the power of the binary GA through optimizing a thinned linear array for the lowest maximum sidelobe level. This example demonstrates the power of using a GA to solve problems with discrete variables. Both examples are presented in minute detail, so the novice can easily follow all the steps.

3.1 PLACING NULLS

The first example is an objective function from Chapter 1: AF_4. This objective function was selected, because it has multiple minima yet has only two variables, so the process can be monitored using plots of the objective function.

Each chromosome contains two variables. The variables have values between zero and one. Since the actual extent of the variables is between 0 and 2π, the objective function must perform the necessary scaling. The binary GA encodes each value using 7 bits. Thus, the binary chromosome contains 14 bits. A uniform random-number generator creates each population. Binary digits are obtained by rounding the output of the random-number generator. Here, we start both GAs at the exact same spots on the objective function

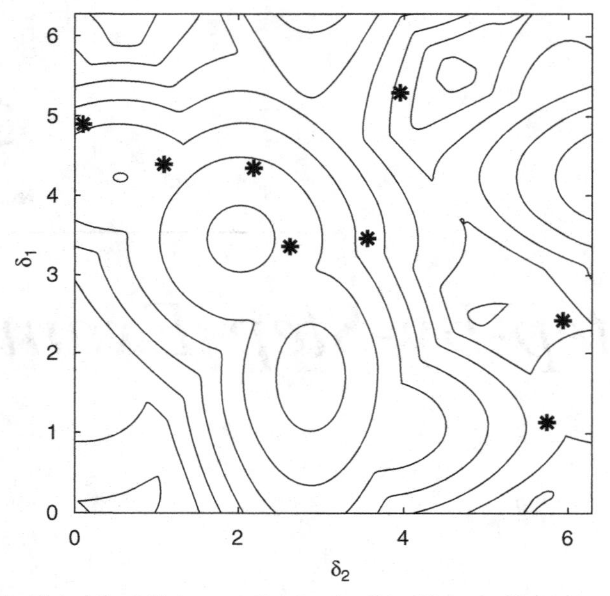

Figure 3.1. *Plot of the initial generation for the GA with and without binary coding.*

TABLE 3.1. Initial Population

GA with Binary Encoding		GA without Binary Encoding		
Chromosome	AF_4	Chromosome		AF_4
01011001011000	1.4502	0.3465	0.6929	1.4502
10010001000110	1.1159	0.5669	0.5512	1.1159
00101101011001	1.3266	0.1732	0.7008	1.3266
00000101100011	1.1597	0.0157	0.7795	1.1597
11110000110001	0.5534	0.9449	0.3858	0.5534
11101000010111	0.6919	0.9134	0.1811	0.6919
10100001101011	0.5362	0.6299	0.8425	0.5362
01101011000100	1.5472	0.4173	0.5354	1.5472

(see Fig. 3.1). The chromosomes are passed to the objective function and the corresponding outputs are listed in Table 3.1. The objective function converts the continuous values into binary form.

The next step sorts the initial population from the lowest to the highest value of the objective function as shown in Table 3.2. Using a 50% discard rate results in discarding the bottom four chromosomes. Note that chromosomes 4 and 5 differ by less than 4% but 4 survives while 5 does not. They do not come from the same area of the objective function, so one area remains active in the population while the other area does not.

Roulette wheel selection will be used to pick two sets of parents that will produce four offspring to replace the discarded chromosomes in the popula-

TABLE 3.2. Initial Population Sorted. (Chromosomes in Italics are Discarded)

GA with Binary Encoding		GA without Binary Encoding		
Chromosome	AF_4	Chromosome		AF_4
10100001101011	0.5362	0.6299	0.8425	0.5362
11110000110001	0.5534	0.9449	0.3858	0.5534
11101000010111	0.6919	0.9134	0.1811	0.6919
10010001000110	1.1159	0.5669	0.5512	1.1159
00000101100011	*1.1597*	*0.0157*	*0.7795*	*1.1597*
00101101011001	*1.3266*	*0.1732*	*0.7008*	*1.3266*
01011001011000	*1.4502*	*0.3465*	*0.6929*	*1.4502*
01101011000100	*1.5472*	*0.4173*	*0.5354*	*1.5472*

tion. The probability of selection is based on the position of the ranked chromosome in the population. As a result,

$$\begin{aligned} p(\text{chromosome } 1) &= \frac{4}{1+2+3+4} = 0.4 \\ p(\text{chromosome } 2) &= \frac{3}{1+2+3+4} = 0.3 \\ p(\text{chromosome } 3) &= \frac{2}{1+2+3+4} = 0.2 \\ p(\text{chromosome } 4) &= \frac{1}{1+2+3+4} = 0.1 \end{aligned} \tag{3.1}$$

In this case, the first chromosome is 4 times more likely to be selected than the last chromosome. Next, generate a uniform random number (r). If $0 \leq r \leq 0.4$, then the first chromosome is selected. If $0.4 < r \leq 0.7$ then the second chromosome is selected, and so on (see Fig. 3.2). For our example, the parent selection parameters are shown in Table 3.3. For the GA with binary encoding, chromosomes 2 and 4 mate with crossover occurring between bits 11 and 12 to produce chromosomes 5 and 6, and chromosomes 3 and 1 mate with crossover occurring between bits 4 and 5 to produce chromosomes 7 and 8. In the GA without binary encoding, chromosomes 2 and 1 mate with crossover occurring at variable 1 to produce chromosomes 5 and 6, and chromosomes 1 and 2 mate with crossover occurring at variable 1 to produce chromosomes 7 and 8. The new crossover values are given by

$$\begin{aligned} 0.9324 &= 0.9449 - 0.03960[0.9449 - 0.6299] \\ 0.6424 &= 0.6299 + 0.03960[0.9449 - 0.6299] \\ 0.7927 &= 0.6299 - 0.5167[0.6299 - 0.9449] \\ 0.7821 &= 0.9449 + 0.5167[0.6299 - 0.9449] \end{aligned} \tag{3.2}$$

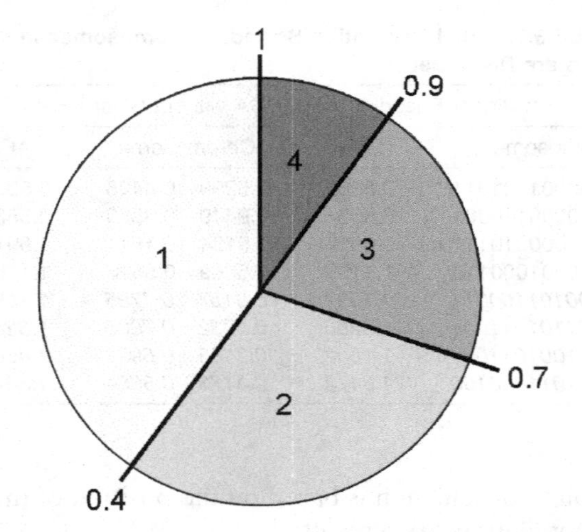

Figure 3.2. *Roulette wheel selection of the top four chromosomes.*

TABLE 3.3. Chromosomes Paired Using Roulette Wheel Selection

GA with Binary Encoding			GA without Binary Encoding		
r	Chromosome	Crossover after Bit	*r*	Chromosome	Crossover at Variable
0.6265	2	11	0.6762	2	1
0.9410	4	11	0.1685	1	1
0.8905	3	4	0.04999	1	1
0.05746	1	4	0.4373	2	1

TABLE 3.4. Population of Generation 1 after Crossover

GA with Binary Encoding		GA without Binary Encoding		
Chromosome	AF_4	Chromosome		AF_4
10100001101011	0.5362	0.6299	0.8425	0.5362
11110000110001	0.5534	0.9449	0.3858	0.5535
11101000010111	0.6919	0.9134	0.1811	0.6919
10010001000110	1.1159	0.5669	0.5512	1.1159
11110000110110	0.5648	0.9324	0.8425	0.7485
10010001000001	1.1426	0.6424	0.3858	1.0555
11101000010011	0.6821	0.7927	0.3858	0.3820
10100001101111	0.5727	0.7821	0.8425	0.3272

Parents and new offspring are shown in Table 3.4. Some of the offspring have lower values of AF_4 than the parents.

Mutation occurs after crossover. A mutation rate of 20% was used for both GAs. Consequently, the GA with binary encoding has

$$\underbrace{14}_{\text{bits}} \times \underbrace{7}_{\text{chromosomes}} \times \underbrace{0.2}_{\substack{\text{mutation}\\ \text{rate}}} = 19.6 \tag{3.3}$$

or 20 bits to be mutated. Only seven chromosomes can be mutated because elitism is used. The mutated (row,column) pairs are given by

$$\begin{matrix} (5,9) & (7,10) & (5,7) & (2,8) & (6,14) & (8,5) & (6,4) & (6,13) & (3,8) & (4,6) \\ (3,6) & (6,4) & (3,8) & (3,9) & (5,11) & (6,3) & (2,2) & (8,1) & (8,10) & (2,2) \end{matrix} \tag{3.4}$$

The GA without binary encoding mutates three variables by replacing them with a uniform random number. These mutated pairs are given by

$$(2,1)(4,2)(7,1) \tag{3.5}$$

Table 3.5 lists the mutated chromosomes and their associated value of AF_4. Several chromosomes improved, while several others lost ground. When the chromosomes are sorted as shown in Table 3.6, new chromosomes moved into the top positions. A major difference in the binary encoding is that all the

TABLE 3.5. Population of Generation 1 after Mutation. (Boldface Indicates a Mutation)

GA with Binary Encoding		GA without Binary Encoding		
Chromosome	AF_4	Chromosome		AF_4
10100001101011	0.5362	0.6299	0.8425	0.5362
1***0***11000***1***110001	0.2619	***0.7157***	0.3858	0.6584
11101***1***0***11***10111	0.6048	0.9134	0.181	0.6919
10010***1***01000110	1.0621	0.5669	***0.4644***	1.2453
111100***1***0***0***1***1***110	0.6588	0.9324	0.8425	0.7485
10***10***00010000***10***	0.8956	0.6424	0.3859	1.0555
111010000***0***0011	0.4016	***0.7164***	0.3858	0.6547
0010***1***0011***1***1111	0.4002	0.7821	0.8425	0.3272

TABLE 3.6. Sorted Results after First Generation

GA with Binary Encoding		GA without Binary Encoding		
Chromosome	AF_4	Chromosome		AF_4
10110001110001	0.2619	0.7821	0.8425	0.3272
00101001111111	0.4002	0.6299	0.8425	0.5362
11101000000011	0.4016	0.7164	0.3858	0.6547
10100001101011	0.5362	0.7157	0.3858	0.6584
11101101110111	0.6048	0.9134	0.1811	0.6919
11110010011110	0.6588	0.9324	0.8425	0.7485
10100001000010	0.8956	0.6424	0.3858	1.0555
10010101000110	1.0621	0.5669	0.4644	1.2453

chromosomes except the top one received at least one mutation while only three chromosomes were mutated in the GA without binary encoding. After the first generation, the binary GA is in the lead.

The second generation begins with the population in Table 3.6 (Figure 3.3). The probabilities of selection for the top four chromosomes do not change. A new set of random numbers are generated to pick the parents and crossover points as shown in Table 3.7. The value of the crossover variable is the same for chromosomes 3 and 4 as well as for chromosomes 1 and 2. As a result, the offspring are exact replicas of the parents. The mutated and sorted population appears in Table 3.8 and Figure 3.4. All the binary encoded chromosomes are unique at this point, but the nonencoded chromosomes have two identical pairs. The binary GA seems to be way ahead at this point.

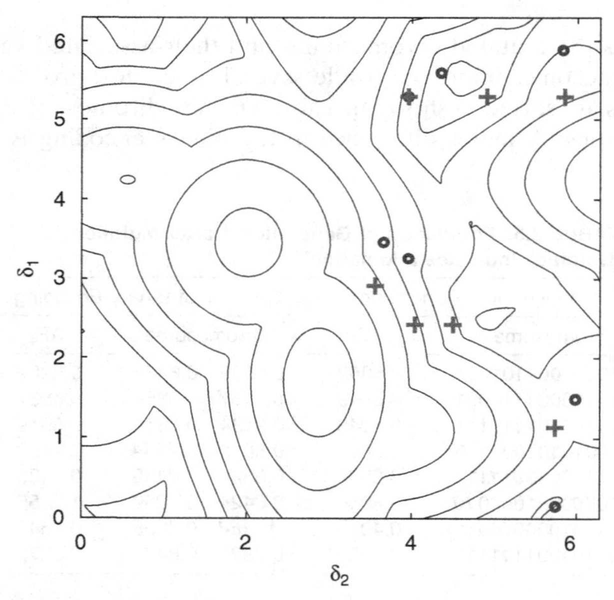

Figure 3.3. *Population after the first generation: o binary coding, + no binary encoding.*

TABLE 3.7. Chromosomes Paired Using Roulette Wheel Selection

GA with Binary Encoding			GA without Binary Encoding		
r	Chromosome	Crossover after Bit	r	Chromosome	Crossover at Variable
0.04064	1	13	0.8062	3	2
0.8391	3	13	0.9494	4	2
0.1394	1	11	0.1743	1	2
0.9086	4	11	0.6071	2	2

TABLE 3.8. Sorted Results after Second Generation

GA with Binary Encoding		GA without Binary Encoding		
Chromosome	AF_4	Chromosome		AF_4
10110011101011	0.2306	0.7821	0.8425	0.3272
10110001110000	0.2503	0.7821	0.8425	0.3272
10110001110001	0.2619	0.7157	0.9473	0.4180
11010101110011	0.4264	0.9223	0.9983	0.4698
11100000001011	0.6396	0.6299	0.8425	0.5362
10100001000011	0.8997	0.6299	0.8425	0.5362
00101001100010	1.1912	0.7164	0.3858	0.6547
10101110010011	1.2526	0.7157	0.3858	0.6584

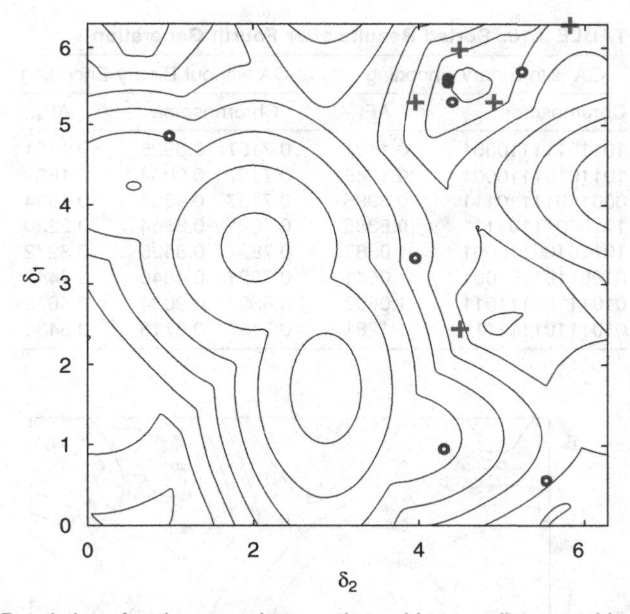

Figure 3.4. *Population after the second generation: o binary coding, + no binary encoding.*

The GAs are run for two more generations. Results are shown in Tables 3.9 and 3.10 as well as in Figures 3.5 and 3.6. We stopped the minimization after four generations, because the GAs were getting very close to the true minimum of 0.11659 at (0.75693,0.87021). Not all the chromosomes ended up in the valley containing the global minimum. Mutations keep the algorithms exploring diverse areas. Figure 3.7 is a plot of the convergence of both GAs. The lower two curves are the best results after each generation. Generation 0 is the initial random population. The upper two curves are the population averages. The best curves either stay the same or go lower than the previous generation. The average curves can increase because some chromosomes may have higher costs.

TABLE 3.9. Sorted Results after Third Generation

GA with Binary Encoding		GA without Binary Encoding		
Chromosome	AF_4	Chromosome		AF_4
10110101110001	0.1746	0.7157	0.9034	0.1634
10110011101011	0.2306	0.7821	0.8864	0.2230
11011101110011	0.5348	0.7821	0.8425	0.3272
00011001110011	0.5827	0.7157	0.9473	0.4180
11110101111000	0.6028	0.9223	0.9983	0.4698
11100001101001	0.6804	0.4647	0.8425	1.0072
10110010110000	0.7523	0.7157	0.2786	1.0146
00110010110000	1.2198	0.5362	0.9473	1.0620

TABLE 3.10. Sorted Results after Fourth Generation

GA with Binary Encoding		GA without Binary Encoding		
Chromosome	AF_4	Chromosome		AF_4
10110111110001	0.1481	0.7157	0.8925	0.1451
10110101110001	0.1746	0.7157	0.9034	0.1634
00011011110111	0.3985	0.7157	0.9034	0.1634
10100011101111	0.5335	0.7821	0.8864	0.2230
10100100110001	1.0387	0.7821	0.8425	0.3272
01001101101001	1.0541	0.7821	0.4340	0.4442
01011101111011	1.0602	0.6397	0.9034	0.5678
01011101111111	1.1281	0.7157	0.6715	0.6430

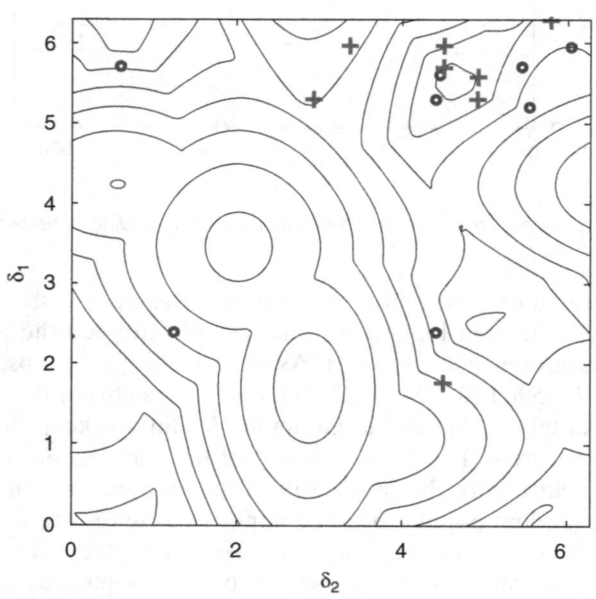

Figure 3.5. *Population after the third generation: o binary coding, + no binary encoding.*

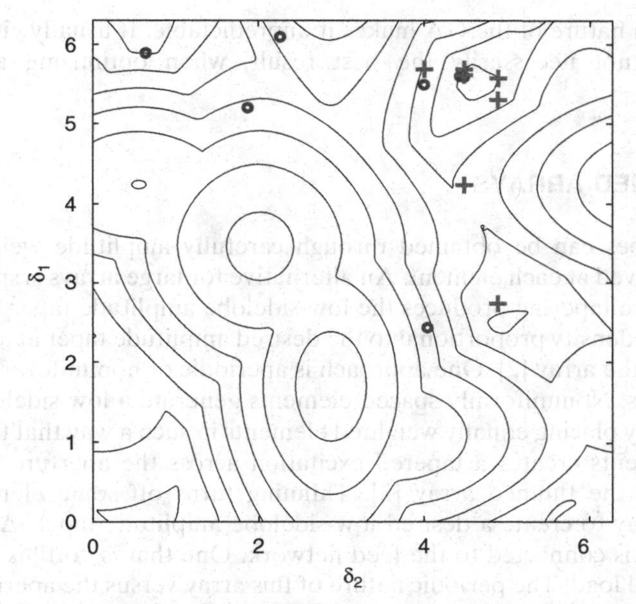

Figure 3.6. Population after the fourth generation: o binary coding, + no binary encoding.

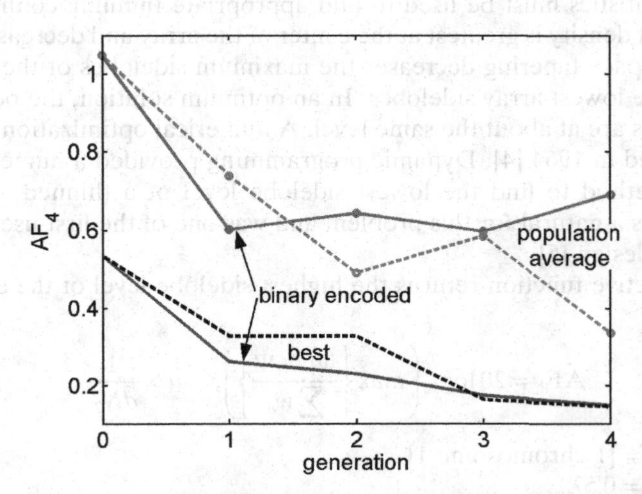

Figure 3.7. Convergence of the GA with (solid lines) and without (dashed lines) binary encoding for optimizing AF_4.

It is important to note that these runs proved successful in a very short period of time. Reinitializing the random-number generator and running the algorithms again will produce different results. Changing the GA parameters, such as population size and mutation rate, will also produce different results.

The random nature of the GA makes it unpredictable. It usually gives a good result, but not necessarily the best result, when optimizing a complex function.

3.2 THINNED ARRAYS

Low sidelobes can be obtained through carefully amplitude weighting the signals received at each element. An alternative for large arrays is space tapering [1]. Space tapering produces the low-sidelobe amplitude taper by making the element density proportional to the desired amplitude taper at a particular location on the array [2]. One approach is aperiodic or nonuniform spacing of the elements. Nonuniformly spaced elements generate a low-sidelobe amplitude taper by placing equally weighted elements in such a way that the spacing of the elements creates a tapered excitation across the aperture. A second approach is the thinned array [3]. Thinning turns off some elements in a uniform array to create a desired low-sidelobe amplitude taper. An element that is "on" is connected to the feed network. One that is "off" is connected to a matched load. The periodic nature of this array versus the aperiodic array facilitates construction of the array.

Since analytical methods do not exist for aperiodic array synthesis, optimization or statistics must be used to find appropriate thinning configurations. The element density is greatest at the center of the array and decreases toward the edges. Space tapering decreases the maximum sidelobes of the array but increases the lowest array sidelobes. In an optimum solution, the peaks of all the sidelobes are at about the same level. A numerical optimization approach was first tried in 1964 [4]. Dynamic programming provided a numerical optimization method to find the lowest sidelobe level of a thinned array. The binary GA is a natural for this problem and was one of the first uses of a GA in antenna design [5].

The objective function returns the highest sidelobe level of the array

$$\mathrm{AF}_{\mathrm{th}} = 20\log_{10}\left(\max\left\{\frac{w\cos\Psi}{\sum w_n}\right\}\right), \quad u > \frac{1}{dN_T} \tag{3.6}$$

where $w = [1 \text{ chromosome } 1]$
$d = 0.5\lambda$
$N_T = 4 + 2N_{\mathrm{cbits}}$
N_{cbits} = number of bits in a chromosome

$$\Psi = \begin{bmatrix} 0.5k\,du_1 & 0.5k\,du_2 & \cdots & 0.5k\,du_M \\ 1.5k\,du_1 & 1.5k\,du_2 & & 1.5k\,du_M \\ \vdots & & \ddots & \vdots \\ (N_T/2-0.5)k\,du_1 & (N_T/2-0.5)k\,du_2 & \cdots & (N_T/2-0.5)k\,du_M \end{bmatrix}$$

Since the array is assumed to be symmetric about its physical center, a cosine is used in the array factor instead of a complex exponential. The center two elements and both edge elements are always turned on. Consequently, the weight vector has its first and last elements equal to one. The center elements are assumed to be on because low-sidelobe amplitude tapers always have a maximum at the center. The edge elements are assumed to be on in order to keep the mainbeam width the same. This is important when searching for the maximum sidelobe level. Dividing by the number of elements turned on normalizes the array factor to its peak. The chromosome is a $1 \times N_{cbits}$ vector full of ones and zeros. Forcing the edge elements to be one establishes a well-defined mainbeam for all possible chromosomes; hence the sidelobe region begins at the first null next to the mainbeam. A good estimate of the location of the first null is $u = 1/(dN_T)$.

An exhaustive search of all possible chromosomes requires $2N_{cbits}$ objective function evaluations. This approach quickly bogs down in computer time. A graph of the lowest sidelobe level as a function of N_T was found through an exhaustive search for $16 \le N_T \le 44$ and is plotted in Figure 3.8. For the most part, the sidelobe level decreases as N_T increases. Calculating AF_{th} for all possible thinning combinations when $N_T = 44$ took approximately 20 min. The computation time approximately doubles every time N_T increments by 2. If this holds true as N_T increases, then to find AF_{th} for $N_T = 62$ would take approximately one week.

A binary GA for an array with $N_T = 62$ has an initial population as given in Table 3.11. The population after sorting appears in Table 3.12. None of the chromosomes are better than a uniform array. Parents are found through tournament selection. Two contenders are randomly selected. The one with

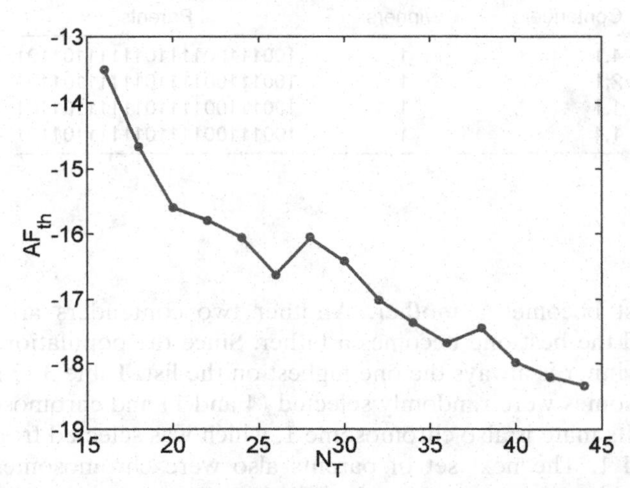

Figure 3.8. *The minimum value of AF_{th} versus N_T was found using exhaustive search.*

TABLE 3.11. Initial Population for Array Thinning

Chromosome	AF_{th}
100111001110000111010000	–9.071
101101000110110000011001	–7.267
010101101010011101010101	–7.252
100111001111011111101101	–10.722
010010111010000101101000	–8.374
010101010010100100100011	–9.557
100101010111100010000111	–8.167
101001000011101001010000	–5.734

TABLE 3.12. Population after Sorting

Chromosome	AF_{th}
100111001111011111101101	–10.722
010101010010100100100011	–9.557
100111001110000111010000	–9.071
010010111010000101101000	–8.374
100101010111100010000111	–8.167
101101000110110000011001	–7.267
010101101010011101010101	–7.252
101001000011101001010000	–5.734

TABLE 3.13. Parents for Generation 1 Were Found Using Tournament Selection

Contenders	Winners	Parents
4,1	1	100111001111011111101101
2,1	1	100111001111011111101101
1,1	1	100111001111011111101101
1,4	1	100111001111011111101101

the best cost becomes a mother. Another two contenders are randomly selected, and the best one becomes a father. Since the population is already sorted, the winner is always the one highest on the list. Table 3.13 shows that two chromosomes were randomly selected (4 and 1) and chromosome 1 was the winner. Its mate is also chromosome 1, which was selected from chromosomes 2 and 1. The next set of parents also were chromosomes 1 and 1. Random masks were generated and applied to the parents in order to create the offspring (see Table 3.14). In this case, all the parents were chromosome 1, so all the offspring are also chromosome 1. No improvements result from these crossovers. A 5% mutation rate results in 9 random bit changes in the population outside of the top chromosome. Table 3.15 lists the mutated chromosomes and their costs. Two of the mutated chromosomes result in AF_{th} values higher than those of the best value in the initial population. Table 3.16 displays the sorted population at the end of generation 1 and the beginning of generation 2.

Generation 2 starts with a new tournament as shown in Table 3.17. This time there is some diversity in the parents. Creating new masks and perform-

TABLE 3.14. Random Masks Are Generated for Uniform Crossover to Create Offspring in Generation 1

Random Mask	Offspring	AF_{th}
001110100110000110010010	100111001111011111101101	−10.722
110001011001111001101101	100111001111011111101101	−10.722
011010100011001110110101	100111001111011111101101	−10.722
100101011100110001001010	100111001111011111101101	−10.722

TABLE 3.15. Population of Generation 1 after Mutation. (Boldface Indicates a Mutation)

Chromosome	AF_{th}
100111001111011111101101	−10.722
01010101001**1**100100100011	−9.6814
1**1**01110011**0**00001110**0**0000	−8.1671
010010111010000101101000	−8.374
1001110**1**1111011111101101	−11.378
10011100111101111110110**0**	−10.857
100111001111011111101101	−10.722
1001**0**1**1**01111**1**11111101101	−9.7262

TABLE 3.16. Population after Sorting Generation 1

Chromosome	AF_{th}
100111011111011111101101	−11.378
100111001111011111101100	−10.857
100111001111011111101101	−10.722
100111001111011111101101	−10.722
100101101111111111101101	−9.726
010101010011100100100011	−9.681
010010111010000101101000	−8.374
110111001100000111000000	−8.167

TABLE 3.17. Parents for Generation 2 Were Found Using Tournament Selection

Contenders	Winners	Parents
3,3	3	100111001111011111101101
3,1	1	100111011111011111101101
1,1	1	100111011111011111101101
3,2	2	100111001111011111101100

TABLE 3.18. Random Masks Are Generated for Uniform Crossover to Create Offspring in Generation 2

Random Mask	Offspring	AF_{th}
000101111011011000001111	100111001111011111101101	–10.722
111010000100100111110000	100111011111011111101101	–11.378
011011000101111001100101	100111001111011111101101	–10.722
100100111010000110011010	100111011111011111101100	–11.378

TABLE 3.19. Population of Generation 2 after Mutation. (Boldface Indicates a Mutation)

Chromosome	AF_{th}
100111011111011111101101	–11.378
10011100111101**0**111101100	–12.12
1001110**1**1111011111101101	–11.378
1001110011110111**0**1101101	–11.679
10011**0**001111011111101101	–9.567
100111**1**111110**0**1111**0**01101	–12.987
100111001111011111**0**01101	–11.962
10011**0**011111011111101100	–10.138

TABLE 3.20. Population after Sorting Generation 2

Chromosome	AF_{th}
100111111111001111001101	–12.987
100111001111010111101100	–12.12
100111001111011111001101	–11.962
100111001111011101101101	–11.679
100111011111011111101101	–11.378
100111011111011111101101	–11.378
100110011111011111101100	–10.138
100110001111011111101101	–9.567

ing the mating results in four new offspring (Table 3.18). Although the offspring are not all the same, they are just replicas of already existing parents. Consequently, mutation will provide the needed population diversity as before. Table 3.19 shows the mutated bits in the population. Table 3.20 ranks the mutated population in preparation for generation 3. After two generations, the peak sidelobe level has decreased by 2 dB in the best chromosome. So far, mutation provides the diversity while crossover results in replication of only some of the good chromosomes. A uniform array still has lower sidelobes at this point.

TABLE 3.21. Final Population after Generation 31

Chromosome	AF_{th}
1111111111111101111001110	−18.35
1111111110111101111001111	−16.652
1111111110111101111001110	−16.087
1111111110111101111001110	−16.087
1111111110111100111001111	−15.556
1111111110111111111001111	−15.118
1110111110111101111001111	−14.231
1111111110011101111001101	−13.664

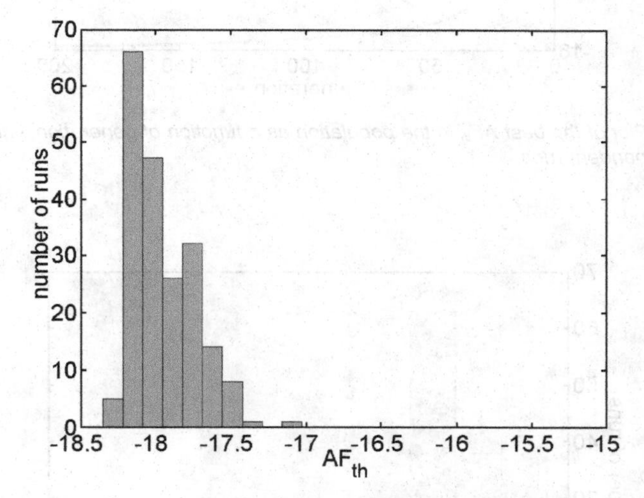

Figure 3.9. *The GA performance was averaged over 200 runs. This plot shows the number of these runs that attained a best value of AF_{th} after 200 generations.*

The GA continues this process. At generation 31, it finds the optimum solution. Table 3.21 presents the final population with their values of AF_{th}. If we assume that seven different chromosomes are mutated every generation (the maximum possible), the number of function evaluations needed to find the minimum in this example is

$$\text{Function evaluations} = 8 + 31 \times 7 = 225 \tag{3.7}$$

The GA does not always perform this well. The random variables produce different results every independent run. Some are good and some are not. To demonstrate the varied performance of the GA, 200 independent runs were made and the results analyzed. Figure 3.9 is a bar graph of the number of runs that attained a given best value for AF_{th}. Over half of the time, the GA found a thinned configuration that had a best result with a peak sidelobe level below

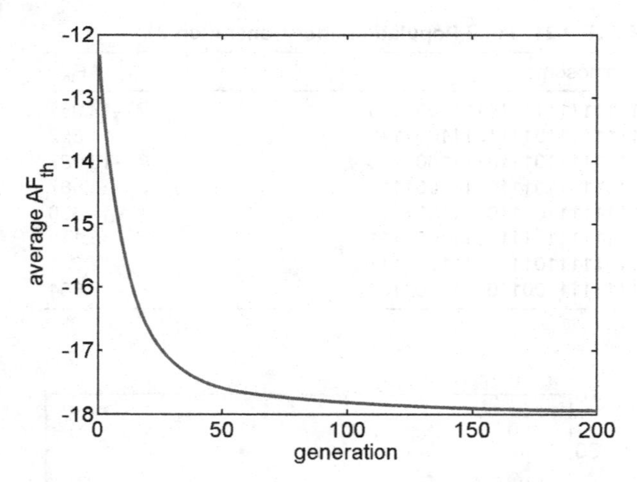

Figure 3.10. Plot of the best AF_{th} in the population as a function of generation when averaged over 200 independent runs.

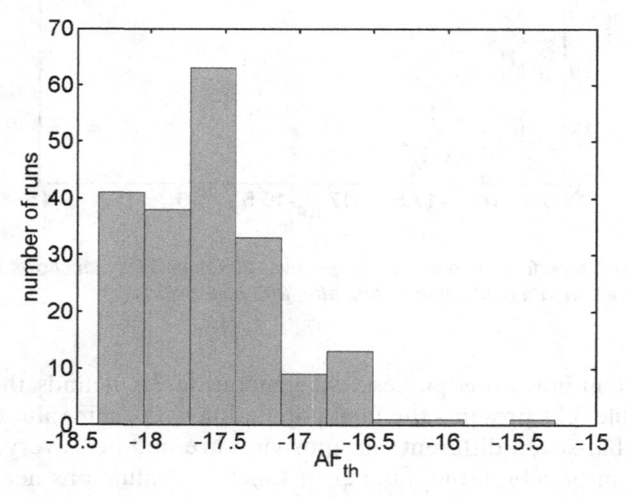

Figure 3.11. Limiting the GA to 50 generations and performing 200 independent results produces this histogram.

−18 dB. Figure 3.10 is a graph of the best result averaged over the 200 independent results for each generation. On the average, the GA starts out fast and then slows down. Decreasing the maximum number of generations to 50 significantly decreases the performance as shown in Figure 3.11. Stopping too soon can deny you the solution you seek. Figure 3.12 shows a small improve-

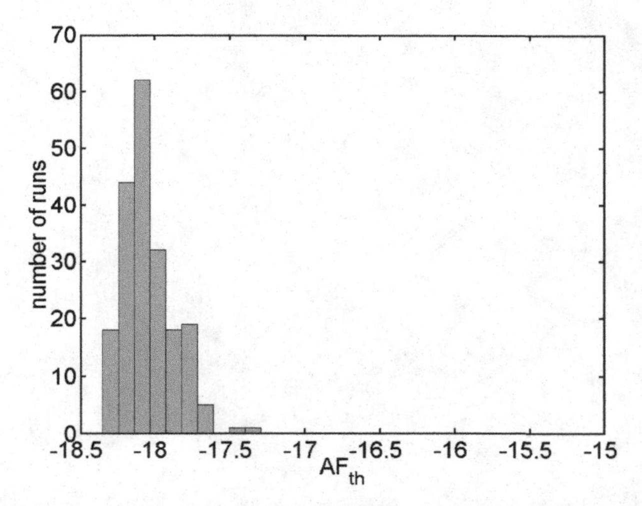

Figure 3.12. *Limiting the GA to 500 generations and performing 200 independent results produces this histogram.*

ment when the number of generations increases to 500. In particular, there are about 20 solutions in Figure 3.11 that are worse than the worst solution in Figure 3.12.

The following chapters cover a wide range of applications of GAs in electromagnetics but do not provide as much detail as this chapter. In most cases, the GAs operate very similar to the ones used in this chapter. Only the cost functions are much more complex and difficult to compute. The Appendix presents some MATLAB files for binary and continuous GAs.

REFERENCES

1. H. Unz, Linear arrays with arbitrarily distributed elements, *IEEE AP-S Trans.* **8**:222–223 (March 1960).
2. R. Harrington, Sidelobe reduction by nonuniform element spacing, *IEEE AP-S Trans.* **9**:187–192 (March 1961).
3. R. E. Willey, Space tapering of linear and planar arrays, *IEEE AP-S Trans.* **10**: 369–377 (July 1962).
4. M. Skolnik, G. Nemhauser, and J. Sherman, III, Dynamic programming applied to unequally spaced arrays, *IEEE AP-S Trans.* **12**:35–43 (Jan. 1964).
5. R. L. Haupt, Thinned arrays using genetic algorithms, *IEEE AP-S Trans.* **42**:993–999 (July 1994).

4

Optimizing Antenna Arrays

An array response based on the weighting and placement of isotropic point sources is called the *array factor* and, for a linear array, is given by

$$\mathrm{AF} = \sum_{n=1}^{N} w_n e^{j\Psi_n} \tag{4.1}$$

where $w_n = a_n e^{j\alpha_n}$ = element weights
Ψ_n = phase due to element position and observation direction
N = number of elements in the array

In many situations, it is important to take into account the type of antenna elements used in the array. The array factor can be converted to a far-field antenna radiation pattern by the following expression

$$\mathrm{FF} = \mathrm{EP} \times \mathrm{AF} \tag{4.2}$$

where EP (element pattern) is an angular description of the far field (FF) of a single antenna in the array, assuming that all the elements in the array have identical patterns. A full-wave solution must be used to take into account coupling between elements and any scattering from the environment, such as a ground plane. One effect of mutual coupling is that EP is different for every element, so (4.2) cannot be used.

The cost function for the GA normally includes some aspect of the array factor or antenna pattern. Including element impedance or feed network

Genetic Algorithms in Electromagnetics, by Randy L. Haupt and Douglas H. Werner

design is also important when practical. Optimizing the amplitude weights is not common, because there are many linear methods available for array factor synthesis based on amplitude weights. Optimizing the amplitude weights is important when certain restrictions on the weights or pattern make linear synthesis difficult. The element phases and positions are contained in the arguments of the exponents, so linear methods do not apply and GAs work well.

Arrays come in many different configurations. A linear array along the x axis has an array factor given by

$$\mathrm{AF} = \sum_{n=1}^{N} w_n e^{jkx_n \sin\theta} \tag{4.3}$$

and a linear array along the z axis has an array factor given by

$$\mathrm{AF} = \sum_{n=1}^{N} w_n e^{jkz_n \cos\theta} \tag{4.4}$$

Amplitude-only synthesis has $a_n = 0$, while phase-only synthesis implies $a_n = 1$. If the elements are equally spaced, then

$$x_n, z_n = (n-1)d \tag{4.5}$$

Symmetry allows some simplification of the array factor equation. Planar and conformal array factors have array factor expressions that satisfy the element locations of the particular array. Steering the array mainbeam is possible by forcing the sum of Ψ_n and the element weight phase to equal zero at the desired steering angle. For instance, a linear array along the x axis has its mainbeam pointing in the direction θ_s when each element has a phase shift given by

$$-kx_n \sin\theta_s \tag{4.6}$$

The use of GAs for solving complex antenna array optimization problems has been and continues to be an extremely active area of research. In fact, the body of published work dealing with this topic of research is considerably larger than any other category related to applications of GAs in electromagnetics. This chapter presents an overview of the various methodologies that have been developed for the optimization of antenna arrays using GAs, along with several illustrative design examples.

4.1 OPTIMIZING ARRAY AMPLITUDE TAPERS

As mentioned before, many excellent analytical techniques exist for synthesizing low-sidelobe amplitude tapers for arrays. A binary GA is appropriate to use when the amplitude weights are quantized [1]. When the quantization is one bit, then the array is thinned. Array thinning will be discussed in section 4.4.1.

Another interesting example is the use of a GA to synthesize low sidelobe levels in almost uniformly excited arrays. It has been shown [2] that the sidelobe level of a uniformly excited array can be lowered significantly by reducing the excitation amplitudes on only a relatively small number of elements throughout the array. A GA method has been introduced [2] for determining which amplitudes to reduce and by how much. The amplitude weights of a symmetric linear array consisting of 50 uniformly excited elements spaced a half-wavelength apart results in the radiation pattern represented by the solid curve plotted in Figure 4.1. The sidelobe level in this case is approximately −13.3 dB. Next, a GA is used to optimize this 50-element array with the objective of achieving a −20 dB sidelobe level by modifying (i.e., reducing) the excitation amplitudes of only 10 elements. The radiation pattern of the optimized array is shown by the dashed curve in Figure 4.1, which has a sidelobe level of −19.0 dB. The corresponding current distribution on the optimized array is shown in Figure 4.2. Sidelobes close to the main beam were lowered at the expense of increasing the sidelobes everywhere else.

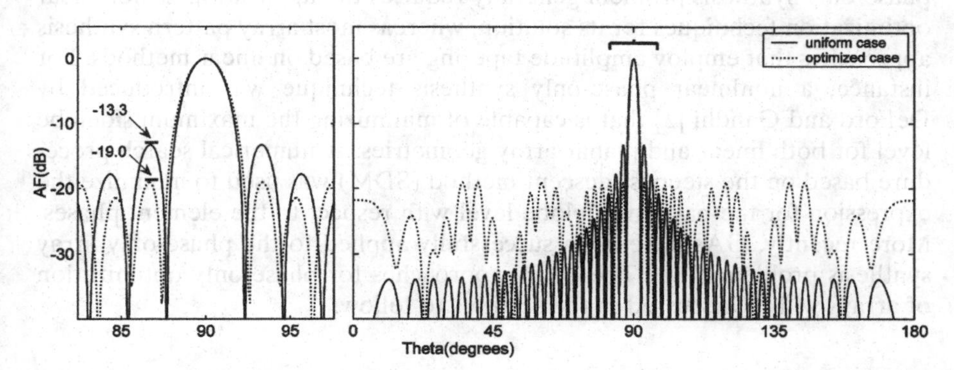

Figure 4.1. *Array factor comparison of two 50-element arrays with 0.5λ periodic spacing between elements. The first array has uniform current excitation and possesses a −13.27 dB sidelobe level. The second array is optimized such that 10 elements of the array have reduced current magnitudes. This array has a sidelobe level of −19.04 dB.*

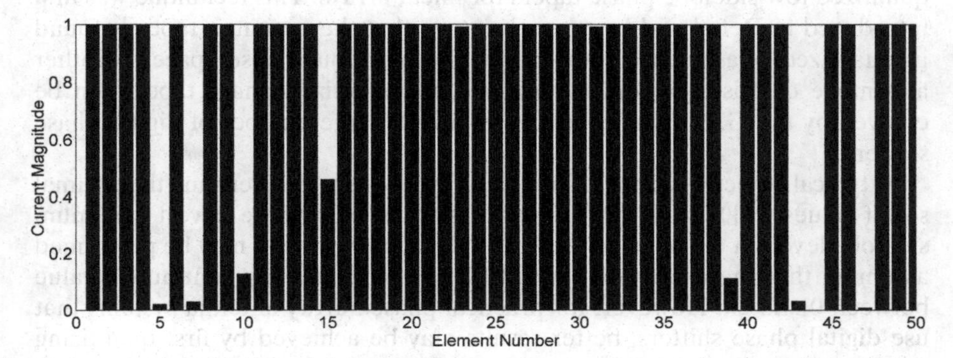

Figure 4.2. *Optimized current magnitude excitation for an almost uniformly excited array with 50 elements and a 0.5λ periodic spacing between elements.*

4.2 OPTIMIZING ARRAY PHASE TAPERS

As discussed in Chapter 1 and in Section 4.1, the conventional approach for synthesizing array far-field radiation patterns with low-sidelobes has been to use some form of amplitude tapering. However, it is also possible to synthesize low-sidelobe radiation patterns by setting the element amplitudes to one and varying the phases. This process of phase tapering is known as *phase-only array synthesis*. One of the main advantages of phase-tapered arrays is the relatively simple feed network that is required compared to amplitude tapered arrays. More specifically, instead of amplitude weighting the signal to each element through a complex feed network, the signal to each element can be phase weighted via the beam steering phase shifters that are already part of the antenna array.

Despite its potential advantages for practical phased array systems, only a relatively small amount of available literature has been devoted to the subject of phase-only array pattern synthesis. This is due primarily to the fact that the phase-only synthesis problem generally requires the application of nonlinear optimization techniques for its solution, whereas most array pattern synthesis approaches that employ amplitude tapering are based on linear methods. For instance, a nonlinear phase-only synthesis technique was introduced by DeFord and Gandhi [3] that is capable of minimizing the maximum sidelobe level for both linear and planar array geometries. A numerical search procedure based on the steepest-descent method (SDM) was used to minimize the expression for the pattern sidelobe level with respect to the element phases. More recently, GAs have been successfully applied to the phase-only array synthesis problem [4–8]. These GA approaches for phase-only optimization of arrays will be discussed in the sections to follow.

4.2.1 Optimum Quantized Low-Sidelobe Phase Tapers

In this section we consider the application of a GA to determine optimum quantized low-sidelobe phase tapers for linear arrays. This technique was first introduced in Ref. 4 and has the advantage that the optimum taper is found in quantized phase space rather than in continuous phase space. Another advantage of this technique is that optimal quantized phase tapers can be evolved by the GA for arrays with a relatively large number of digital phase shifters.

A typical objective of phase-only array synthesis is to determine the optimal set of values $\{\alpha_n\}_{n=1}^{N}$ such that the array factor results in the lowest maximum sidelobe level. In the most general case, the array synthesis may be performed assuming that the excitation current phases can have any continuous value between 0 and 2π. However, for practical phased array antenna systems that use digital phase shifters, better results may be achieved by first quantizing the excitation phases and then performing the optimization. An example is presented next that illustrates how a GA can be used to determine an optimum quantized phase taper for a linear array.

Let us start by considering a uniformly excited and equally spaced linear array of N elements as illustrated in Figure 4.3. If we assume that the individual array elements are short dipole antennas oriented along the x axis, then their element patterns can be approximated by

$$\mathrm{EP}(\theta) = \cos\theta \tag{4.7}$$

The phase settings for a digital phase shifter with B bits may be expressed in the form

$$\alpha_n = [b_1^n + b_2^n 2^{-1} + \cdots + b_B^n 2^{(B-1)}]\pi \tag{4.8}$$

where the coefficients $b_1^n, b_2^n, \ldots, b_B^n$ represent the binary sequence associated with a specific phase setting for the nth element of the array. Moreover, in the context of a GA, this binary encoding of the phase provides a convenient way to express the genes. For example, suppose that a phase shifter on the first element of an array with a 3-bit accuracy is set to 90°, then the gene in this case would be

$$gene = [b_1^1 b_2^1 b_3^1] \tag{4.9}$$

A chromosome is formed by placing several genes together in a sequence. If we consider a four-element array with phase settings of 90°, 0°, 270°, and 90°, then it has a chromosome representation of

$$chromosome = [b_1^1 b_2^1 b_3^1 b_1^2 b_2^2 b_3^2 b_1^3 b_2^3 b_3^3 b_1^4 b_2^4 b_3^4] = [010000110010] \tag{4.10}$$

Hence, an individual chromosome contains the binary encodings for one possible quantized phase taper.

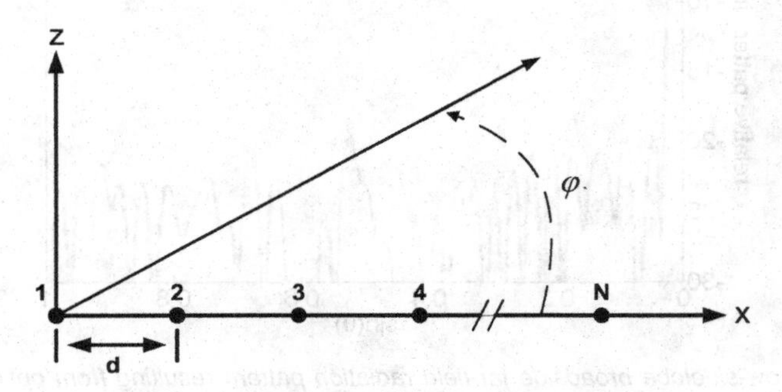

Figure 4.3. Geometry for an N-element linear array.

Next we consider using a binary GA to evolve the optimal phase tapers that produce the lowest possible maximum relative sidelobe level for a 70-element array with 3-bit phase shifters (i.e., $N = 70$ and $B = 3$). During the evolutionary process, the sidelobe performance of each candidate array is evaluated and ranked according to its fitness. The optimum phase taper that results in a −19.7 dB maximum relative sidelobe level for this array is shown in Figure 4.4, while the corresponding far-field radiation pattern is graphed in Figure 4.5.

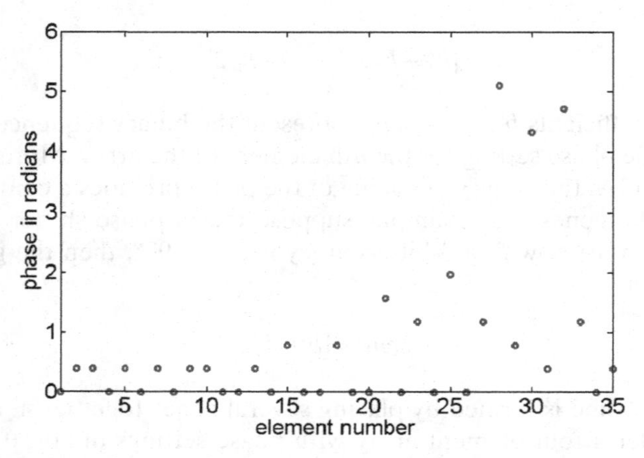

Figure 4.4. *Plot of optimum phase weights as determined from a GA.*

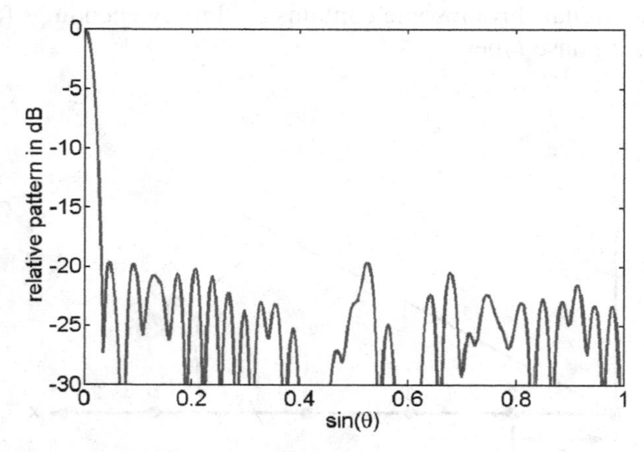

Figure 4.5. *Low-sidelobe broadside far-field radiation pattern resulting from optimized phase taper on a 70-element uniformly excited array.*

4.2.2 Phase-Only Array Synthesis Using Adaptive GAs

It was shown in Section 4.2.1 that it is possible to lower the maximum sidelobe level of a uniformly excited array through phase-only adjustment of the element excitations by using a GA. In this section we will demonstrate that it is also possible to use a GA to lower the sidelobes via phase-only optimization for an array with some fixed (i.e., built-in) amplitude taper. A GA with an adaptive mutation rate is used to speed convergence for phase-only array synthesis problems.

The time-consuming cost function evaluations required in many electromagnetic synthesis problems has prompted research into ways to reduce the number of iterations and function evaluations required for GAs. It has been observed that the best choice for GA parameters at the beginning of an optimization, when the solution has not yet coalesced, usually differs from the best parameters to use toward the end of the optimization, when the final answer is being refined. This observation has led to the development of adaptive GA schemes, where two broad categories of supervisory parameter control have been identified [9]. The first category, often referred to as *self-adaptive parameter control*, involves encoding algorithm parameters into the candidates and allowing them to evolve as the optimization proceeds. On the other hand, the second category involves adjusting parameters using time-varying quantities such as iteration number, population diversity, solution quality, or relative improvement.

Adaptive GAs have only recently been introduced to the electromagnetics community. Most of the emphasis to date has been placed on the development and application of adaptive GA schemes that considerably speed up the convergence for array synthesis problems [5–7]. Boeringer and Werner [5] introduced an adaptive GA for phase-only array synthesis that works by toggling between a small set of parameters during the optimization process to maximize relative improvement. More recently, a real-valued GA has been presented [6,7] that is capable of simultaneously adapting several parameters such as mutation rate, mutation range, and number of crossovers. This adaptive GA was shown to outperform its static counterparts when used to synthesize the phased array weights (amplitude-only, phase-only, or complex) required to satisfy specified far-field sidelobe constraints.

Let us consider the phase-only array synthesis technique presented in Ref. 5, which is based on an adaptive mutation parameter toggling GA. The application chosen to illustrate this technique is determination of the optimal phase-only array weights required to best meet a specified far-field sidelobe requirement, which includes the presence of several 50 dB notches in the radiation pattern. Suppose that we have a linear array consisting of 100 antenna elements uniformly spaced a half-wavelength apart along the z axis. In this case the expression for the far-field radiation pattern (4.2), based on the form of the array factor given in (4.4), reduces to

$$\mathrm{FF}(\theta)=\sqrt{\sin^{\nu}\theta}\sum_{n=1}^{N=100}a_n e^{j[(n-1)\pi\cos\theta+\alpha_n]} \tag{4.11}$$

It will also be assumed for this example that the elements have a 30 dB Taylor taper [10]. Finally, the element pattern is a function of ν, where $\nu = 0$ would represent ideal isotropic array elements and $\nu > 0$ represents directive array elements. Various values for ν are cited in the literature for different radiating element types; for example, $\nu = 1.5$ [11] or $\nu = 1.6$ [12] have been shown to provide good approximations for the patterns of dipole antenna array elements. Moreover, the value of ν is often chosen to match the half-power beamwidth of the radiating element under consideration [13,14]. An ideal value of $\nu = 1.0$ will be assumed here for the purposes of this example.

A continuous GA is used to optimize the excitation phases α_n. The cost function is the sum of the squares of the far-field magnitude above the specified sidelobe envelope. This penalizes sidelobes that are above the envelope, while neither penalty nor reward is given for sidelobes that fall below the specification.

For the adaptive toggling GA considered here, an initial population of 100 random sets of phase weights is generated and scored. The scoring process assigns more importance or weight to those solutions that better meet the notched sidelobe specification. For the generations that follow, 20 pairs of parents are chosen by tournament, whereby each parent is selected as the best of five randomly chosen from the best ten candidates. A simple single-point crossover scheme is employed to produce two children from each of the 20 sets of parents. These 40 children have mutations applied according to the adaptive toggling procedure described below. On completion of this process, the resulting 40 children are evaluated for their performance and assigned a corresponding value of fitness. Of the 140 total individuals (100 parents plus 40 children), the best-scoring 100 survive to the next generation, and the process repeats itself. The termination criterion in this case is to take the best-scoring individual after 1000 generations as the final phase-only distribution for the array.

The mutation operation within the GA is accomplished by replacing the phase of an individual radiating element with a randomly chosen value. Moreover, the mutation rate and range of the mutation process are adaptively controlled during the course of the optimization. The mutation rate in this case is defined to be the probability that any given radiating element will be assigned a new phase weight, while the mutation range governs how far a mutated element phase weight may be from its original (premutation) value. For the adaptive toggling scheme considered first, the algorithm has the freedom to choose between a mutation rate of 2%, 4%, or 6%. On the other hand, if the mutation range is r (where $0 \le r \le 1$), the original value is x, and the allowable weight limits are $x_{\min}$ and $x_{\max}$, then the mutated value is chosen with uniform probability from the range

$$x_{\text{new}} \in \left\{ \max\left(x - r\frac{x_{\max} - x_{\min}}{2}, x_{\min}\right), \quad \min\left(x + r\frac{x_{\max} - x_{\min}}{2}\right), x_{\max} \right\} \quad (4.12)$$

Note that since the range determined from (4.12) is centered on its original value x, smaller mutation ranges imply a more local search, while larger mutation ranges are better suited for introducing diversity into the population. First we will limit the adaptive toggling algorithm to choosing a mutation range from one of three possible values, namely, 0.009, 0.09, or 0.9.

To begin the adaptive toggling process, the algorithm will start off by alternating between 2% and 4% mutation rates as well as between 0.9 and 0.09 mutation ranges. Hence, there are four possible combinations, each of which is tested 8 times, yielding a total of 32 generations. On completion of these first 32 generations, the 16 results obtained using a 2% mutation rate and the 16 results obtained using a 4% mutation rate are compared. If the 4% case provides better average cost function improvement, then, for the next 32 generations the algorithm will toggle between mutation rates of 4% and 6%. Otherwise, if the average cost function did not improve, then the algorithm will continue to toggle between 2% and 4% mutation rates. An analogous procedure is followed for the mutation range, where the 16 results for a 0.9 mutation range are compared to the 16 results for a 0.09 mutation range. If the 0.09 case provides greater average cost function improvement, then for the next 32 generations the algorithm will toggle between mutation ranges of 0.09 and 0.009. On the other hand, if the average cost function did not improve, then the algorithm will continue to toggle between a mutation range of 0.9 and 0.09. Choosing to update the parameters every 32 generations provides some averaging of the results, which helps mitigate the inherent noisiness that would otherwise be associated with improvements in the cost function.

An example of a far-field radiation pattern with three 50 dB notches that was synthesized using the adaptive mutation parameter toggling GA scheme described above is shown in Figure 4.6. The 30 dB Taylor weights as well as the optimized phase weights are also shown in Figure 4.6, as the left and right insets, respectively. The discrete mutation rate and mutation range parameter values chosen by the adaptive toggling GA for this example are shown in Figure 4.7. We see from Figure 4.7a that the mutation range tends downward as the solution is refined, which is consistent with the intuitive notion that diversity should be greatest at the beginning and less toward the end of the evolutionary process. Finally, Figure 4.8 compares the convergence performance of the adaptive toggling algorithm (solid line) to the nine static cases (dashed lines) corresponding to holding constant each of the possible combinations of mutation rate and mutation range. The comparisons made in Figure 4.8 clearly demonstrate that the adaptive toggling technique has the fastest convergence.

The implementation of this method can be easily generalized to include toggling between more than three choices, which can produce even further improvement in the results. Extending the concept outlined above from 3

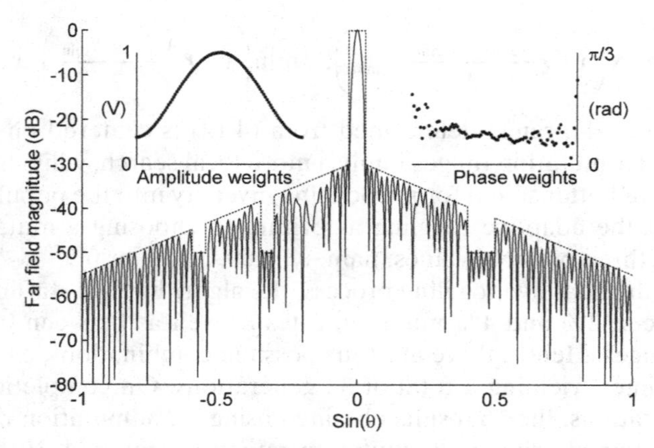

Figure 4.6. Optimized far-field pattern with sidelobe notches. Left inset—amplitude weights (constrained to be 30 dB Taylor weights in this case); right inset—optimized phase weights.

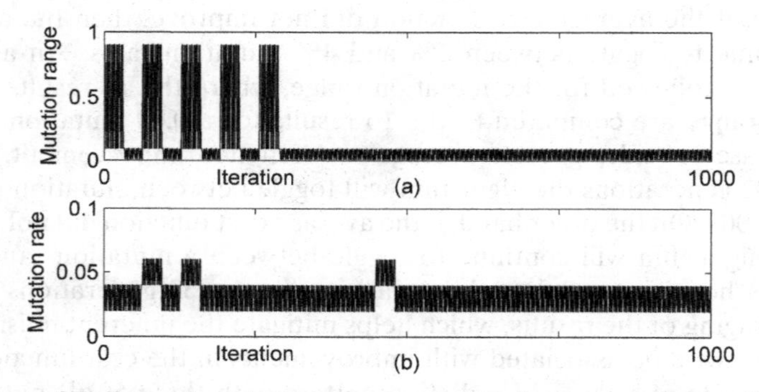

Figure 4.7. Parameter values chosen by adaptive toggling genetic algorithm.

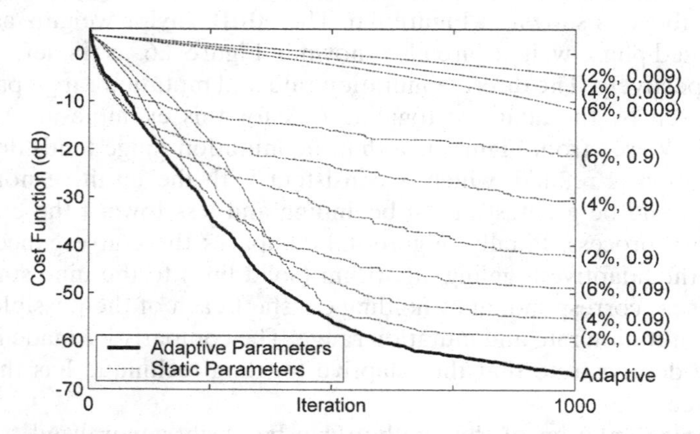

Figure 4.8. Adaptive toggling of parameters outperforms static parameter choices. The values in parentheses for each static parameter combination are the mutation rate and range, respectively.

parameter choices to 10 parameter choices, the same example is synthesized more quickly, or more accurately for an equivalent number of iterations. In this case the possible mutation rates are 10 equally spaced values ranging from 2% to 6% and the mutation ranges are 10 logarithmically spaced values ranging from 0.009 to 0.9, so for both parameters the range of values is the same as previously considered but there are finer gradations. The optimized far-field pattern that results after the same number of iterations is shown in Figure 4.9, where the three 50 dB notches are somewhat cleaner than those of Figure 4.6 obtained using just three parameter choices. The newly optimized phase weights are shown in the right inset of Figure 4.9. Both patterns meet the sidelobe requirements and clearly show the specified notches. The discrete mutation rate and mutation range parameter values chosen by the adaptive toggling GA for this example are shown in Figure 4.10. We see from Figure

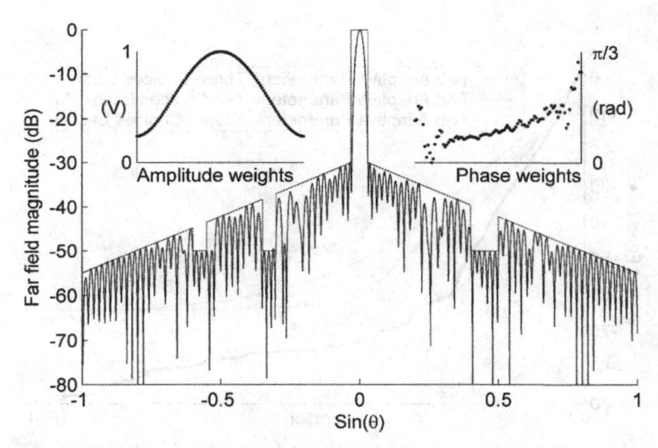

Figure 4.9. Optimized far-field pattern with sidelobe notches, using 10 parameter choices for slightly better results than those of Figure 4.6 using three parameter choices. Left inset—amplitude weights (constrained to be 30 dB Taylor weights in this case); right inset—optimized phase weights.

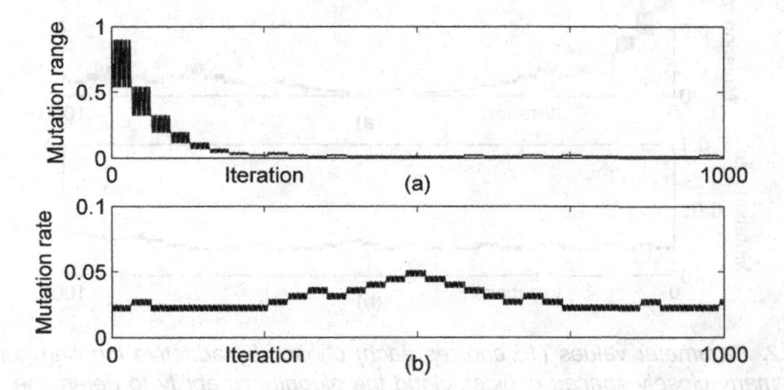

Figure 4.10. Parameter values (10 choices each) chosen by adaptive toggling genetic algorithm.

4.10a that again the mutation range tends downward as the solution is refined. Figure 4.11 compares the convergence performance of the adaptive toggling algorithm using 10 choices (thick solid line) to the adaptive toggling algorithm using 3 choices as previously described (thin solid line) and 15 choices equally spaced along the same range (dashed line). The comparisons made in Figure 4.11 clearly demonstrate that the adaptive toggling technique with 10 choices has the fastest convergence. Having more choices is not always better; as illustrated by the case with 15 choices, if there are too many choices, then they will necessarily be closely spaced, and the algorithm can stagnate because the performance of one choice may not be consistently discernable from the nearby neighboring choices and the algorithm cannot readily choose the best direction to adjust the parameters. This aimless wandering can be seen in the parameter choices taken when there are 15 possibilities as shown in Figure 4.12.

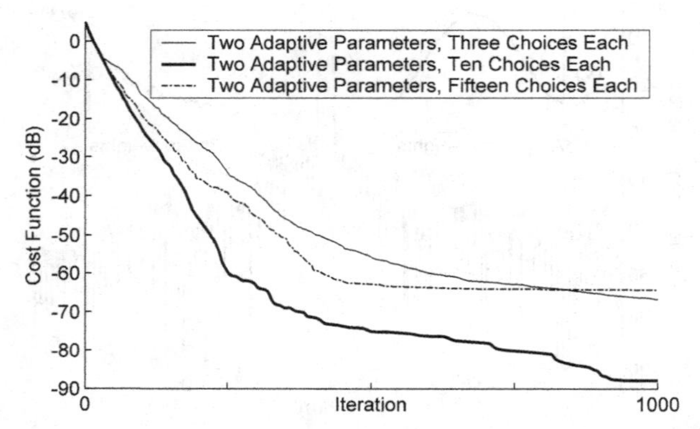

Figure 4.11. *Adaptive toggling algorithm converges faster with 10 parameter choices than either 3 choices or 15 choices.*

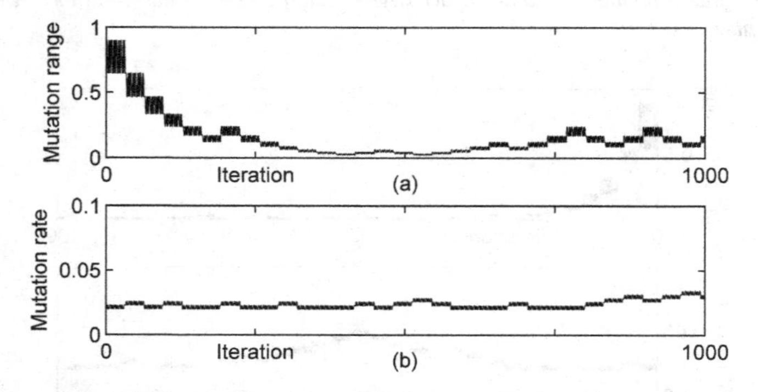

Figure 4.12. *Parameter values (15 choices each) chosen by adaptive toggling genetic algorithm. Too many closely spaced choices cloud the algorithm's ability to determine which way to adjust the various parameters, especially during the latter half of this optimization, where the convergence has stagnated as shown above in Figure 4.11.*

4.3 OPTIMIZING ARRAYS WITH COMPLEX WEIGHTING

So far, we've presented approaches to controlling the array pattern using either amplitude or phase weights. This section presents examples of the more powerful method of using complex amplitude weights to achieve the desired antenna array pattern.

4.3.1 Shaped-Beam Synthesis

GAs have been used for shaped-beam radiation pattern synthesis of arrays by optimizing the complex weighting [6–8,15–19]. For example, a technique was reported in Ref. 15, where a GA evolved a set of amplitude and phase weights for a linear array to synthesize a narrow, flattop beam design, given a limited number of amplitude and phase states. It was also shown that the GA optimization was able to produce better results by operating on the states directly than could be achieved by optimizing linear variable weights and rounding off to the nearest available digital state.

A simple and versatile GA methodology has been introduced [16] for radiation pattern synthesis of antenna arrays with an arbitrary geometric configuration. This GA approach directly represents the array excitation weighting vectors in terms of complex-valued chromosomes, as opposed to using a more conventional binary coding scheme. In other words, the chromosomes are represented directly by complex weighting vectors of the form

$$C = [c_1 c_2 \cdots c_n \cdots c_N] \tag{4.13}$$

where each c_n is a gene that corresponds to the complex weighting of the nth radiating element in the array and N is the length of the weighting vector. The technique also employs a decimal linear crossover rather than a binary crossover, which has the advantage of not entailing the need to choose a crossover location or locations as would be required in a binary crossover. Several examples are provided in Ref. 16 that demonstrate the effectiveness of this synthesis method for linear and circular arrays.

Another technique that utilizes a GA for the synthesis and optimization of shaped-beam radiation patterns is presented in Ref. 17. This approach is based on a modification of the Woodward synthesis method with an additional sidelobe minimization. In this case the optimization of synthesized array excitations is achieved by using a combination of array polynomial complex root swapping and the GA. Similarly, a technique that combines Schelkunoff's method with a GA for the synthesis of linear arrays with complex weighting and arbitrary radiation patterns has been reported [18]. A radiation pattern synthesis technique for an $M \times N$ planar array with rectangular cells is also discussed in Ref. 18, where a GA is used to determine the optimal excitation amplitude and phase for each element in the array.

The synthesis of amplitude-only, phase-only, and complex weighting for a specified far-field sidelobe envelope has been demonstrated [8,19] for a GA as well as for particle swarm optimization, with good performance obtained by both methods. The application considered in Refs. 8 and 19 is the determination of complex phased array weights to best meet a specified far-field requirement that includes a 60 dB notch on one side. To illustrate the utility of this synthesis technique, let's consider a linear array of equally spaced antenna elements along the x axis with elements having $\cos^{\nu}\theta$ element patterns, where ν can be any real number greater than or equal to zero. The cost measure to be minimized is the sum of the squares of the excess far-field magnitude that exceeds the specified sidelobe envelope. This penalizes sidelobes above the envelope, while it provides neither a penalty nor a reward for sidelobes that fall below the specification. Figure 4.13 shows an example of a synthesized far-field radiation pattern with a 60 dB sidelobe notch obtained using the GA, with the final amplitude and phase weights inset. The array used in this example consists of 100 half-wavelength spaced antenna elements, each assumed to have an element pattern with $\nu = 0.6$.

As mentioned in Section 4.2.2, an efficient real-valued adaptive GA has been developed and applied [6,7] to the synthesis of phased array radiation patterns using amplitude-only, phase-only, and complex weights. To accomplish this, an adaptive supervisory algorithm is employed that periodically updates the parameters of the GA to maximize relative improvement. The interaction of the supervisory algorithm with the GA is illustrated in Figure 4.14. When compared to conventional static parameter implementations, computation time is saved in two ways: (1) the algorithm converges more rapidly and (2) the need to tune the parameters by hand (generally done by repeatedly

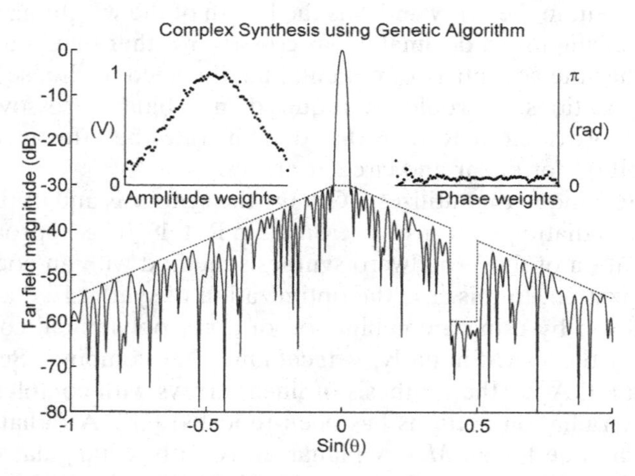

Figure 4.13. Complex synthesis of a 60 dB sidelobe notch using a GA (Boeringer and Werner, © IEEE, 2004 [19]).

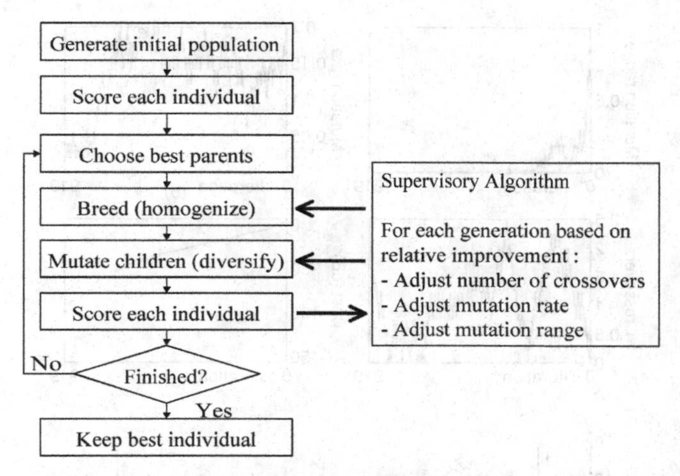

Figure 4.14. *Flowchart of GA with adaptive supervisory algorithm (Boeringer et al., © IEEE, 2005 [6]).*

running the code with different parameter choices) is significantly reduced. As an example, the same synthesis problem considered in Figure 4.13 has been performed using the adaptive GA technique, the results of which are shown in Figure 4.15. The progress of the mutation range, mutation rate, and number of crossovers as the solution adaptively evolves is shown in Figures 4.15a, 4.15b, and 4.15c, respectively. From Figure 4.15a we see that the mutation range adapts downward as the solution becomes refined, which is consistent with the intuitive notion that diversity should be highest at the beginning and less at the end of the evolutionary process. Finally, the cost performance for this synthesis example is shown in Figure 4.15d.

Two additional examples are now considered to further demonstrate the versatility of the adaptive GA approach to the synthesis of shaped-beam radiation patterns [6,7]. Figure 4.16 shows the outcome of the adaptive GA with two different array sizes (40 and 60 elements) for the same flattop beam goals. Figure 4.17 compares optimization with directive versus isotropic elements ($\nu = 0.6$ and $\nu = 0$) for some arbitrary pattern goals, with excellent results obtained in both cases. The optimized amplitude and phase weights are shown as insets in Figures 4.16 and 4.17.

The array failure correction problem represents another interesting and useful application of the GA to array synthesis [20–24]. A GA approach has been introduced [20] for failure correction in digital beamforming of arbitrary arrays, where the weights of the array are represented directly by a vector of complex numbers. The GA in this case employs a decimal linear crossover scheme so that no binary coding and decoding is necessary. In addition, the utility of the array failure correction approach has been demonstrated [20] by considering an example of a 32-element linear array with single-, double-, and triple-element failures. A technique that combines the GA with a fast Fourier

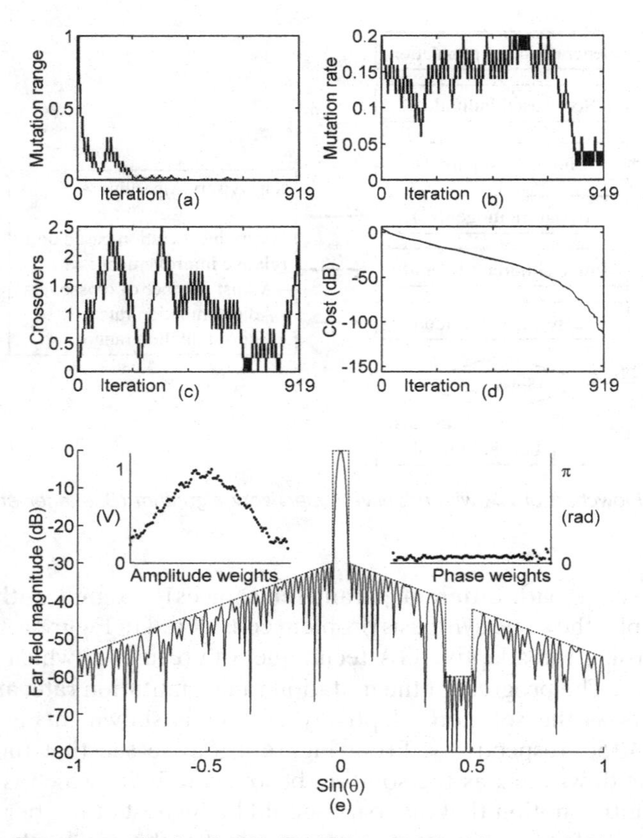

Figure 4.15. Complex synthesis of a 60 dB sidelobe notch using an adaptive GA (Boeringer et al., © IEEE, 2005 [6]): (a) algorithm's choice of mutation range; (b) algorithm's choice of mutation rate; (c) algorithm's choice of number of crossovers; (d) cost performance; (e) resulting far-field pattern with desired sidelobe constraints. Left inset—resulting amplitude weights; right inset—resulting phase weights.

transform (FFT) for array failure correction is presented in Ref. 21. The FFT is used to speed up the evaluation of the far-field array pattern, which consequently speeds up the convergence of the GA. Array failure correction techniques have also been proposed on the basis of amplitude-only optimization using a GA [22,23]. Finally, a GA optimization technique has been described [24] for the diagnosis of array faults from far-field amplitude-only data.

4.3.2 Creating a Plane Wave in the Near Field

Accurate far-field antenna pattern measurements require separating the transmit antenna and the antenna under test (AUT) by a distance sufficient to ensure that the field amplitude and phase variations across the test aperture

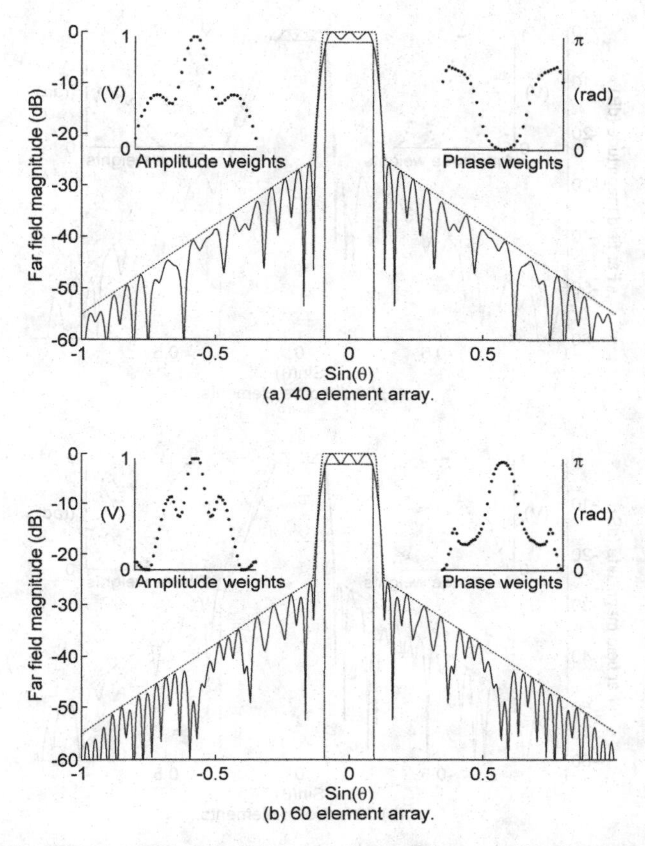

Figure 4.16. *Flattop beam with limited ripple synthesized for two different array sizes using an adaptive GA (Boeringer et al., © IEEE, 2005 [6]): (a) 40-element phased array; (b) 60-element phased array with the same pattern goals. Insets: optimized amplitude and phase weights for each case.*

are small. The IEEE defines the far-field [25] in terms of a maximum phase deviation. For uniform apertures the maximum phase variation is $\pi/8$ radians, which results in a far-field distance defined by

$$R_{ff} \geq \frac{2D^2}{\lambda} \tag{4.14}$$

where D is the largest dimension of the aperture. The ratio of the maximum field amplitude to the minimum field amplitude at a distance prescribed by (4.14) is approximately

$$\frac{8R_{ff}^2}{8R_{ff}^2 + D^2} \tag{4.15}$$

In practice, (4.15) is almost equal to one, since R_{ff} is much larger than D.

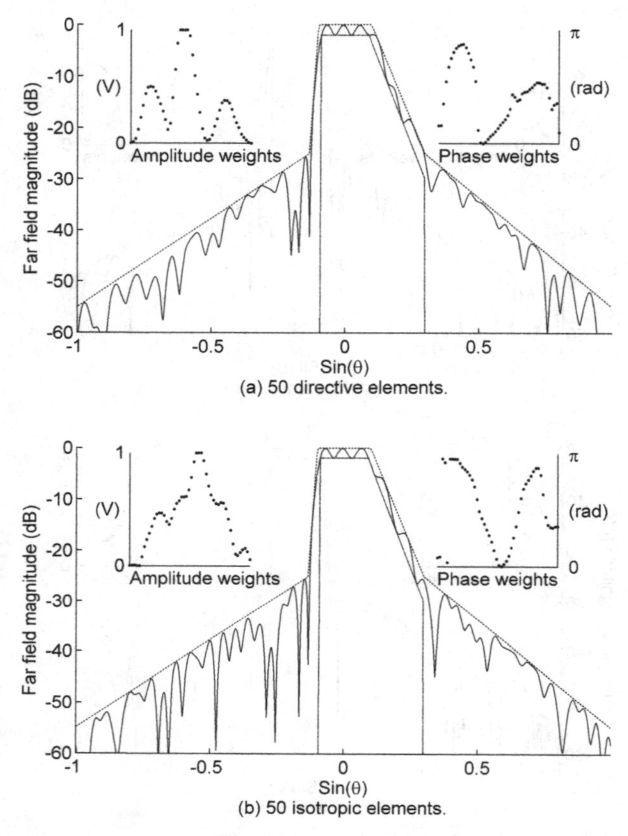

Figure 4.17. Arbitrary beam with limited ripple synthesized for two different types of array elements using an adaptive GA (Boeringer et al., © IEEE, 2005 [6]): (a) 50-element phased array of directive sources ($v = 0.6$); (b) 50-element phased array of isotropic sources ($v = 0$). Insets: optimized amplitude and phase weights for each case.

If the distance between the transmit antenna and AUT does not satisfy (4.14), then some other approach, such as a compact range, is needed to generate a plane wave at the AUT. The compact range reflector is large, making applications that require moving the plane-wave location impractical. One possibility of generating the plane wave in the near-field is to build an array that projects a plane wave at a specified distance. Projection differs from focusing in that focusing requires the fields to add in phase at a specific point in space [26], while projecting requires that there be a minimum variation in the field phase over a specified plane. Minimizing the amplitude and phase over a plane in the near-field implies an optimization or least-squares approach [27]. Additional background information on creating a plane wave in the near-field is presented in Ref. 28.

A GA has been used to create a plane wave with linear arrays [29,30] and planar arrays [31,32]. These approaches are summarized in the next two examples. A linear array of line sources that are infinite in the y direction has a relative radiation pattern at the point (x_m, z_p) given by

$$E_m = \sum_{n=1}^{N} w_n H_0^{(2)}(kR_{mn}) \tag{4.16}$$

where $H_0^{(2)}(\ ) =$ zero-order Hankel function of the second kind
$R_{mn} = \sqrt{(x_n - x_m)^2 + (z_n - z_p)^2}$
$(x_n, z_n) =$ element locations

The objective function for minimizing the phase and amplitude oscillations is given by

$$cost = c_1 \left| \frac{\max \angle E_m - \min \angle E_m}{P} \right| + (1 - c_1) \frac{\max|E_m| - \min|E_m|}{\min|E_m|} \tag{4.17}$$

for all sample points. In the cost function, the first term compares the maximum phase deviation across the AUT to a constant, P. For this example, $P = \pi/8$. The second term compares the ratio of the maximum amplitude value to the minimum value across the AUT to the ratio of the maximum amplitude to the minimum amplitude across that same AUT due to a line source. A third term can be added to keep the optimal solution out of a null, although this term is not necessary if the array is assumed to have symmetric phase weighting.

The first example optimizes the weights of a 9 element linear array along the x axis with an element spacing of $d = 0.5\lambda$ to produce a "plane-wave" across an AUT that has a width of 10λ at a distance of 20λ. Figure 4.18 shows the optimized amplitude and phase weights for the array, and Figures 4.19 and 4.20 show the amplitude and phase values of the projected field. The optimized field amplitude is an improvement over the uniform array, but the optimized field phase is about the same as that of the uniform array. Optimizing the element locations as well as the amplitude and phase tapers provides improvement to the projected plane-wave at close and far distances. Tradeoffs exist in the ripples in the phase and the ripples in the amplitude of the projected plane-wave. Adjusting the z location of the elements improves the quality of the projected plane-wave but makes the array configuration more sensitive. Since out of plane elements would produce some blockage of the elements in the rear, adjusting the x spacing of the elements is the recommended procedure and the most practical to implement.

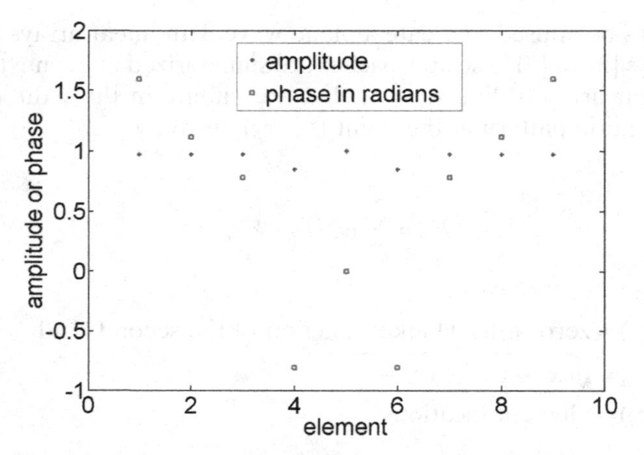

Figure 4.18. *Optimized amplitude and phase weights of a nine-element array (d = 0.5λ) for an AUT of 10λ at a distance of 20λ.*

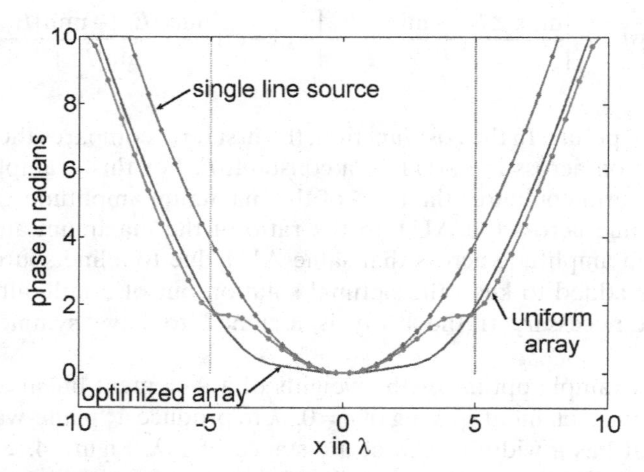

Figure 4.19. *Field phase of an optimized nine-element array (d = 0.5λ) across an AUT of 10λ at a distance of 20λ.*

The electric field of the planar array of isotropic point sources shown in Figure 4.21 is given by

$$E(x_m, y_m, z_p) = \sum_{n=1}^{N} w_n \frac{e^{jkR_{mn}}}{4\pi R_{mn}} \tag{4.18}$$

where R_{mn} is the distance from element n to the field point (x_m, y_m, z_p) on the plane-wave. In this example, a 6×6 element array with a square lattice having

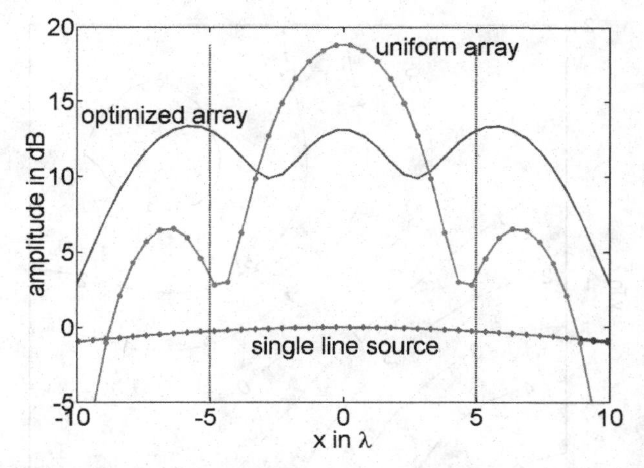

Figure 4.20. *Field amplitude of an optimized nine-element array (d = 0.5λ) across an AUT of 10λ at a distance of 20λ.*

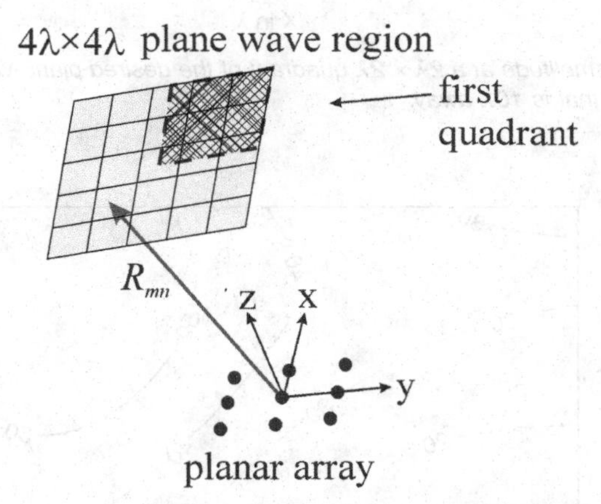

Figure 4.21. *The planar transmit array generates an approximate plane-wave in the near field.*

a spacing of $d = 1.0\lambda$ projects a plane-wave over a $4\lambda \times 4\lambda$ area that is at $z_p = 10\lambda$. At the normal far-field distance of $z_p = 64\lambda$, the phase variation is 22.5° and the amplitude variation is 0.33 dB. The amplitude and phase plots at $z_p = 10\lambda$ for the array with uniform weighting appear in Figures 4.22 and 4.23, respectively. The maximum phase variation is 60°, and the maximum amplitude variation is 8.6 dB.

The objective function is given by (4.17) with $P = \pi$. The continuous variable GA used a population size of 80 with a 1% mutation rate, 50% crossover rate, single-point crossover, and ran for 30,008 function evaluations. In the

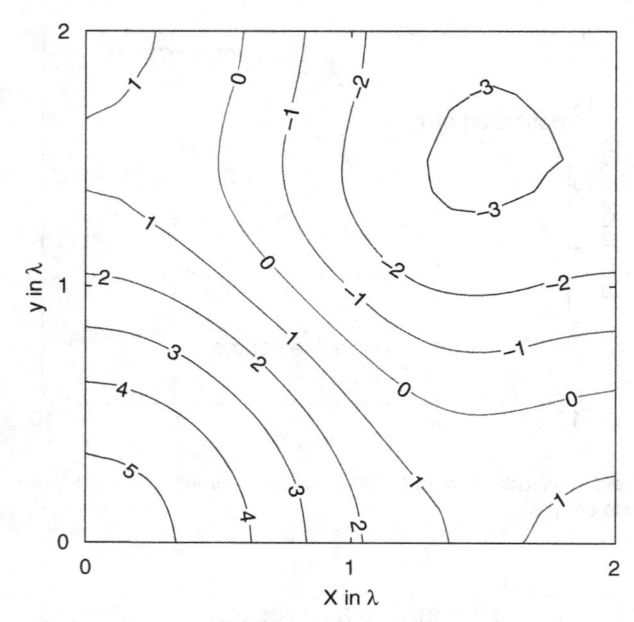

Figure 4.22. *Field amplitude at a $2\lambda \times 2\lambda$ quadrant of the desired plane-wave region due to a 6×6 uniform array that is 10λ away.*

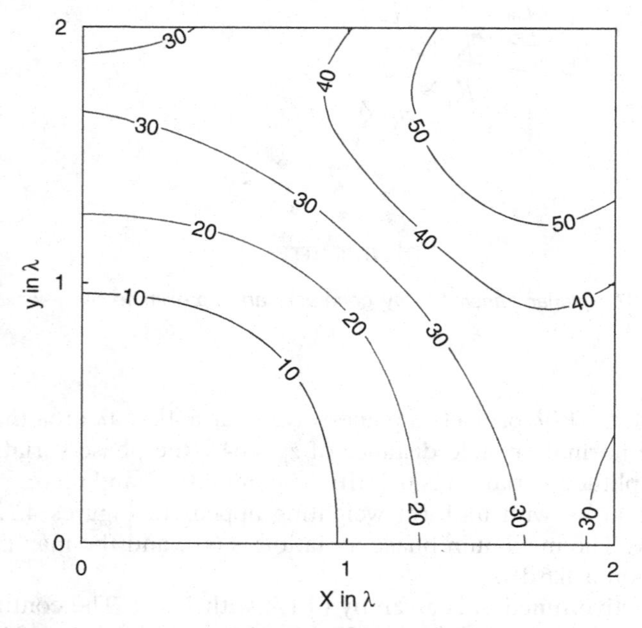

Figure 4.23. *Field phase at a $2\lambda \times 2\lambda$ quadrant of the desired plane wave region due to a 6×6 uniform array that is 10λ away.*

TABLE 4.1. Optimized Weights Found by the GA

x-Element Spacing →	0.5λ		1.5λ		2.5λ	
y-Element Spacing ↓	Normalized Amplitude	Phase (deg)	Normalized Amplitude	Phase (deg)	Normalized Amplitude	Phase (deg)
0.5λ	1.00	0	0.29	29	0.67	−18
1.5λ	0.61	4	0.34	−140	0.59	58
2.5λ	0.39	15	0.61	4	0.29	108

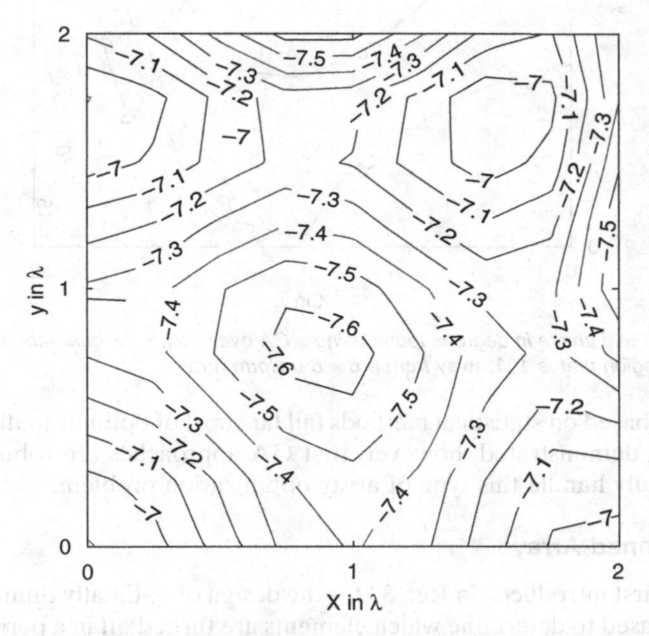

Figure 4.24. Field amplitude in dB found using a GA over a $2\lambda \times 2\lambda$ quadrant of the desired plane-wave region that is 2λ away from a 6×6 uniform array.

end, the GA found the weights in Table 4.1, which produce the fields shown in Figures 4.24 and 4.25. The maximum amplitude variation is 0.75 dB, and the maximum phase variation is 32.4°.

4.4 OPTIMIZING ARRAY ELEMENT SPACING

A considerable amount of discussion in the literature has been devoted to the problem of using GAs to optimize array element spacing. This is primarily because this class of array synthesis problems has been extremely challenging for traditional optimization methods due to the large number of variables that are typically involved. Moreover, conventional aperiodic array synthesis

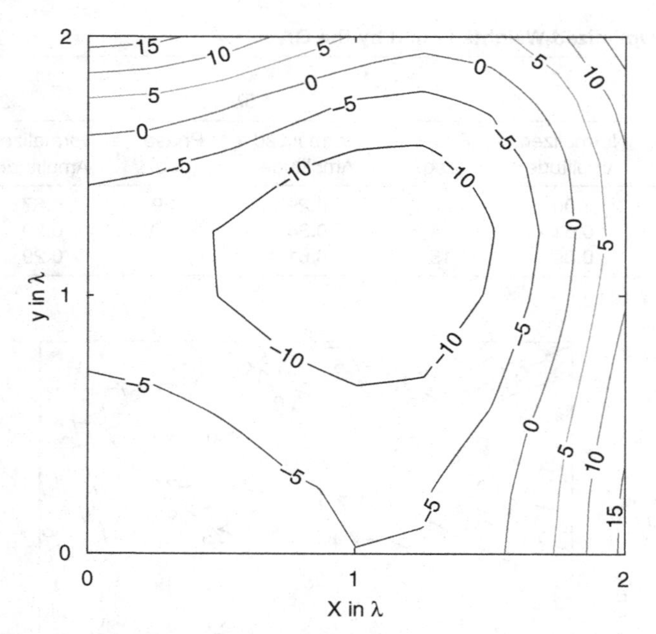

Figure 4.25. Field phase in degrees found using a GA over a $2\lambda \times 2\lambda$ quadrant of the desired plane-wave region that is 10λ away from a 6×6 uniform array.

techniques based on statistical methods fall far short of optimal configurations. It has been demonstrated, however, that GA approaches are robust enough to successfully handle this type of array optimization problem.

4.4.1 Thinned Arrays

GAs were first introduced in Ref. 33 for the design of optimally thinned arrays. The GA is used to determine which elements are turned off in a periodic array to yield the lowest maximum relative sidelobe level. In other words, a thinned array has discrete parameters, where a single bit represents the element state as "on" = 1 or "off" = 0. Thinned designs for both 200-element linear arrays and 200-element planar arrays were considered in Ref. 33. An example of a 40-element GA thinned array with 0.3λ spacing has also been presented [34]. These arrays were thinned by a GA to obtain sidelobe levels of less than −20 dB. To illustrate the thinning process, we consider a design example for a uniformly excited broadside 400-element planar array of isotropic sources spaced a half-wavelength apart in a square lattice. A GA is used to thin this array in order to achieve the lowest possible maximum relative sidelobe level. The cost function is given by

$$cost = \max\left\{\sum_{n=1}^{10}\sum_{m=1}^{10} a_{mn}\cos[k(m-0.5)d_y u]\cos[(k(n-0.5)d_x v)]\right\}$$
$$\text{for} \quad \sqrt{u^2+v^2} > c_1 \tag{4.19}$$

where $u = \sin\theta\cos\phi$
$v = \sin\theta\sin\phi$
c_1 = start of mainbeam

This cost function assumes symmetry about the x and y axes, and the element closest to the origin has an amplitude of 1. Thus, a chromosome has 99 variables. The radiation pattern for the thinned array evolved using the GA is shown in Figure 4.26, which has a sidelobe level of –22.6 dB. The resulting thinned array (60% filled) is also shown in Figure 4.27. An element that is "on" is represented by a white square, while one that is "off" is represented by a black square. Figure 4.28 shows the convergence of the GA with a population size of 8, a mutation rate of 0.15, and uniform crossover. Only very small

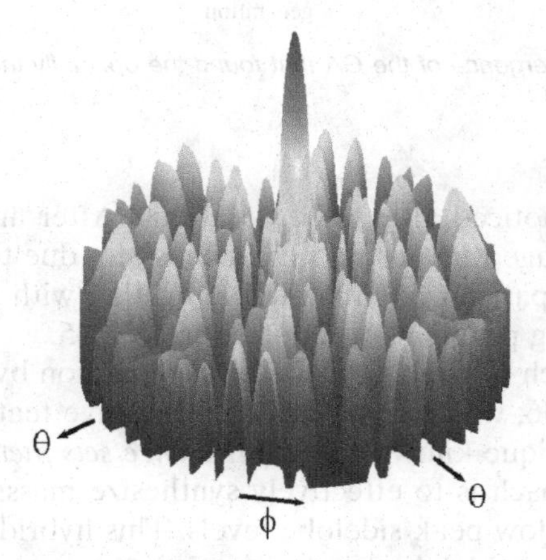

Figure 4.26. The array factor for the optimally thinned planar array. The maximum sidelobe level is –22.6 dB.

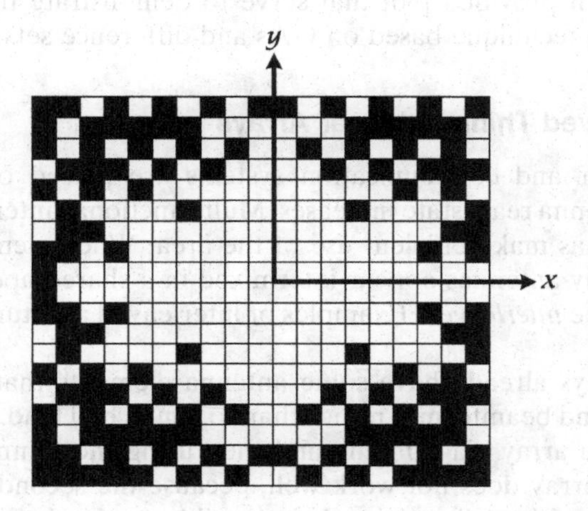

Figure 4.27. Optimally thinned planar array. White squares indicate that the element has an amplitude of one; black squares, zero.

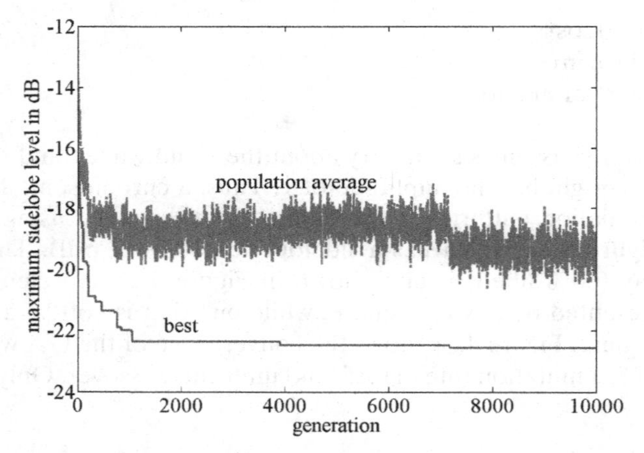

Figure 4.28. *Convergence of the GA that found the optimally thinned planar array.*

improvements are noticed after 1000 iterations. After an initial sharp drop, the population average remains relatively constant, due to the high mutation rate. Finally, a comparison of simulated annealing with GAs for linear and planar array thinning problems is reported in Ref. 35.

A hybrid approach to peak sidelobe level reduction by array thinning was introduced in Ref. 36, which combines the attractive features of GAs with a combinatorial technique known as the *difference sets method*. The main goal of this hybrid approach is to effectively synthesize massively thinned arrays that have relatively low peak sidelobe levels. This hybrid approach has been shown to yield better sidelobe level control than is possible with a conventional GA. Several design examples of massively thinned linear and planar arrays have been provided [36] that serve to demonstrate the utility of the hybrid synthesis technique based on GAs and difference sets.

4.4.2 Interleaved Thinned Linear Arrays

As more sensor and communication systems are placed on vehicles, the demand for antenna real estate increases. Multifunctional antennas and shared aperture antennas make efficient use of the area. When elements dedicated to different array antennas appear intermixed in a shared aperture, then the array is said to be *interleaved*. Examples of interleaved apertures are reported in Refs. 37–41.

Thinned arrays already have some antenna elements that could be connected to a second beamformer rather than to a matched load. Optimizing the sidelobes of one array via thinning and then using the thinned elements to form a second array does not work well, because the second array has very high sidelobes and low directivity. It is possible to simultaneously optimize

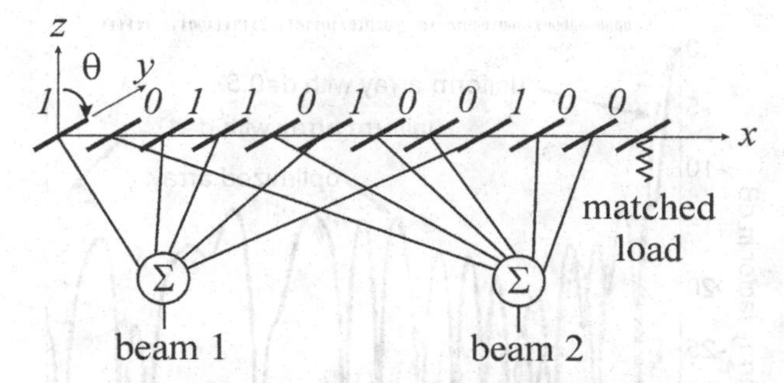

Figure 4.29. *Diagram of the interleaved array concept.*

two thinned arrays in either a full or partial interleave configuration [42]. Figure 4.29 shows how two arrays could be interleaved.

A fully interleaved array has every element in the aperture attached to a beamformer. Assume all the elements are identical and equally spaced. Elements in the arrays are antisymmetric in that when one array has an element on/off, the other array has it off/on. Hence, we may define

$$w = [w_1, w_2, \ldots, w_N, 1-w_N, \ldots, 1-w_2, 1-w_1] \tag{4.20}$$

and

$$w' = 1 - w \tag{4.21}$$

Adding w and w' yields a vector of all ones, which implies that the aperture efficiency η_{ap} is 100%:

$$\eta_{ap} = \frac{\text{number of elements in aperture turned on}}{\text{total number of elements in aperture}} \tag{4.22}$$

This arrangement ensures that each interleaved array has half of the elements and an aperture efficiency of 50% where aperture efficiency is

$$\eta_{ar} = \frac{\text{number of elements in array turned on}}{\text{total number of elements in array}} \tag{4.23}$$

There are many different ways to fully interleave two arrays over the same aperture area. For example, suppose that a $(2N = 60)$-element equally spaced $(d = 0.5\lambda)$ array is the starting aperture. One approach is to divide the aperture in two, so that $w_n = 1$ for $n = 1, \ldots, N$ and $w_n = 0$ otherwise. The array on the left side forms one beam, and the array on the right side forms the other beam. The pattern of one array is shown by the dashed line in Figure 4.30. Another approach is to interleave the two arrays where every other (alternate) element belongs to one array and the remaining belongs to the other array, or $w_{2n} = 1$

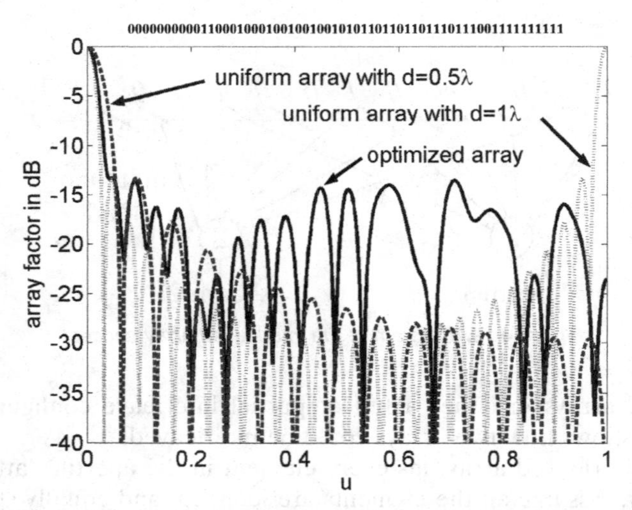

Figure 4.30. *The array factors for an optimized fully interleaved array, a side-by-side array (d = 0.5λ), and an alternate-element interleaved array (d = 1.0λ) when there are 60 total elements with 30 turned on for one beam.*

and $w_{2n-1} = 0$ for $n = 1, \ldots, N$. The pattern of one array ($d = 1.0\lambda$) is the dotted line in Figure 4.30. As seen in Figure 4.30, the side-by-side arrays have the advantage of lower average sidelobe levels, while the interleaved arrays have the advantage of narrower beamwidth. Now, if a GA picks the element assignments, then the array factor with the solid line is found. The optimized weight assignments are shown at the top of Figure 4.30. This optimized array has the narrow beamwidth of the full aperture and has the same peak sidelobe level of a uniform array. Thus, it is a compromise between the side-by-side uniform array and the every-other-element (alternate-element) interleaved array.

It is also possible to interleave a sum-and-difference array while optimizing both for low sidelobes. In this case, the thinning is assumed to be symmetric about the center of the array. Half of the elements in the difference array receive 180° phase shifts. A better approach is to define the aperture size and element spacing and let a GA decide which elements should be on/off for the sum and difference arrays. The resulting optimized sum and difference patterns for a 0.6λ-spaced dipole array of 120 elements are shown in Figures 4.31 and 4.32, respectively. The peak sidelobe level for the sum pattern is −13.1 dB and for the difference pattern, −13.3 dB. Efficiency for the sum array is $\eta_{ar} = 0.45$ and for the difference array, $\eta_{ar} = 0.55$. The array of dipoles was modeled using the method of moments, so mutual coupling is included.

Fully interleaving two arrays limits the peak sidelobe reduction. One way to increase sidelobe control at the expense of aperture efficiency is to shift one array half an aperture to the right and then interleave the two arrays. The heavily populated center of the first array overlaps the lightly populated edge

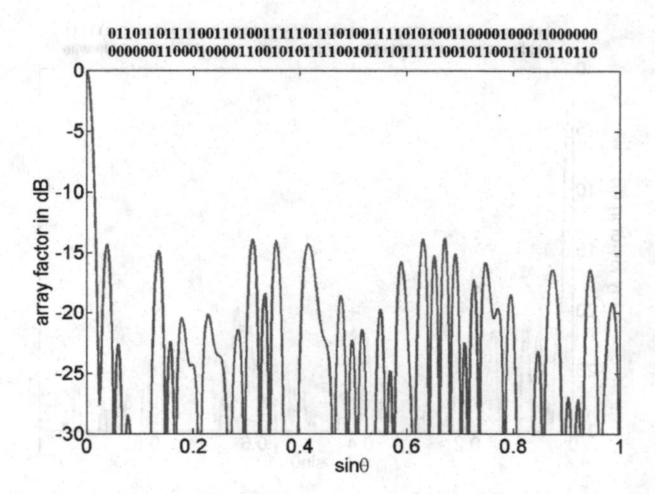

Figure 4.31. *Sum array factor for a 120-element aperture in which sum and difference arrays are interleaved.*

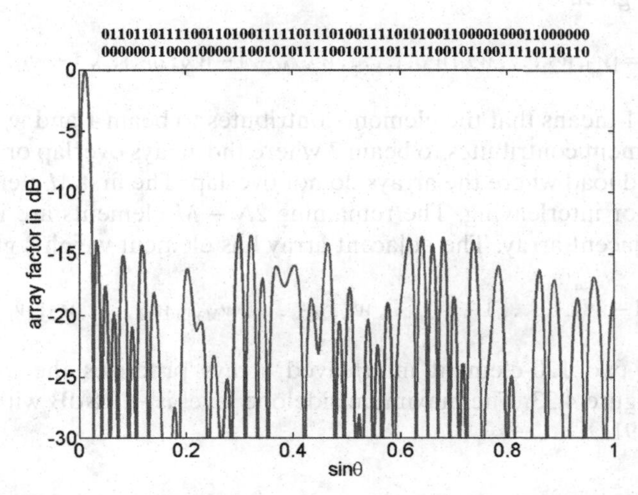

Figure 4.32. *Difference array factor for a 120-element aperture in which sum and difference arrays are interleaved.*

of the second array. Aperture efficiency is less than 100%. Elements not connected to one of the matched array feed networks are terminated in a matched impedance.

Another approach to interleaving is to partially interleave two arrays so that they are interleaved over 50% of their elements. Now, the right half of one array is interleaved with the left half of the other array. The center third

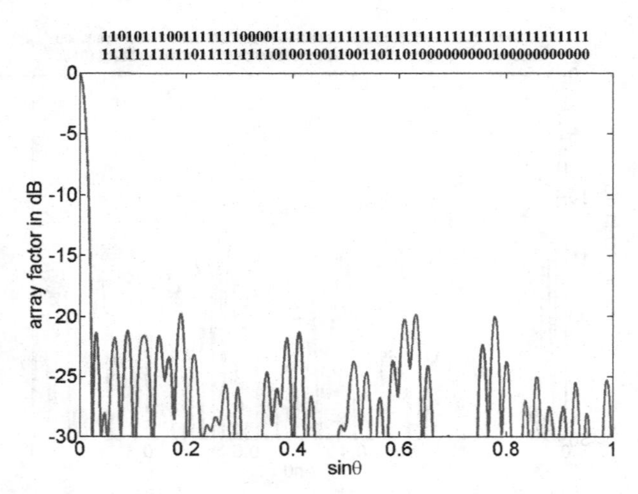

Figure 4.33. *Array factor for a 120-element array using asymmetric partial interleaving.*

of the aperture is 100% efficient, while the end thirds are thinned. The element weights are given by

$$w = [w_1, w_2, \ldots, w_M, w_{M+1}, \ldots, w_{N+M/2}, 1 - w_{N+M/2}, \ldots, 1 - w_{M+1}] \tag{4.24}$$

where $w_n = 1$ means that the element contributes to beam 1 and $w_n = 0$ means that the element contributes to beam 2 where the arrays overlap or is attached to a matched load where the arrays do not overlap. The first M elements have no overlap or interleaving. The remaining $2N - M$ elements are interleaved with the adjacent array. The adjacent array has element weights given by

$$[1 - w_{M+1}, \ldots, 1 - w_{N+M/2}, w_{N+M/2}, \ldots, w_{M+1}, w_M, \ldots, w_2, w_1] \tag{4.25}$$

Optimizing two 120-element interleaved arrays produces the array factor shown in Figure 4.33. The maximum sidelobe level is –19.8 dB with $\eta_{ar} = 0.83$ and $\eta_{ap} = 0.91$.

4.4.3 Array Element Perturbation

A method has been presented [43] for pattern nulling that works by perturbing the element positions within a uniformly spaced linear array. To achieve the best possible results, the perturbations of the array elements from their original positions were treated as optimization variables within a GA. This technique has been shown to be useful for null steering, where nulls with specified depths can be placed in the directions of undesirable interfering signals. Two examples are presented in Ref. 43, where perturbed array patterns have been determined for a 20-element array of isotropic sources by using a GA. In the first case the elements are perturbed along the array axis to produce two nulls at 14.5° and 40.5°, while in the second case the elements are perturbed in a direction normal to the array axis to produce two nulls at 26° and 32.5°.

Finding the element spacings of a uniformly weighted array that produced the lowest possible sidelobe levels was first investigated using a GA in Ref. 44. We illustrate this technique by considering an example where a GA optimizes the sidelobe level of a 19-element nonuniformly spaced linear array lying along the x axis. In this case the spacings of the elements are quantized to 3 bits. The array factor cost function is given by

$$sll = \max\left\{1 + 2\sum_{n=1}^{9} \cos[k \sin\theta(x_{n-1} + d_0 + \Delta_n)]\right\} \quad \text{for } \theta > \theta_{MB} \tag{4.26}$$

where $x_0 = 0$

$d_0 = \lambda/2$

Δ_n = variable spacing between 0 and $\lambda/2$

θ_{MB} = first null next to the mainbeam

The GA had a population size of 8 and mutation rate of 0.15. After 1000 generations, a Nelder–Mead algorithm polished the results. Figure 4.34 shows the resulting array factor with a maximum sidelobe level of −20.6 dB. The 19 round dots in the figure represent the relative element spacings of the array elements. This method of using a GA to optimize nonuniformly spaced linear arrays is also outlined in a review article [45].

If an array is optimally thinned for low sidelobes at boresite, then scanning the mainbeam will cause the sidelobes to go up (assuming directional elements). Sidelobes can be optimized for the entire scan range as shown in

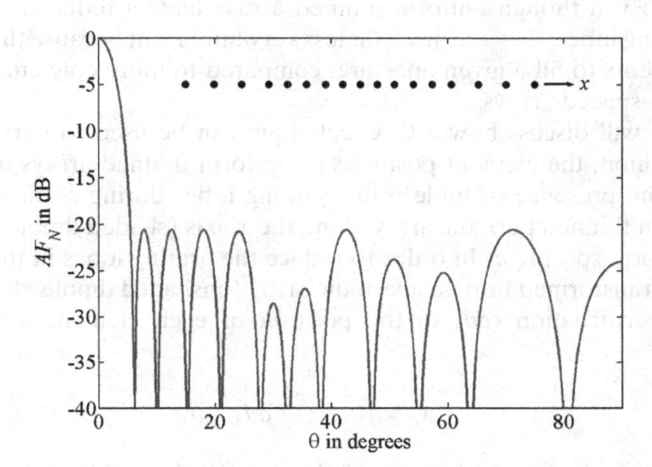

Figure 4.34. *Array factor of nonuniformly spaced array. Maximum sidelobe level is −20.6 dB. The optimum relative element spacings are shown along the x axis.*

Ref. 33. If the array elements have a spacing greater than $\lambda/2$, then grating lobes could emerge, depending on the scan angle. This problem has been addressed [46–48] by considering a different representation of the thinned array. Instead of removing elements from a fully populated half-wavelength spaced array, the thinned array is based on a periodic linear phased array with an interelement spacing greater than a half-wavelength. In this case the GA optimization parameter is a perturbation added to each interelement spacing in such a way as to create an aperiodic array that maintains low sidelobes during scanning. In addition, the GA places restrictions (i.e., an upper bound and a lower bound) on the driving-point impedance of each element in the array to ensure that they are well behaved during scanning. A GA approach was also developed [47,48] for the purpose of evolving an optimal set of the simplest possible matching networks to be used in conjunction with impedance constrained thinned arrays.

The maximum angle that a uniformly spaced thinned linear phased array can be scanned from broadside before a grating lobe will appear is given by [49]

$$\theta_{\max} = \sin^{-1}\left(\frac{\lambda}{d} - 1\right) \tag{4.27}$$

where $\theta_{\max}$ is the maximum steering angle from broadside ($\theta = 0°$) and d is the uniform element spacing. From this equation it can be seen that half-wavelength-spaced arrays will have a complete theoretical scan range of 90° from broadside. However, as the uniform spacing of the array is increased beyond a half-wavelength, the scan range of the array begins to be severely reduced. For example, with a uniform spacing of just 0.8 wavelength the maximum scan angle of the array without grating lobes is only 14.5° from broadside. Even though uniform thinned arrays have a reduced scan range due to grating lobes, they are nevertheless very important because they require fewer elements to fill a given aperture, compared to more conventional half-wavelength-spaced arrays.

Next we will discuss how a GA technique can be used to perturb, in an optimal fashion, the element positions of uniform thinned arrays in order to eliminate the presence of undesirable grating lobes during scanning. Figure 4.35 shows a thinned periodic array along the x axis (shaded dipole elements) with a uniform spacing d. In order to reduce the grating lobes in the periodic array, it is transformed into an aperiodic array (unshaded dipole elements) by adding a perturbation (δd_n to the position of each element in the array where

$$x_n = d(n-1) + \delta d_n \tag{4.28}$$

An eight-element thinned array of half-wave dipoles with a uniform spacing of 0.8 wavelengths and a uniform current excitation is used as the starting

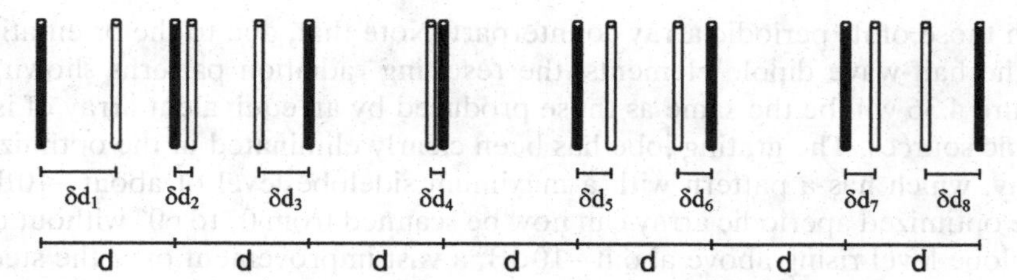

Figure 4.35. *An illustrative example of a thinned periodic array with a uniform spacing d (shaded dipole elements) compared to an aperiodic array formed by perturbing the element locations of the periodic thinned array (unshaded dipole elements) (Bray et al., © IEEE, 2002 [48]).*

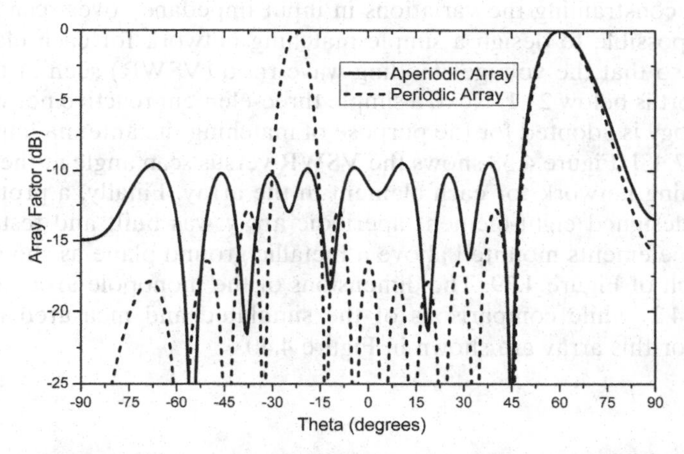

Figure 4.36. *Far-field radiation pattern obtained from the GA optimized eight-element aperiodic array compared with that of its periodic array counterpart.*

point for GA optimization. It has already been established using (4.27) that this array would have a maximum scan angle of 14.5° from broadside without generating grating lobes. The goal of the GA for this design will be to optimize the element position perturbations so that the mainbeam can be scanned to 60°, well beyond the scan angle limitation of the periodic array. This is accomplished by representing the phased array as a list of element position perturbations, which is the chromosome to be optimized by the GA. In addition, the real part of the driving-point impedance for each dipole element in the optimized aperiodic array was limited to between 20 and 250 Ω over the entire scanning range of the array (in this case, for all angles between 0° and 60°). Imposing these additional restrictions within the GA will guarantee that a set of matching networks can be designed for the array that have a minimal number of components as well as practical component values.

Figure 4.36 compare the results for the far-field radiation pattern obtained from the GA optimized eight-element aperiodic array of half-wave dipoles with those of its periodic array counterpart. Note that, due to the orientation of the half-wave dipole elements, the resulting radiation patterns shown in Figure 4.36 will be the same as those produced by an equivalent array of isotropic sources. The grating lobe has been clearly eliminated in the optimized array, which has a pattern with a maximum sidelobe level of about –10 dB. The optimized aperiodic array can now be scanned from 0° to 60° without the sidelobe level rising above about –10 dB, a vast improvement over the steerability of the thinned periodic array. Moreover, the input resistance for each half-wave dipole element in the eight-element aperiodic array is plotted in Figure 4.37 as a function of scan angle between 0° and 90°. Note that these values of input resistance are well behaved and stay within the desired minimum and maximum bounds imposed by the GA of 20 and 250 Ω, respectively. By constraining the variations in input impedance over scan range, it becomes possible to design a simple matching network for each element in the array so that the voltage standing-wave ratio (VSWR) seen at the input of each port is below 2 : 1. Next, a simple three-element reactive network with a Π topology is adopted for the purpose of matching the antenna impedances to 50 Ω [47,48]. Figure 4.38 shows the VSWR versus scan angle at the input to the matching network for each element in the array. Finally, a prototype of this GA designed eight-element aperiodic array was built and tested using monopole elements mounted above a metallic ground plane as shown in the photograph of Figure 4.39. The dimensions of the monopole array are listed in Table 4.2, while comparisons of the simulated and measured radiation patterns for this array are shown in Figure 4.40.

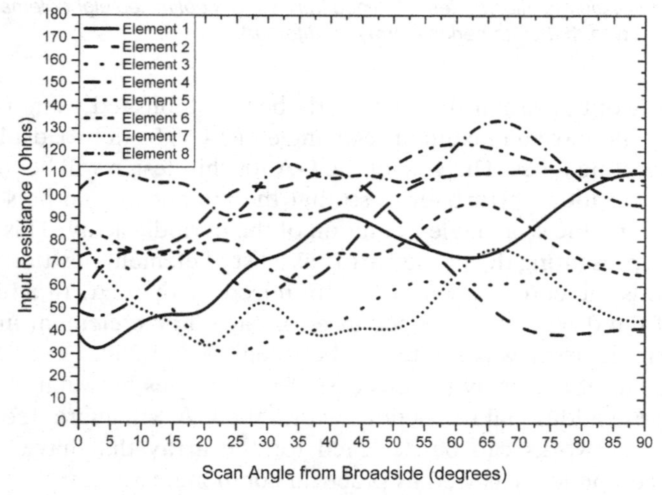

Figure 4.37. Input resistance for each half-wave dipole element in the eight-element aperiodic array plotted versus scan angle from broadside.

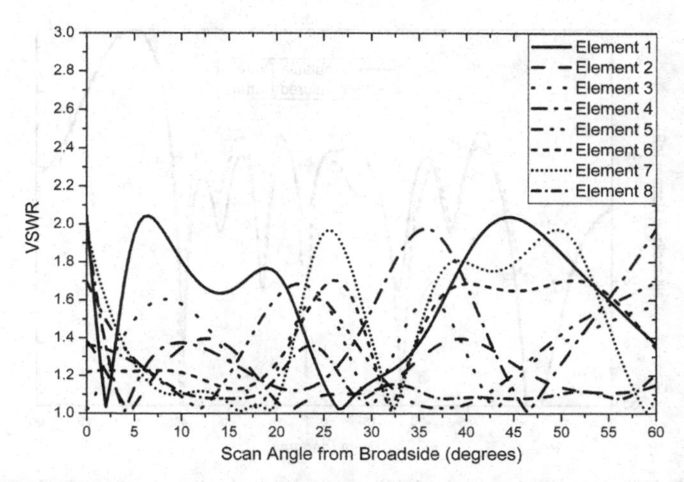

Figure 4.38. *VSWR versus scan angle at the input to the GA optimized matching network for each element in the array.*

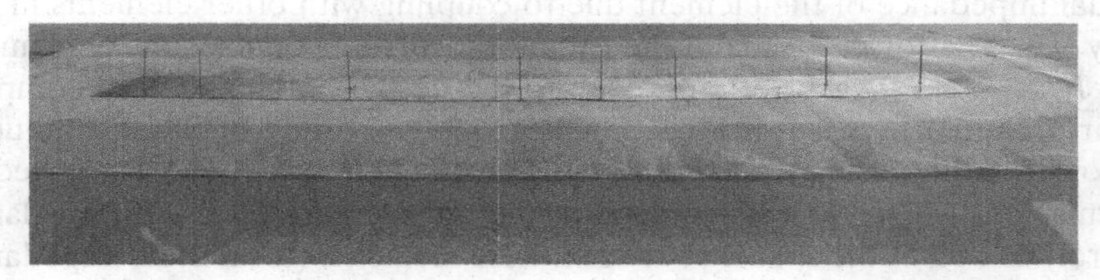

Figure 4.39. *Photograph of an optimized aperiodic eight-element monopole array designed for 1296.1 MHz mounted above a metallic ground plane.*

TABLE 4.2. Positions of Monopole Elements for Operation at 1296.1 MHz

Element	1	2	3	4	5	6	7	8
Centimeters	0	13.8	35.4	46.1	57.6	82.4	104.9	113.6
Wavelengths	0	0.598	1.532	1.992	2.490	3.561	4.537	4.910

The driving-point impedance for the nth element of an N-element linear array may be expressed in the form

$$Z_n = Z_{n1}\left(\frac{I_1}{I_n}\right) + Z_{n2}\left(\frac{I_2}{I_n}\right) + \ldots + Z_{nN}\left(\frac{I_N}{I_n}\right) \tag{4.29}$$

where $n = 1, 2, \ldots, N$. Hence, the driving-point impedance of an individual array element is a function of the self-impedance of the element (i.e., Z_{nn}), the

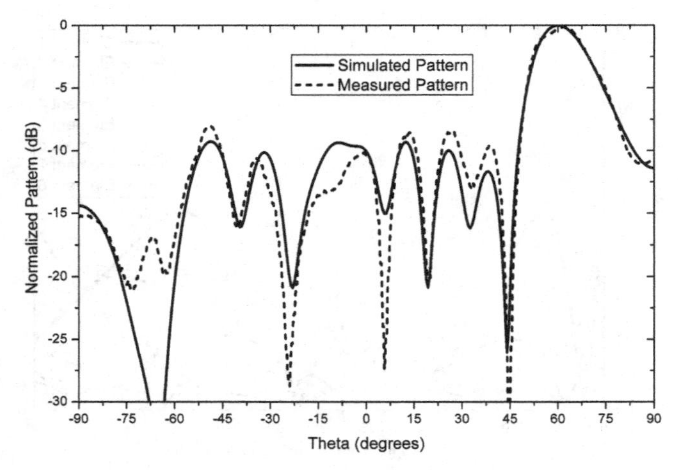

Figure 4.40. *Simulated and measured radiation patterns for the array shown in Figure 4.39.*

mutual impedance of the element due to coupling with other elements in the array (i.e., Z_{nm} for $m \neq n$), and the excitation currents on each array element (i.e., $I_1, I_2, \ldots, I_N$). A novel approach to the design optimization of compact linear phased arrays is introduced in Ref. 50. The utility of the technique is demonstrated by using fractal dipoles as array elements. A GA is applied to optimize the shape of each individual fractal element (for self-impedance control) as well as the spacing between these elements (for mutual impedance control) in order to obtain compact array configurations with dramatically improved driving-point impedance versus scan angle performance.

In Ref. 51, a design methodology is presented for the optimization of aperiodic planar arrays of periodic planar subarrays with an emphasis on the application to satellite mobile earth stations. In this case the steepest-descent method was used to optimize the arrangement of subarrays in order to eliminate grating lobes as well as reduce sidelobes within a certain scan volume. An example of an aperiodic array consisting of 16 subarrays designed to operate in the 14 GHz band, which is capable of tracking a geostationary satellite up to a maximum of 2° from broadside, was provided in Ref. 51.

4.4.4 Aperiodic Fractile Arrays

A new family of fractal arrays, known as *fractile arrays*, has been introduced [52,53]. A *fractile array* is defined as any array which has a fractal boundary contour that tiles the plane. Fractals are a class of geometric shapes capable of being divided into parts that are similar to the whole. Fractal geometries are a consistent theme found in nature. Terrain such as coastlines, mountain ranges, and river systems can all be described using fractal geometries. In addition, the shapes of vascular and bronchial systems in animals and the overall shape of trees and plants are fractal in nature. It is possible to create deterministic fractal designs by repeatedly applying a simple geometric pattern called a *generator*. However, natural objects are rarely fully deterministic and often have at least some degree of random properties.

Fractal tiles, or fractiles, represent a unique subset of all possible tile geometries that can be used to cover the plane without gaps or overlaps. An example of a fractile is shown in Figure 4.41. The unique geometrical properties of fractiles have been exploited [52–56] to develop a new design methodology for modular broadband antenna arrays. Another important property of fractile arrays is that their self-similar tile geometry can be exploited to develop a rapid iterative procedure for calculating the far-field radiation patterns corresponding to these arrays, which is much faster than conventional approaches based on the discrete Fourier transform (DFT).

A considerable amount of research has been devoted to the analysis and design of a specific type of fractile array that is based on the Peano–Gosper family of space-filling curves [52–54]. The elements of the array are uniformly distributed along a Peano–Gosper curve, which leads to a planar array configuration with a regular hexagonal lattice on the interior that is bounded by an irregular closed Koch fractal curve around its perimeter. The far-field radiation characteristics of the uniformly spaced Peano–Gosper fractile array are compared with the conventional square and hexagonal array in Ref. 54. It was shown that the Peano–Gosper array has the same desirable grating lobe conditions as the hexagonal array, because the elements are arranged in the same equilateral triangular lattice on the interior of both arrays. It was also shown, however, that the Peano–Gosper array has a considerably lower overall

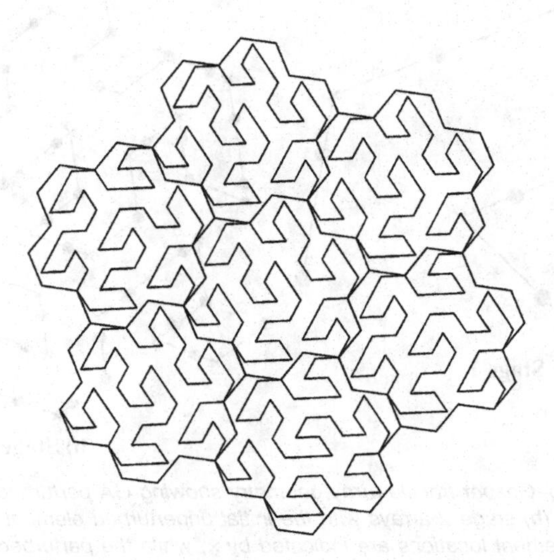

Figure 4.41. *Stage 3 Gosper island fractile.*

sidelobe level than does the hexagonal array even though both arrays are uniformly excited. This highly desirable property of Peano–Gosper arrays is a direct consequence of their irregular Koch fractal boundaries.

Grating lobes will occur when the mainbeam of the Peano–Gosper fractile array is steered from broadside for a uniform element spacing of one wavelength or greater. In Refs. 55 and 56 a GA is applied to optimize the Peano–Gosper fractile array geometry with a goal to eliminate the presence of grating lobes during scanning. A binary GA is employed along with two-point crossover and mutations. The GA perturbs the element locations along the stage 1 curve to produce a radiation pattern with no grating lobes within a specified conical scan volume and the lowest sidelobe level possible for a predetermined higher-order stage array. By optimizing the stage 1 array in this manner, it is possible to preserve the beneficial properties of the fast iterative beamforming algorithm, which can then be used to considerably speed up the convergence of the GA. The end result of this GA optimization procedure is a Peano–Gosper array with an irregular boundary contour and a nonuniform or aperiodic arrangement of elements on the interior.

There are a total of eight elements in the stage 1 array, where the positions of the uniformly spaced elements are indicated in Figure 4.42a by an "x." However, the GA will output only six perturbed element locations as the two endpoint element positions are held fixed. The genetically optimized stage 1 Peano–Gosper curve, shown in Figure 4.42a, with perturbed element locations denoted by an "•" can then be used to create higher-order stage arrays and

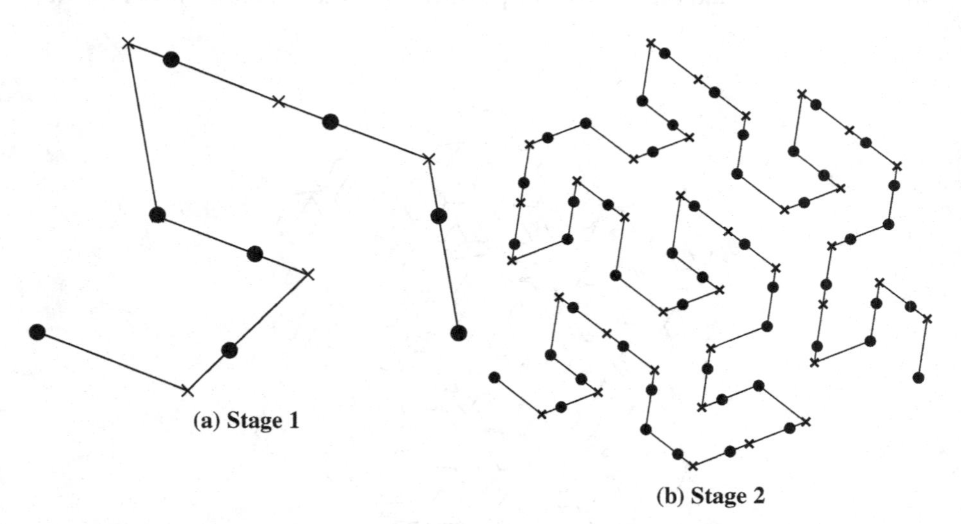

Figure 4.42. Peano–Gosper fractile array geometry showing GA perturbed element locations for (a) stage 1 and (b) stage 2 arrays with the initial unperturbed element spacing set to 2λ. The unperturbed element locations are indicated by $\times$, while the perturbed element locations are indicated by •.

evaluate their corresponding radiation characteristics efficiently via an iterative approach. Figure 4.42b shows the stage 2 Peano–Gosper fractile array with GA perturbed element locations. For the optimized Peano–Gosper fractile array design shown in Figure 4.42, a restriction was placed on the GA such that the spacing between any two consecutive elements in the array would be no less than one wavelength. Moreover, the optimization starts with an initial uniform element spacing of 2λ and the mainbeam steered to $\theta = 30°$ and $\varphi = 180°$. The final GA-optimized array has a minimum spacing between elements of 1.13λ. Also, the average element spacing for the same array was found to be 1.68λ. Figure 4.43 shows the $\varphi = -86.9°$ plane cut while varying θ. There are four grating lobes present in Figure 4.43a when the elements are equally spaced along the stage 3 Peano–Gosper fractile array. The GA was used to eliminate these grating lobes by perturbing the element spacing as is demonstrated in Figure 4.43b. In the case of the GA-optimized stage 4 Peano–Gosper fractile array, the maximum sidelobe level over the entire visible region and for scan angles up to and including 30° from broadside was found to be −10.12 dB. Finally, we note that for a stage 4 Peano–Gosper fractile array the iterative beamforming algorithm is 19.5 times faster than the conventional beamforming technique based on a DFT.

4.4.5 Fractal–Random and Polyfractal Arrays

One of the most promising concepts in optimizing element spacing is the use of layouts based on fractal–random geometries. Fractal–random geometries can be used to create objects that more closely resemble natural structures. Fractal–random structures can be created in a manner similar to deterministic fractals; however, the generator pattern must first be randomly selected from

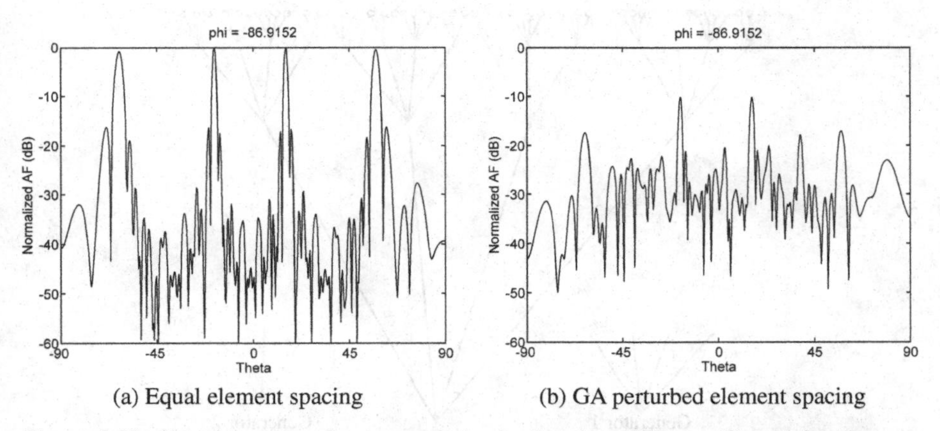

(a) Equal element spacing (b) GA perturbed element spacing

Figure 4.43. *Plot of the normalized array factor for the stage 3 Peano–Gosper fractile array with initial element spacing* $d_{min} = 2\lambda$ *and* $\varphi = -86.9152°$*: (a) equal element spacing; (b) GA perturbed element spacing.*

a set of multiple generators and then applied [57]. We visualize how multiple generators can be employed to construct fractal–random arrays by using the fractal–random tree geometry. Figure 4.44 shows a 51-element fractal–random array constructed from two generators, one 4-branch generator, and one 3-branch generator. Each generator is scaled according to its level in the tree structure. Finally, the positions of the antenna elements are represented by the tips of the tree's topmost branches.

The properties of fractal–random arrays were originally studied by Kim and Jaggard in 1986 [57]. One of the key benefits of these array layouts is that they act as an intermediary between purely periodic arrays and random arrays, possessing physical and electrical attributes of both. The fractal nature of these arrays makes them partially deterministic; however, the random selection of generators also introduces a degree of randomness into these arrays. The deterministic properties of the array keep the peak sidelobe levels lower than a purely random array.

The primary difficulty with using fractal–random arrays in optimization problems is that the array layouts cannot be exactly recreated from the generators alone. An additional parameter must be used to specify the combination of the generators selected. The number of combinations can be extremely large. For instance, 345,600 different three-level fractal–random arrays can be created from the two generators shown in Figure 4.44. For higher-level fractal–random arrays the number of combinations increases exponentially. To overcome this problem, a special subclass of fractal–random arrays has been introduced [58], called *polyfractal arrays*, which are particularly suitable for use in GA optimizations. Polyfractal arrays are constructed from multiple generators as are fractal–random arrays; however, the generator attached to

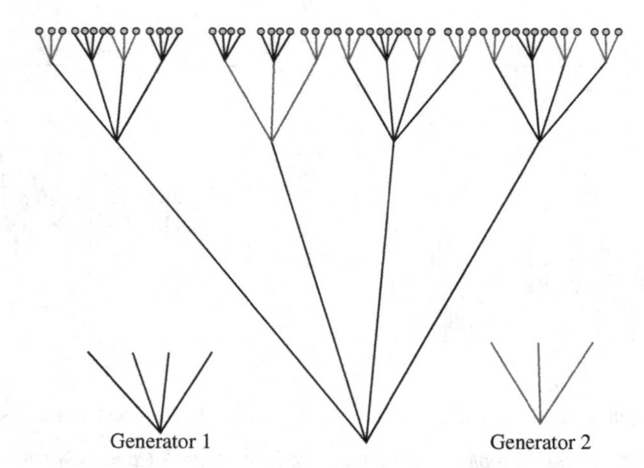

Figure 4.44. *Fractal–random tree analogy for the construction of a 51-element linear fractal-random array. The element positions are represented by the tips of the topmost branches.*

the end of each branch is not selected randomly but instead in accordance with the branch's *connection factor* [58]. Each branch of each generator has a connection factor associated with it. In addition, one global connection factor is introduced to specify which generator is used for the base of the tree. An example shown in Figure 4.45 describes how a 46-element polyfractal array can be constructed from the same two generators and an accompanying set of connection factors. By using the connection factors, a polyfractal array can be recreated with ease.

Polyfractal arrays are well suited for optimization techniques not only for their ability to be reproduced but also for the speed in which their radiation characteristics can be evaluated [58]. The fractal properties of polyfractal arrays create subarrays in which the layouts and therefore the radiation patterns are identical. To take full advantage of this self-similarity, the topmost subarrays in the fractal tree are first evaluated. Next, the radiation patterns of the immediately lower-stage subarrays are evaluated using the radiation patterns of the upper-stage subarrays as elemental radiation patterns. In this way the polyfractal array can be treated as an array of arrays. Finally, this procedure is repeated until the bottom level is reached, at which point the radiation pattern of the entire array is found. This recursive procedure can evaluate radiation patterns many times faster than the conventional methods based on the DFT. An illustration of this procedure is shown in Figure 4.46 for a 46-element polyfractal array. Note that in this figure the fractal tree structure of the array is shown inverted.

An example of a genetically optimized polyfractal array is now presented. The example was evolved from a population size of 500 members for over 400 generations. The resulting array consists of 256 elements, has a peak sidelobe level of −18.84 dB, and has a half-power beamwidth of 0.203°. The average

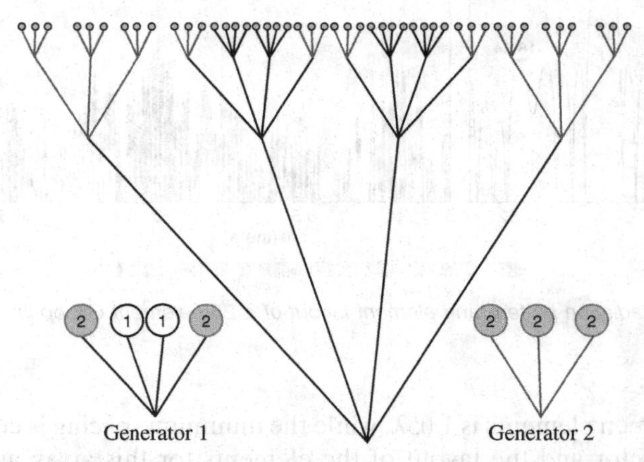

Figure 4.45. *Fractal–random tree analogy for the construction of a 46-element polyfractal array. The connection factors are illustrated by the numbers above the branches of the four- and three-branch generators.*

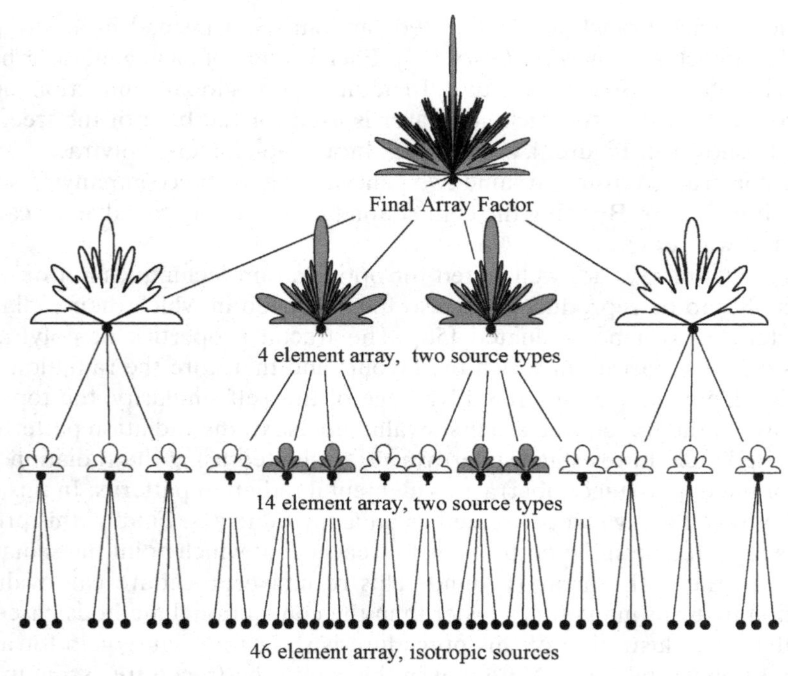

Figure 4.46. *Illustration of the recursive beamforming algorithm for a 46-element polyfractal array.*

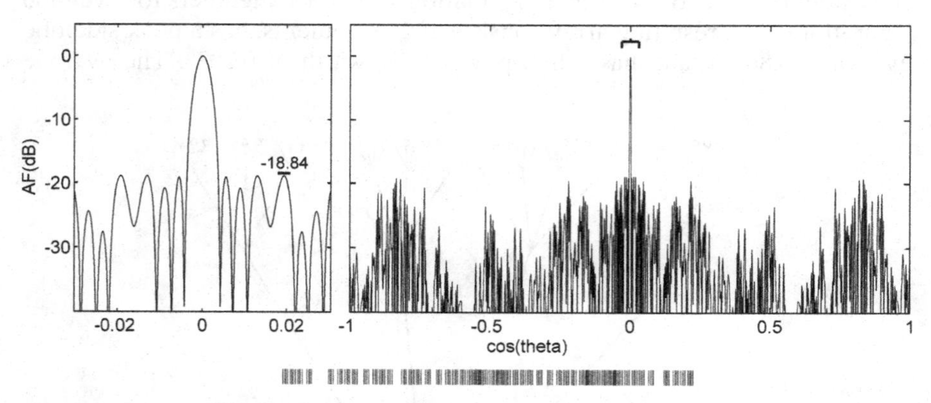

Figure 4.47. *Radiation pattern and element layout of a 256-element GA optimized polyfractal array.*

spacing between elements is 1.05λ, while the minimum spacing is equal to 0.5λ. The array factor and the layout of the elements for this array are shown in Figure 4.47. The performance properties of this array are summarized in Table 4.3. The recursive beamforming algorithm significantly reduces the time

TABLE 4.3. Performance Characteristics for a 256-Element Polyfractal Array Evolved Using a GA

Minimum Spacing	Average Spacing	Number of Elements	SLL (dB)	Half-power Beamwidth
0.5λ	1.05λ	256	−18.84	0.203°

required to calculate the radiation pattern of the polyfractal array. For the radiation pattern of the 256-element array with a resolution of 0.001°, the calculation is performed 454% faster using the recursive method versus a DFT-based method. Because of this speed increase, optimizations for large arrays that would normally take weeks now only take a matter of days or even hours.

4.4.6 Aperiodic Reflectarrays

Nonuniform element spacings can also be exploited in reflectarray design in order to achieve performance improvements such as reduced sidelobe levels. To this end, a technique for the design of unequally spaced microstrip reflectarrays has been presented [59]. An optimization approach was used for deriving the reflectarray element positions that is a variant of the GA known as the *differential evolution algorithm*. The technique was used to design a nonuniformly spaced 18 × 18-element reflectarray for operation at 10 GHz. A prototype of the reflectarray was fabricated with measured radiation patterns compared to simulations.

4.5 OPTIMIZING CONFORMAL ARRAYS

A particularly important area of practical interest is the application of robust optimization techniques to the design of conformal arrays. This is a very challenging problem in many respects. Nevertheless, significant progress in this area has recently been made. Conformal array synthesis techniques were presented in Ref. 60 for the design of one-dimensional circular arc arrays as well as for rectangular arrays on circular cylinders of infinite length. These conformal array synthesis methods were based on using simulated annealing (SA) techniques to determine the optimal set of current excitations (both amplitudes and phases) required to best approximate a desired far-field radiation pattern. Examples considered in Ref. 60 include the synthesis of a flat-topped shaped beam with a −30 dB maximum sidelobe level and a sum pattern with a −34 dB sidelobe level in addition to three nulls on either side of the mainbeam, each obtained via SA optimization of a conformal array with 25 half-wavelength spaced axial dipoles arranged along a 120° circular arc. Another example discussed in Ref. 60 concerns using SA to synthesize a $\csc^2(\theta)$-shaped radiation pattern for an 8 × 8 cylindrical–rectangular array that is conformal to the surface of a 5λ radius circular cylinder with half-wavelength spacing between elements. Finally, we note that GA techniques could just as easily be used in place of the SA approaches for these types of array synthesis problems.

A hybrid domain decomposition/reciprocity procedure has been presented [61] that allows the radiation patterns of microstrip patch antennas mounted on arbitrarily shaped three-dimensional metallic platforms to be computed accurately as well as efficiently. *Domain decomposition* in this context refers to the process of separating the composite problem into two parts, each of which can be solved with much less computational complexity than the original. The first problem involves computing the magnetic currents $\vec{M}_1$ in the aperture of the microstrip patch antenna, while the second involves computing the tangential magnetic fields $\vec{H}_2$ induced on the platform due to plane-wave scattering in the absence of the patch antenna. Once these two separate calculations have been performed, the results are used in conjunction with the reciprocity theorem to determine the far-field radiation pattern of the microstrip patch antenna when placed on the arbitrarily shaped platform. In its standard form the reciprocity theorem states that [62]

$$\iiint_{V_2} \left(\vec{E}_1 \cdot \vec{J}_2 - \vec{H}_1 \cdot \vec{M}_2\right) dV = \iiint_{V_1} \left(\vec{E}_2 \cdot \vec{J}_1 - \vec{H}_2 \cdot \vec{M}_1\right) dV \tag{4.30}$$

In order to make use of the reciprocity theorem for this procedure, it is assumed that source 1 is a microstrip patch antenna and source 2 is an ideal dipole current source located at (x_0, y_0, z_0) in the far-field of source 1 such that

$$\vec{J}_2 = \delta(x - x_0)\delta(y - y_0)\delta(z - z_0)\hat{p} \tag{4.31}$$

$$\vec{M}_2 = 0 \tag{4.32}$$

where $\vec{p} = Il\hat{u}$ is the dipole moment and $\hat{u}$ represents the unit vector tangent to the dipole. The microstrip patch antenna is assumed to be embedded in a cavity between a dielectric substrate and superstrate layer. The fields in the aperture of the cavity due to the presence of the microstrip patch antenna are denoted by $\vec{E}_{ap}$, which through equivalence may be transformed into surface magnetic currents denoted by

$$\vec{M}_1 = -\hat{n} \times \vec{E}_{ap} \tag{4.33}$$

The incident plane wave radiation produced by the far-zone source $\vec{J}_2$ causes an electric current to be induced on the surface of the arbitrarily shaped perfect electric conductor (PEC) platform, denoted here as $\vec{J}_s$. This current may be transformed into a tangential magnetic field $\vec{H}_2$ on the surface of the platform using the equivalence principle, such that

$$\vec{H}_2 = \hat{n} \times \vec{J}_s \tag{4.34}$$

We also note that $\vec{E}_2 = 0$ since $\vec{M}_s = 0$. Hence, under these conditions, the general reciprocity expression given in (4.30) reduces to

$$\vec{E}_1(\theta, \phi) \cdot \hat{p} = -\iint_S (\vec{H}_2 \cdot \vec{M}_1) dS' \tag{4.35}$$

where S represents the surface of the microstrip antenna aperture. This provides a useful expression for determining the far-zone electric fields radiated by a cavity-backed microstrip patch antenna mounted on a metallic platform of arbitrary shape [61].

As an example, let us consider the practical case of a cylindrical metallic platform with a circular cross section and finite length. Suppose that a uniformly spaced array of microstrip patch antenna elements is mounted along the length of the circular–cylindrical platform as shown in Figure 4.48. The

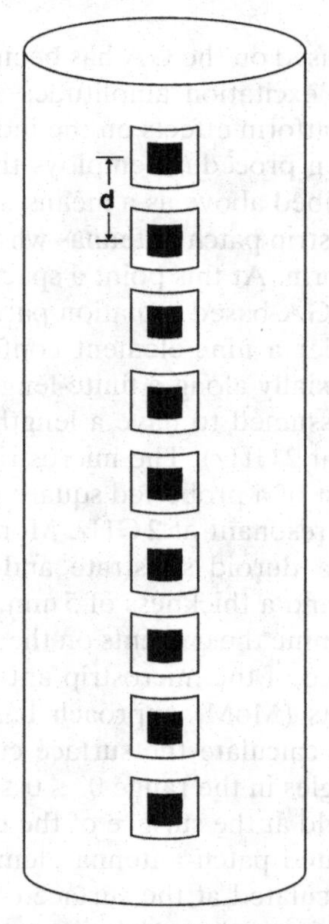

Figure 4.48. *Geometry of a nine-element microstrip patch antenna array mounted conformally to the surface of a finite-length metallic circular cylinder.*

cylindrical mounting platform is assumed to be oriented along the z axis. In this case, the far-field radiation pattern produced by a linear array of N elements may be expressed in the form

$$\mathrm{FF}(\theta, \varphi) = \sum_{n=0}^{N-1} w_n \mathrm{EP}_n(\theta, \varphi) e^{jnkd\cos\theta} \tag{4.36}$$

where $\mathrm{EP}_n(\theta, \varphi)$ are the individual array element patterns. If the cylindrical mounting platform is assumed to be infinitely long, then all the individual radiation patterns of the antenna elements in the array will be identical, provided that the effects of mutual coupling are neglected.

If the cylindrical platform is truncated (i.e., is finite in length), then the assumption that the radiation patterns for each element in the array are identical is no longer valid [61]. For instance, the radiation patterns for elements near the ends of the cylinder will be different from those produced by the elements near the center of the cylinder. In general, each individual array element pattern may be different and the simple expression based on a progressive phase shift (i.e., $\alpha_n = -nkd\,cos\,\theta_s$) can no longer be used to steer the mainbeam.

A synthesis technique based on the GA has been introduced [61] for determining the optimal set of excitation amplitudes and/or phases required in order to compensate for platform effects on the individual array element patterns. The GA optimization procedure employs the domain decomposition/reciprocity approach described above as a means of efficiently evaluating the radiation patterns of microstrip patch antennas when placed at arbitrary locations on a cylindrical platform. At this point a specific design example will be presented to illustrate the GA-based radiation pattern synthesis technique.

Suppose that we consider a nine-element conformal array of microstrip patch antennas mounted axially along a finite-length metallic (PEC) circular cylinder. The cylinder is assumed to have a length of 60 cm and a radius of 7.5 cm (a half-wavelength at 2 GHz). The microstrip patch antenna elements used in this example consist of a probe-fed square patch that is 4.915 cm on a side and is designed to be resonant at 2 GHz. Moreover, the patch antennas are sandwiched between a duroid substrate and superstrate, each with a dielectric constant of 2.33 and a thickness of 5 mm.

The first step is to determine the currents on the surface of the finite-length PEC cylinder in the absence of the microstrip antenna array. To accomplish this, a method of moments (MoM) approach based on the fast multipole method (FMM) is used to calculate the surface currents due to plane-wave incidence from far-field angles in the range $0° \le \theta \le 360°$. The tangential components of the magnetic field at the surface of the cylinder can then be found from (4.34). Next, an isolated patch antenna element is considered and the aperture fields $\vec{E}_{\mathrm{ap}}$ are calculated at the surface of the dielectric superstrate layer using a suitable full-wave analysis technique such as MoM, finite-element method (FEM), finite-element boundary integral (FEBI), or finite-difference time domain (FDTD). The corresponding magnetic surface currents $\vec{M}_1$ are then determined using (4.33). Finally, the results obtained from (4.33) and (4.34) can be combined together via the reciprocity expression given in (4.35) to find the far-field radiation pattern produced by an individual microstrip patch antenna element when placed at any arbitrary location on the circular–cylindrical platform.

Once the individual element patterns for each of the microstrip patch antennas in the cylinder-mounted array have been determined, the far-field radiation pattern of the array may be found using (4.36) for a specific set of excitation currents (amplitudes and phases). For the synthesis problem, a desired far-field radiation pattern is specified and the set of excitation currents is sought that produces the "best" approximation to this desired pattern. Here we will consider a radiation pattern synthesis technique for the cylinder-mounted nine-element microstrip patch antenna array described above that is based on a GA. The radiation pattern chosen as the objective for the GA is an ideal cosine pattern defined by [63]

$$f(\theta) = \begin{cases} \cos\left[\dfrac{\pi}{2}\left(\dfrac{\theta-\theta_s}{\delta}\right)\right], & \text{for} \quad \theta \in [\theta_a, \theta_b] \subseteq [-\pi/2, \pi/2] \\ 0, & \text{elsewhere} \end{cases} \tag{4.37}$$

where θ_s is the desired steering angle (measured from the z axis), δ is the half-power beamwidth, and θ_a and θ_b are limits on the extent of the mainbeam.

Two examples will now be considered where the design parameters for the desired ideal cosine pattern (4.37) are chosen to be $\theta_s = 60°$ (mainbeam steered to 30° from broadside), $\delta = 16°$, $\theta_a = 44°$, and $\theta_b = 76°$. In the first case a GA approach was used to evolve the optimal set of complex weightings (i.e., current amplitudes and phases) for the array that would be required to best approximate this radiation pattern, while in the second case a phase-only optimization was performed by the GA. The synthesized radiation patterns for these two cases are indicated in Figure 4.49 by the solid curve and the dashed curve respectively.

In Ref. 64 a general methodology based on the GA was developed to optimize positioning for any number of radiators on the surface of an aircraft. Moreover, the technique was demonstrated for positioning of antennas that operate in the VHF (very high-frequency) band (150–300 MHz) on a realistic model of a Boeing 747-200 aircraft. A novel aggregate objective function is used in the GA that incorporates surface wave coupling predicted via uniform theory of diffraction (UTD), variance of the coupling values, and far-field radiation pattern analysis. Simple GA, micro-GA, and GA population seeding techniques were all used and compared with an exhaustive GA/local search benchmark. The problem scenarios that were considered include one, three, and eight movable antennas.

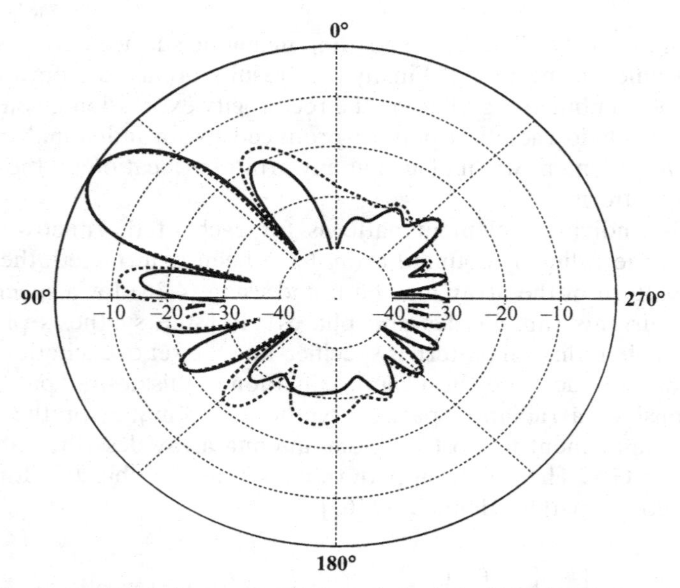

Figure 4.49. Comparison between magnitude and phase (solid line) and phase-only (dashed line) GA optimization of the radiation pattern for a nine-element array mounted along the cylinder axis (Allard et al., © IEEE, 2003 [61]).

A robust synthesis technique has been introduced [65] that uses a GA to evolve optimal low-sidelobe radiation patterns for spherical arrays (i.e., arrays that conform to the surface of a sphere). The far-field radiation pattern produced by a spherical array may be expressed in the form [66,67]

$$\mathrm{FF}(\theta, \varphi) = \sum_{n=1}^{N} w_n EP_n(\theta, \varphi) e^{-jkR(\cos\xi_n - \cos\xi_0)} \tag{4.38}$$

where

$$\xi_n = \sin\theta \sin\theta_n \cos(\varphi - \varphi_n) + \cos\theta \cos\theta_n \tag{4.39}$$

$$\xi_0 = \sin\theta_0 \sin\theta_n \cos(\varphi_0 - \varphi_n) + \cos\theta_0 \cos\theta_n \tag{4.40}$$

$$R = \text{radius of the sphere} \tag{4.41}$$

$$(\theta_n, \varphi_n) = \text{angles for element } n \tag{4.42}$$

Radiation pattern synthesis for two types of spherical arrays is considered in Ref. 65. The first type is called a *spherical–planar array*, where the spherical array layout is generated by molding a planar array to the surface of a sphere. The second type is called a *spherical–circular array*, where a series of circular arrays are created on the surface of a sphere. In addition to this, a GA approach for adaptive nulling with spherical arrays has been reported [68].

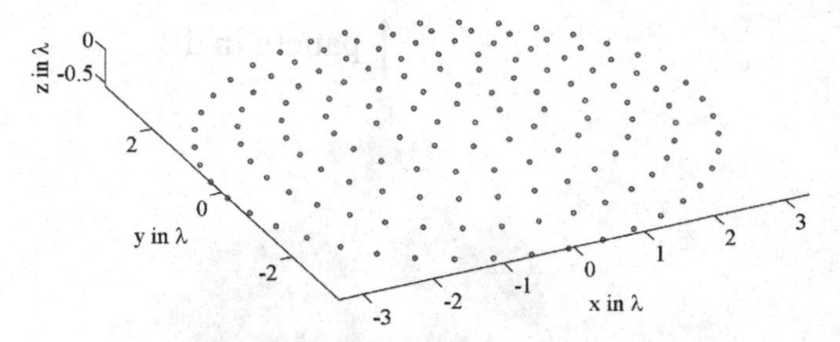

Figure 4.50. *Model of the spherical circular array.*

As an example of optimizing the array factor of a spherical array, consider a spherical circular array with one center element and six rings. This array lies on a sphere of radius 10λ. The spacing between rings when projected on the x–y plane is 0.557λ. A model of the array appears in Figure 4.50. When this array has uniform weights, then the maximum relative sidelobe level is –2.61 dB. Placing a phase taper on the array that compensates for the z distance between the ring elements and the center element results in a relative sidelobe level of –17.5 dB. This sidelobe level reduces to –17.9 dB after applying a 30 dB, $\bar{n} = 2$ Taylor taper.

In order to further reduce the sidelobe levels, a hybid GA is used. In this case, a continuous variable GA is linked with a Nelder–Mead downhill simplex algorithm. The center element is assumed to have an amplitude of 1 and a phase of 0. Thus, a single chromosome has 12 variables: six for amplitude and six for phase. The cost function returns the maximum relative sidelobe level. After running the algorithm, the resulting amplitude weights for the rings are

$$amplitude = [0.73 \quad 0.73 \quad 0.66 \quad 0.51 \quad 0.25 \quad 0.36]$$

and the corresponding phase weights are

$$phase = 2\pi[0.24 \quad 0.19 \quad 0.32 \quad 0.36 \quad 0.57 \quad 0.71]$$

These weights produce a maximum array factor sidelobe level of –26.7 dB. Figure 4.51 is a plot of the resulting array factor.

A method for the generation of optimal distribution sets for single-ring cylindrical arc arrays was presented in Ref. 69. In this case a GA is used to optimize the amplitude and phase coefficients of the conformal cylindrical arc array in order to achieve a desired radiation pattern goal. A design procedure for microstrip patch antenna arrays with aperture-coupled elements that has application to point-to-multipoint radio links at 25 GHz is described in Ref. 70. The array synthesis has been performed through an optimization scheme based on the GA. In Ref. 71 a GA technique is applied to optimize the design

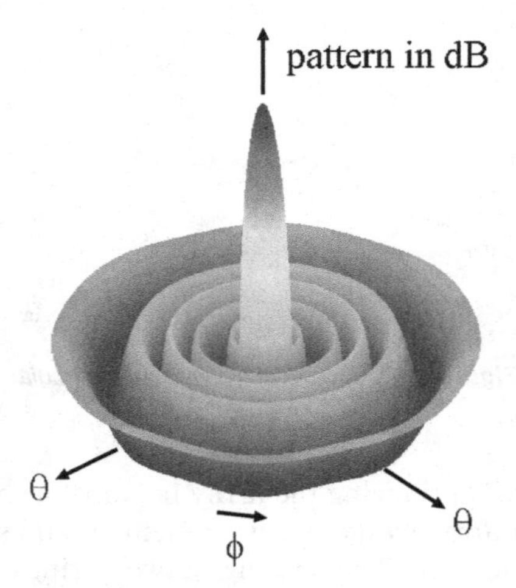

Figure 4.51. Optimized array factor for the spherical circular array.

of a novel array prototype with seven microstrip patch elements in a hexagonal arrangement. Furthermore, this hexagonal array was shown to be capable of configuring its radiation pattern adaptively in real time and is thereby suitable for wireless communications applications. Finally, an efficient methodology based on an accelerated hybrid evolutionary GA has been introduced [72] for printed array pattern synthesis that takes into account the effects of coupling between the radiating elements. Numerical design results were presented that illustrate the advantages of the proposed technique when compared to conventional synthesis methods.

4.6 OPTIMIZING RECONFIGURABLE APERTURES

Reconfigurable aperture (RECAP) antennas are attractive, because they can provide a high degree of versatility in their performance when compared to more traditional antennas. For example, electronic reconfiguration can be used to change the bandwidth of operation or steer the radiation pattern of the antenna. Another advantage of reconfigurable apertures is that they can be designed to have similar performance to conventional phased arrays, but with only a single feed point rather than multiple feeds.

A novel RECAP antenna based on switched links between electrically small metallic patches has been described in [73–75]. The antenna can be reconfigured to meet different performance goals by simply changing the switches that are open and closed in an appropriate way. A GA was used to determine the optimal switch configuration (i.e., which should be open and which should be closed) that is required to achieve a particular goal, such as maximum gain over a specified bandwidth. A finite-difference time-domain (FDTD) method was used in conjunction with the GA to provide rigorous

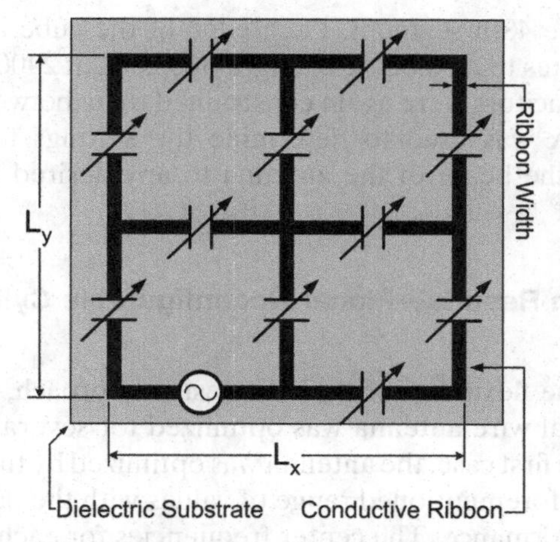

Figure 4.53. Planar reconfigurable ribbon antenna geometry.

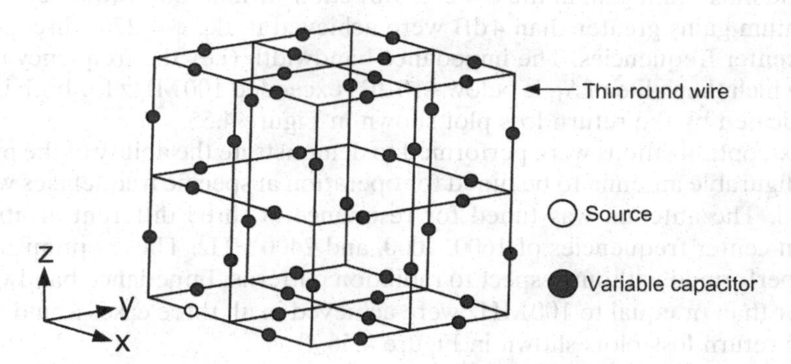

Figure 4.54. Volumetric reconfigurable cylindrical wire antenna geometry.

the cylindrical wires replaced by conducting ribbons. A thin finite-size dielectric substrate was also added, as a physical antenna requires a substrate on which to print the conducting ribbons. A dielectric substrate also provides a surface on which supporting components can be fabricated and/or tuning elements can be mounted. The source and variable capacitor locations are identical to those of the cylindrical wire version of the planar reconfigurable antenna. The antenna geometry is shown in Figure 4.53.

4.6.3 Design of Volumetric Reconfigurable Antennas

The volumetric reconfigurable antenna introduced here is based on the cubical geometry shown in Figure 4.54, which is composed of 48 individual wire segments on the surface of the cube. A variable capacitor was placed at the center of 47 of these wire segments, while the antenna feed was assumed to be located at the center of the 48th segment. Each edge of the cube antenna measures 3.5 cm, which equates to an electrical length of 0.280λ at 2400 MHz. The values of the variable capacitors were again constrained to lie between 0.1 and 1.0 pF. In this case a GA was used to determine the settings for each capacitor required to steer the beam of the antenna to any desired location in three-dimensional space.

4.6.4 Simulation Results—Planar Reconfigurable Cylindrical Wire Antenna

To demonstrate the flexibility of the new design approach, the planar reconfigurable cylindrical wire antenna was optimized for several different performance goals. In the first case, the antenna was optimized by tuning the capacitor values within the aforementioned range of values with the objective of achieving dual-band performance. The center frequencies for each band were specified as 2000 and 2400 MHz. The antenna was optimized for minimum return loss and maximum gain in the $\phi = 270°$ direction within both frequency bands. Maximum gains greater than 4 dB were achieved in the $\phi = 270°$ direction at both center frequencies. The impedance bandwidth (i.e., the frequency range over which the value of S_{11} is below −10 dB) exceeded 100 MHz for both bands as indicated by the return loss plot shown in Figure 4.55.

Next, optimizations were performed to demonstrate the ability of the planar reconfigurable antenna to be tuned for operation at specific frequencies within a band. The antenna was tuned for resonance at three different arbitrarily chosen center frequencies of 1600, 2000, and 2400 MHz. These optimizations were performed without respect to radiation patterns. Impedance bandwidths greater than or equal to 100 MHz were achieved in all three cases as indicated by the return loss plots shown in Figure 4.56.

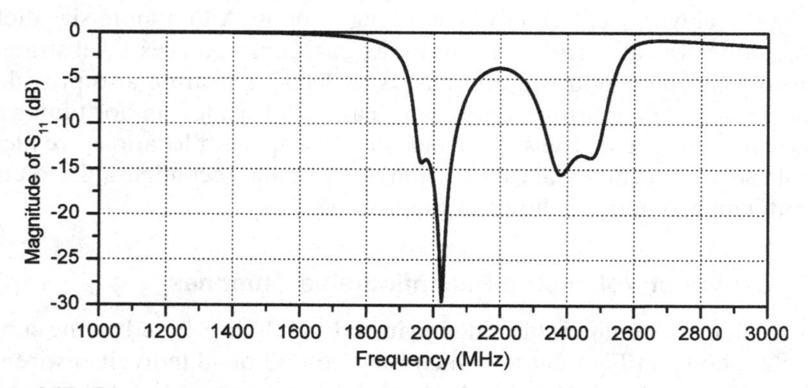

Figure 4.55. Return loss of the planar reconfigurable cylindrical wire antenna tuned for dual-band performance.

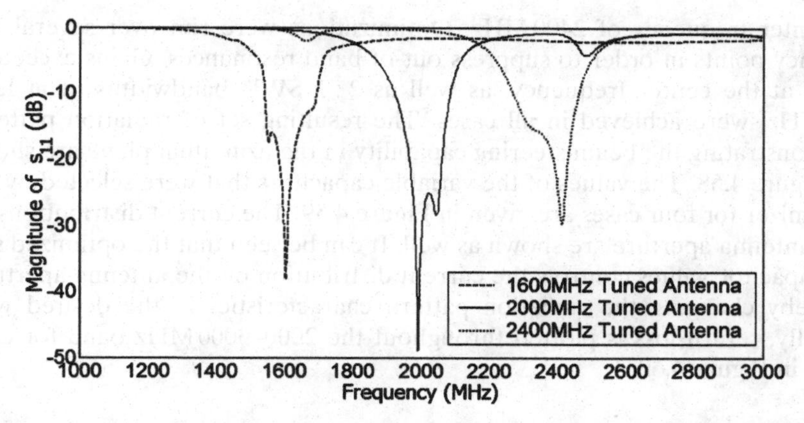

Figure 4.56. *Tuning of the planar reconfigurable cylindrical wire antenna for resonance in three frequency bands.*

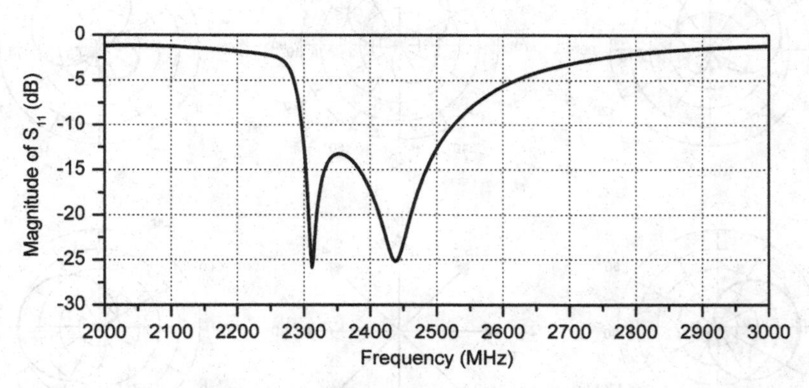

Figure 4.57. *Tuning of the planar reconfigurable cylindrical wire antenna for broadband operation.*

Another optimization was performed in order to achieve a relatively large impedance bandwidth irrespective of radiation patterns. To do so, the GA optimizer was configured to minimize return loss in the frequency range of 2300–2500 MHz, and to suppress out-of-band resonances. An impedance bandwidth of 200 MHz was obtained, as shown in Figure 4.57.

A final set of optimizations were performed on the planar reconfigurable cylindrical wire antenna to demonstrate its beam steering capabilities. The antenna was optimized for maximum gain in eight directions in the azimuthal plane. Return loss was also simultaneously minimized in each case assuming a center frequency of 2400 MHz. Optimizations were run over several frequency points in order to suppress out-of-band resonances. Gains exceeding 5 dB at the center frequency, as well as 2 : 1 SWR bandwidths of at least 50 MHz, were achieved in all cases. The resulting set of radiation patterns demonstrating the beam steering capability in the azimuthal plane are shown in Figure 4.58. The values of the variable capacitors that were selected by the optimizer for four cases are given in Figure 4.59. The current distributions on the antenna aperture are shown as well. It can be seen that the optimized sets of capacitor values controls the current distribution on the antenna aperture, thereby changing the radiation pattern characteristics in the desired way. Finally, return loss is plotted throughout the 2000–3000 MHz band for each case in Figure 4.60.

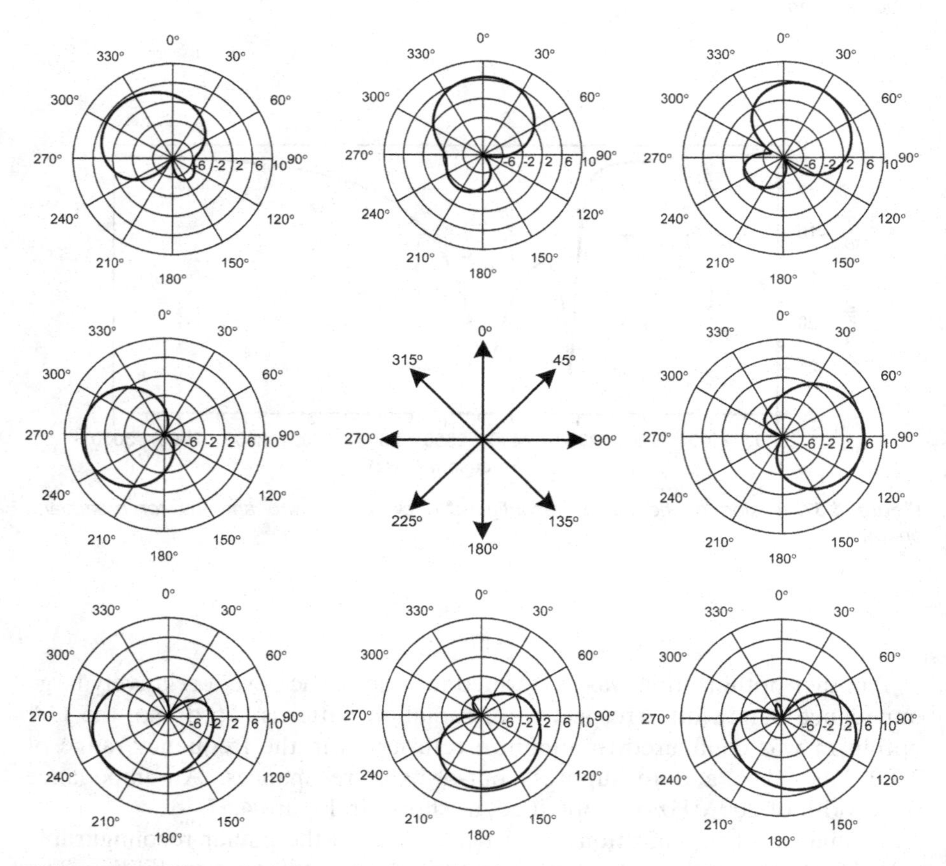

Figure 4.58. *Gain (dB) in the azimuthal plane of the planar reconfigurable cylindrical wire antenna tuned for maximum gain in eight different directions.*

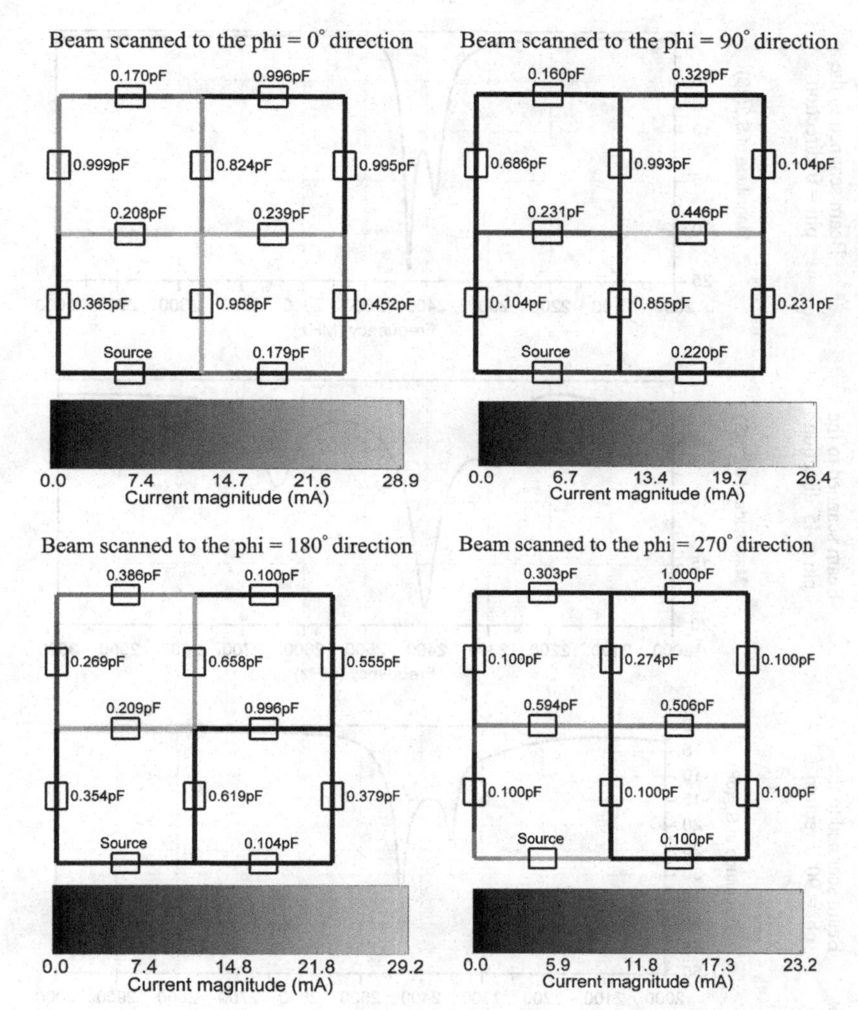

Figure 4.59. *Current distributions and capacitor values of the planar reconfigurable cylindrical wire antenna optimized for maximum gain in four selected directions.*

4.6.5 Simulation Results—Volumetric Reconfigurable Cylindrical Wire Antenna

The volumetric reconfigurable cylindrical wire antenna shown in Figure 4.54 was optimized to steer the mainbeam in the x, y, $-y$, and z directions. Return loss was also simultaneously minimized in each case assuming a center frequency of 2400 MHz. Optimizations were performed with the additional

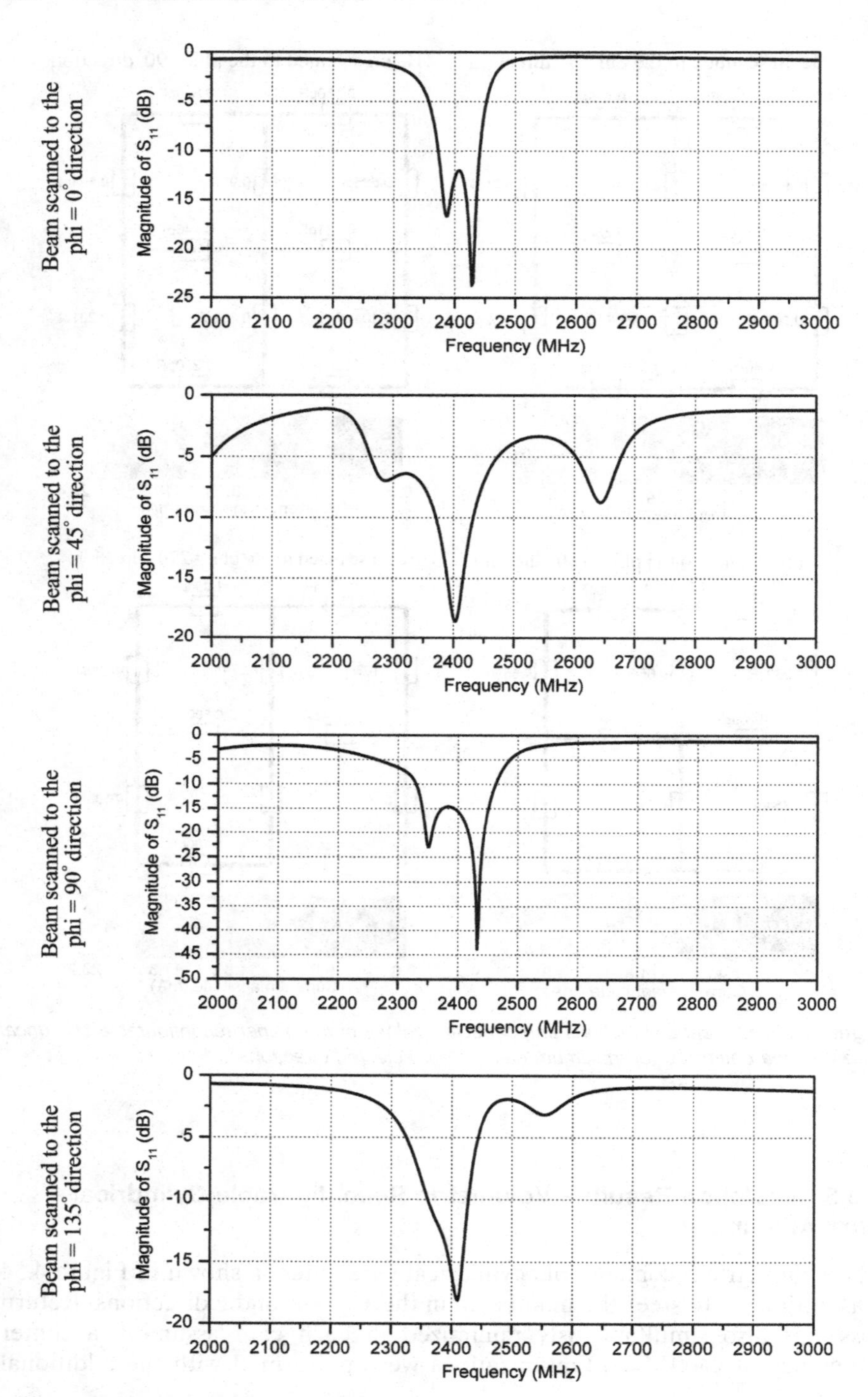

Figure 4.60. *Return loss of the planar reconfigurable cylindrical wire antenna when optimized for maximum gain in eight different directions.*

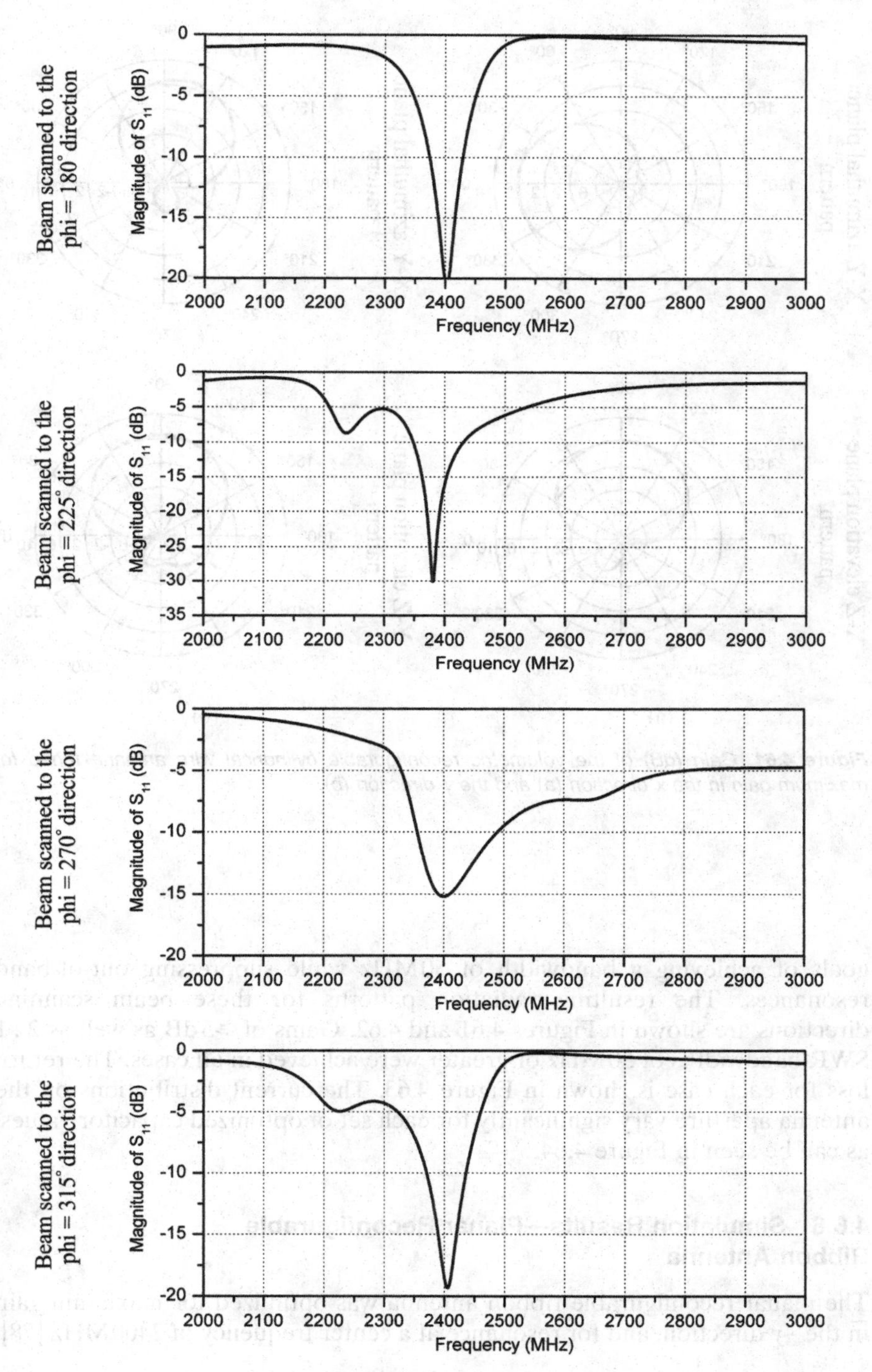

Figure 4.60. *Continued*

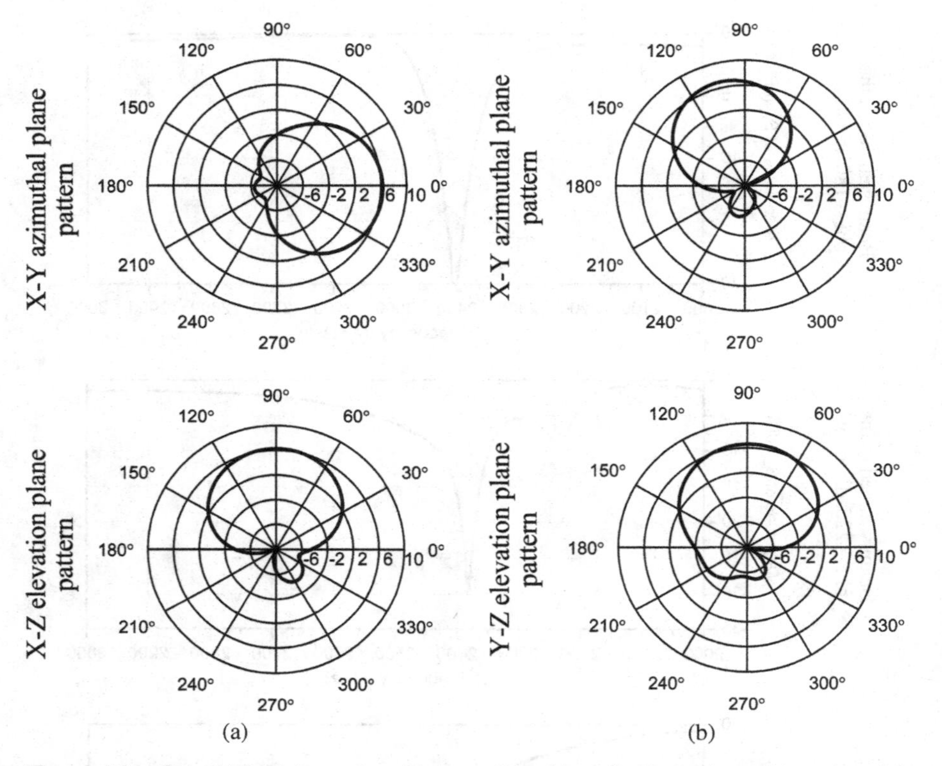

Figure 4.61. Gain (dB) of the volumetric reconfigurable cylindrical wire antenna tuned for maximum gain in the x direction (a) and the y direction (b).

goals of achieving a bandwidth of 50 MHz while suppressing out-of-band resonances. The resulting radiation patterns for these beam scanning directions are shown in Figures 4.61 and 4.62. Gains of ⩾5 dB as well as 2 : 1 SWR bandwidths of 50 MHz or greater were achieved in all cases. The return loss for each case is shown in Figure 4.63. The current distributions on the antenna aperture vary significantly for each set of optimized capacitor values, as can be seen in Figure 4.64.

4.6.6 Simulation Results—Planar Reconfigurable Ribbon Antenna

The planar reconfigurable ribbon antenna was optimized for maximum gain in the $-y$ direction, and for resonance at a center frequency of 2400 MHz [78].

The performance of this antenna was evaluated via full-wave method of moments simulations. The length and width of the antenna were set to 2.9 cm, and the ribbon width used was 1.0 mm. Capacitor values were again constrained to the range of 0.1–1.0 pF. A finite dielectric substrate was added with a relative dielectric constant of 3.8 and dimensions 3.2 × 3.2 × 0.1524 cm. The gain at the center frequency in the $-y$ direction was approximately 5 dB, as shown in Figure 4.65. A bandwidth of approximately 75 MHz was achieved, as indicated in Figure 4.66. Three dimensional views of the radiation pattern superimposed on the antenna model are shown in Figure 4.67. These results indicate that performance similar to that of the planar cylindrical wire geometry can be achieved with the ribbon geometry for this particular optimization goal.

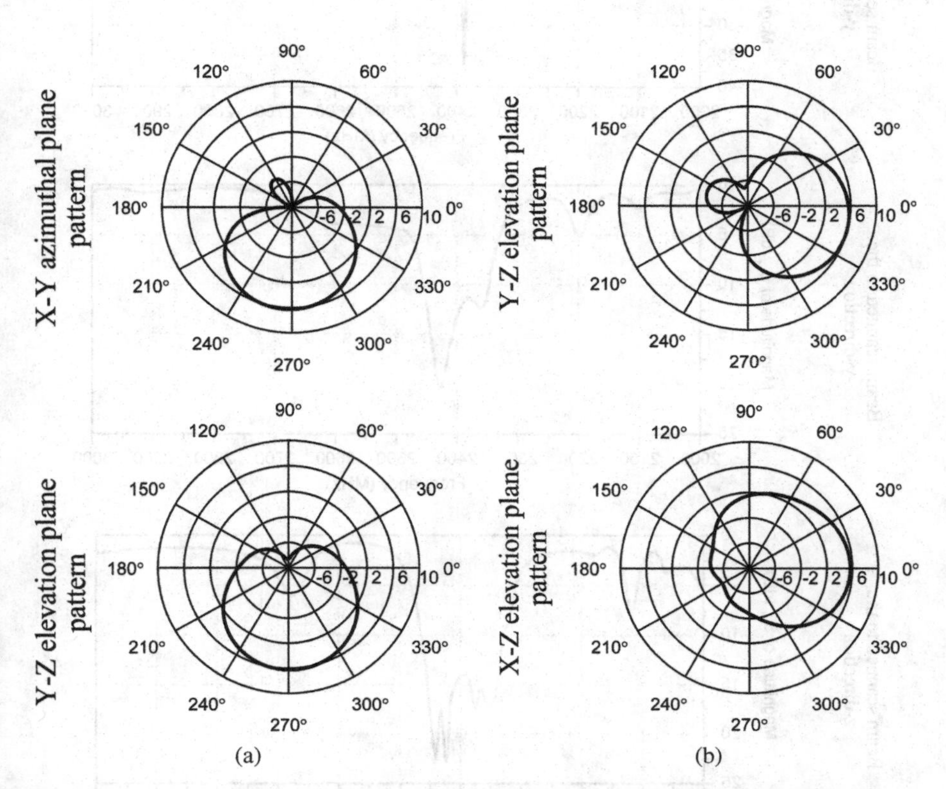

Figure 4.62. *Gain (dB) of the volumetric reconfigurable cylindrical wire antenna tuned for maximum gain in the –y direction (a) and the z direction (b).*

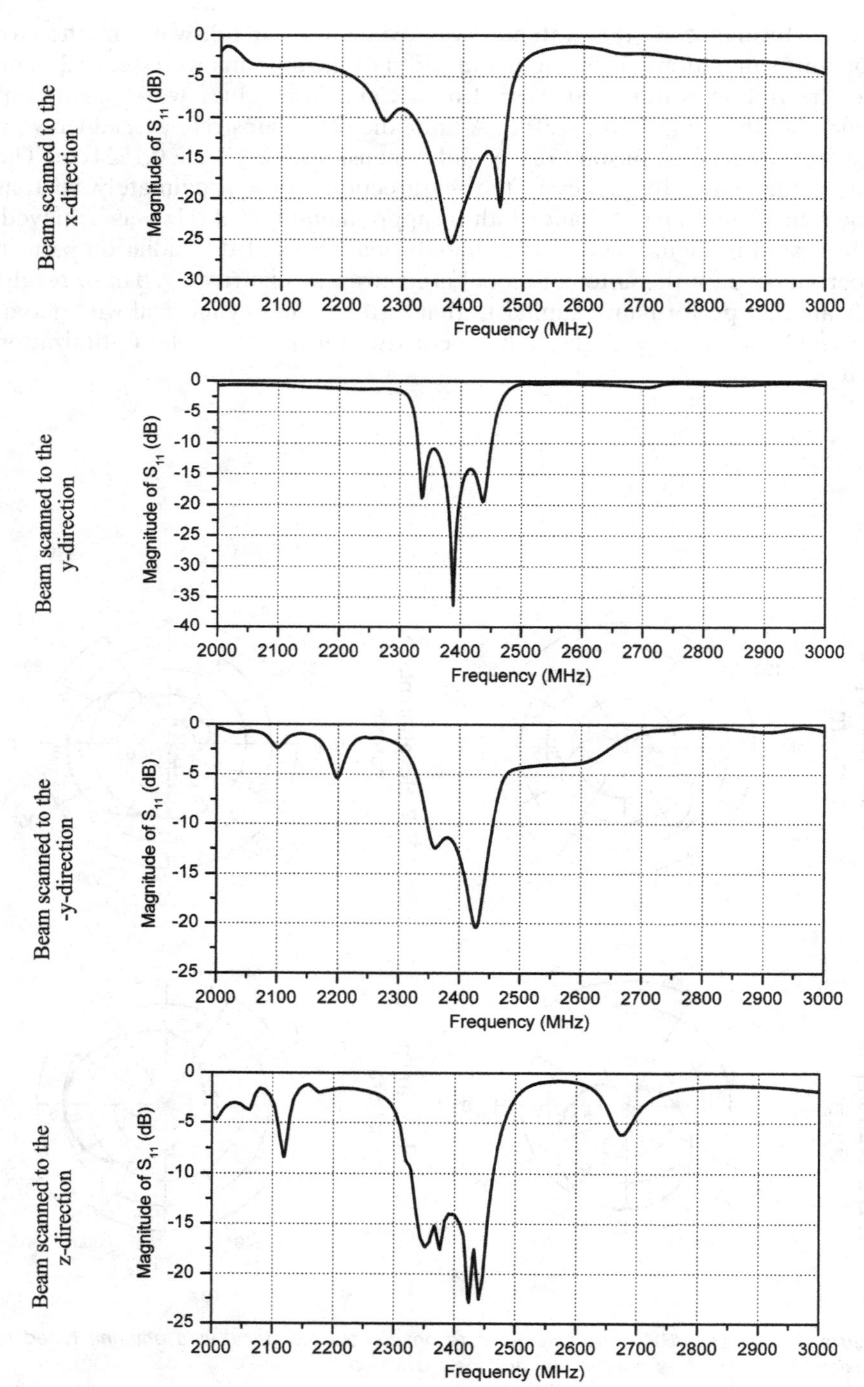

Figure 4.63. *Return loss of the volumetric reconfigurable cylindrical wire antenna when optimized for maximum gain in four different directions.*

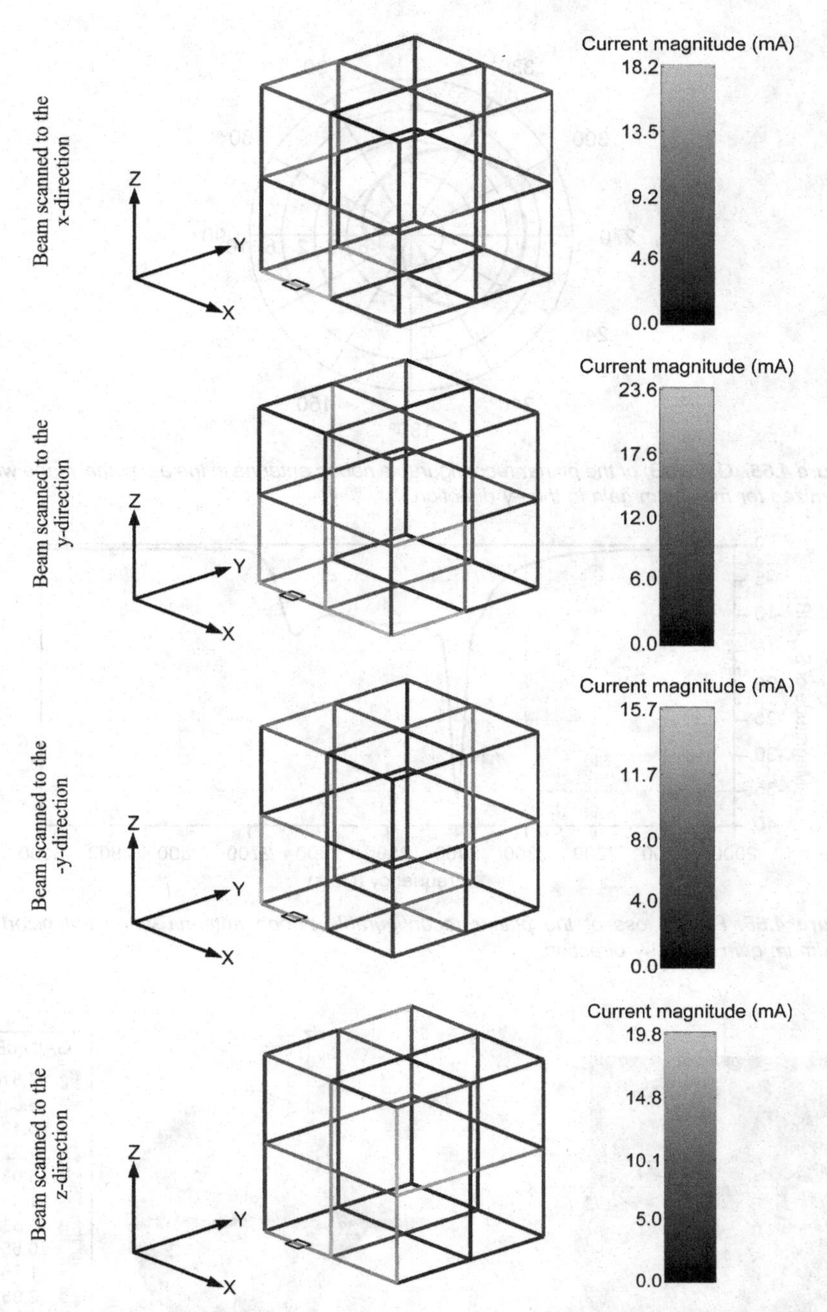

Figure 4.64. *Current distributions for the volumetric reconfigurable cylindrical wire antenna optimized for maximum gain in four different directions.*

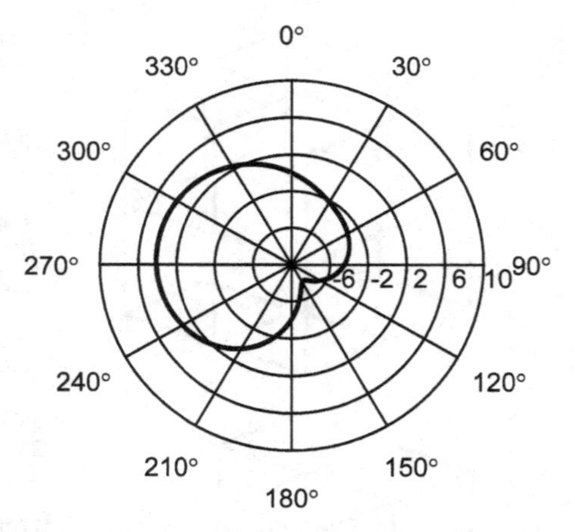

Figure 4.65. Gain (dB) of the planar reconfigurable ribbon antenna in the azimuthal plane when optimized for maximum gain in the −y direction.

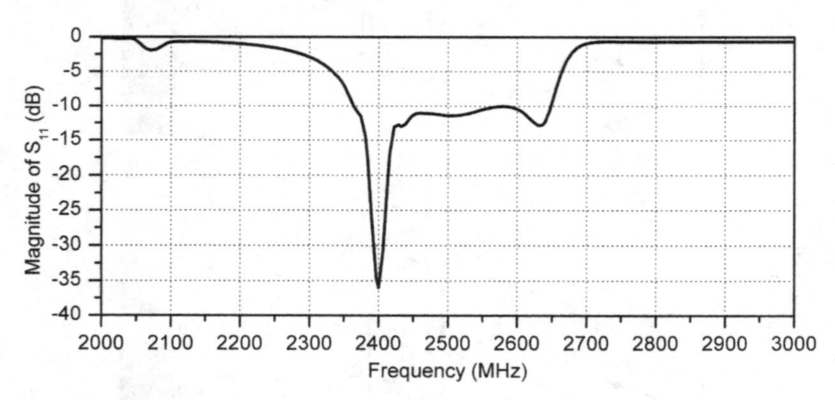

Figure 4.66. Return loss of the planar reconfigurable ribbon antenna when optimized for maximum gain in the −y direction.

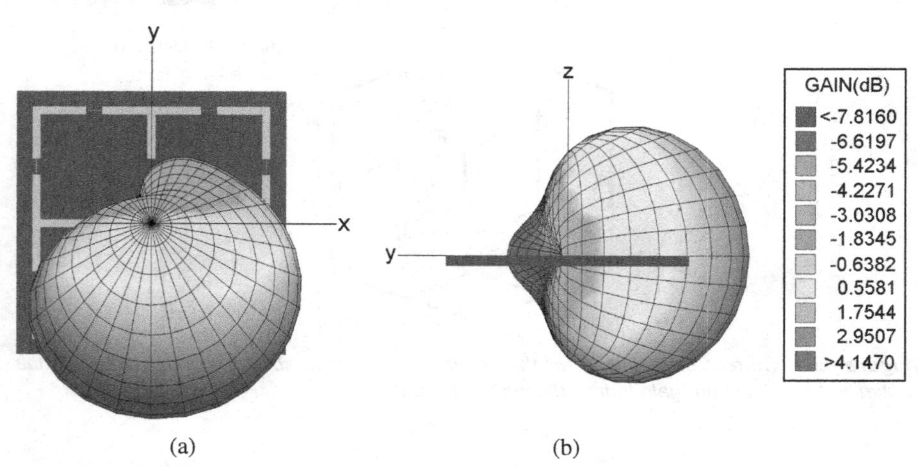

Figure 4.67. Radiation pattern of the planar reconfigurable ribbon antenna when optimized for maximum gain in the −y direction: (a) top view; (b) side view.

REFERENCES

1. R. L. Haupt, Synthesizing low sidelobe quantized amplitude and phase tapers for linear arrays using genetic algorithms, *Proc. Inte. Conf. Electromagnetics in Advanced Applications*, Torino, Italy, Sept. 1995, pp. 221–224.
2. P. Lopez, J. A. Rodriguez, F. Ares, and E. Moreno, Low sidelobe level in almost uniformly excited arrays, *IEE Electron. Lett.* **26**(24):1991–1993 (Nov. 2000).
3. J. F. DeFord and O. P. Gandhi, Phase-only synthesis of minimum peak sidelobe patterns for linear and planar arrays, *IEEE Trans. Anten. Propag.* **36**(2):191–201 (Feb. 1988).
4. R. L. Haupt, Optimum quantized low sidelobe phase tapers for arrays, *IEE Electron. Lett.* **31**(14):1117–1118 (July 1995).
5. D. W. Boeringer and D. H. Werner, Adaptive mutation parameter toggling genetic algorithm for phase-only array synthesis, *IEE Electron. Lett.* **38**(25):1618–1619 (Dec. 2002).
6. D. W. Boeringer, D. H. Werner, and D. W. Machuga, A simultaneous parameter adaptation scheme for genetic algorithms with application to phased array synthesis, *IEEE Trans. Anten. Propag.* **53**(1):356–371 (Jan. 2005).
7. D. W. Boeringer and D. H. Werner, Genetic algorithms with adaptive parameters for phased array synthesis, *Proc. 2003 IEEE Inte. Sympo. Antennas and Propagation and URSI North American Radio Science Meeting*, Columbus, OH, June 2003, Vol. 1, pp. 169–172.
8. D. W. Boeringer and D. H. Werner, A comparison of particle swarm optimization and genetic algorithms for a phased array synthesis problem, *Proc. 2003 IEEE Int. Symp. Antennas and Propagation and URSI North American Radio Science Meeting*, Columbus, OH, June 2003, Vol. 1, pp. 181–184.
9. A. E. Eiben, R. Hinterding, and Z. Michalewicz, Parameter control in evolutionary algorithms, *IEEE Trans. Evol. Computa.* **3**(2):124–141 (1993).
10. T. T. Taylor, Design of line-source antennas for narrow beamwidth and low side lobes, *IRE Trans. Anten Propaga.* **3**:16–28 (1955).
11. R. C. Hansen, Dipole array scan performance over a wide band, *IEEE Trans. Anten. Propag.* **47**(5):956–957 (May 1999).
12. S. Mano, M. Ono, and Y. Takeichi, Effects of the element pattern in the directive properties of a scanning planar array, *IEEE Trans. Anten. Propag.* **AP-22**:169–172 (March 1974).
13. H. E. King and J. L Wong, Directivity of a uniformly excited N × N array of directive elements, *IEEE Trans. Anten. Propag.* **23**(3):401–404 (May 1975).
14. D. H. Werner, D. Mulyantini, and P. L. Werner, A closed-form representation for the directivity of nonuniformly spaced linear arrays with arbitrary element patterns, *IEE Electron. Lett.* **35**(25):2155–2157 (1999).
15. J. Michael Johnson and Y. Rahmat-Samii, Genetic algorithms in electromagnetics, *IEEE Anten. Propag. Mag.* **39**(4):7–21 (Aug. 1997).
16. K. K. Yan and Y. Lu, Sidelobe reduction in array-pattern synthesis using genetic algorithm, *IEEE Trans. Anten. Propag.* **45**(7):1117–1122 (July 1997).
17. K. Markus and L. Vaskelainen, Optimization of synthesized array excitations using array polynome complex root swapping and genetic algorithms, *IEE Proc. Microw. Anten. Propag.* **145**(6):460–464 (Dec. 1998).

18. D. Marcano and F. Duran, Synthesis of antenna arrays using genetic algorithms, *IEEE Anten. Propag. Mag.* **42**(3):12–20 (June 2000).
19. D. W. Boeringer and D. H. Werner, Particle swarm optimization versus genetic algorithms for phased array synthesis, *IEEE Trans. Anten. Propag.* **52**(3):771–779 (March 2004).
20. B. K. Yeo and Y. Lu, Array failure correction with a genetic algorithm, *IEEE Trans. Anten. Propag.* **47**(5):823–828 (May 1999).
21. L. L. Wang and D. G. Fang, Combination of genetic algorithm and fast Fourier transform for array failure correction, *Proc. 6th Int. Symp. Antennas, Propagation and EM Theory*, Oct. 28–Nov. 1, 2003, pp. 234–237.
22. A. Taskin and C. S. Gurel, Antenna array pattern optimization in the case of array element failure, *Proc. 33rd European Microwave Conf.* Munich, Germany, 2003, pp. 1083–1085.
23. J. A. Rodriguez, F. Ares, E. Moreno, and G. Franceschetti, Genetic algorithm procedure for linear array failure correction, *IEE Electron. Lett.* **36**(3):196–198 (Feb. 2000).
24. O. M. Bucci, A. Capozzoli, and G. D'Elia, Diagnosis of array faults from far-field amplitude-only data, *IEEE Trans. Anten. Propag.* **48**(5):647–652 (May 2000).
25. *IEEE Standard Definitions of Terms for Antennas*, IEEE Press, New York, 1993.
26. R. C. Hansen, Focal region characteristics of focused array antennas, *IEEE Trans. Anten. Propag.* **33**(12):1328–1337 (Dec. 1985).
27. D. A. Hill, A numerical method for near-field array synthesis, *IEEE Trans. Electromagn. Compat.* **27**(4):201–211 (Nov. 1985).
28. J. E. Hansen, Plane-wave synthesis, in *Spherical Near-Field Antenna Measurements*, Peter Peregrinus, London, 1988, Chapter 7.
29. R. L. Haupt, Generating a plane wave with a linear array of line sources, *IEEE Trans. Anten. Propag.* **51**(2):273–278 (Feb. 2002).
30. C. C. Courtney, D. E. Voss, R. Haupt, and L. LeDuc, The theory and architecture of a plane wave generator, *Proc. 2002 AMTA Conf.* Cleveland, OH, Nov. 2002.
31. N. N. Jackson and P. S. Excell, Genetic-algorithm optimization of an array for near-field plane wave generation, *Appl. Comput. Electromagn. Soc. J.* **15**(2):61–74 (July 2000).
32. R. Haupt, Generating a plane wave in the near field with a planar array antenna, *Microw. J.* **46**(9):152–158 (Aug. 2003).
33. R. L. Haupt, Thinned arrays using genetic algorithms, *IEEE Tran. Anten. Propag.* **42**(7):993–999 (July 1994).
34. J. M. Johnson and Y. Rahmat-Samii, Genetic algorithms in engineering electromagnetics, *IEEE Anten. Propag. Mag.* **39**(4):7–21 (Aug. 1997).
35. C. A. Meijer, Simulated annealing in the design of thinned arrays having low sidelobe levels, *Proc. 1998 South African Symp. Communications and Signal Processing* (COMSIG '98), Sept. 1998, pp. 361–366.
36. S. Caorsi, A. Lommi, A. Massa, and M. Pastorino, Peak sidelobe level reduction with a hybrid approach based on GAs and difference sets, *IEEE Trans. Anten. Propag.* **52**(4):1116–1121 (April 2004).

37. J. Hsiao, Analysis of interleaved arrays of waveguide elements, *IEEE Trans. Anten. Propag.* **19**(6):729–735 (Nov. 1971).
38. J. Boyns and J. Provencher, Experimental results of a multifrequency array antenna, *IEEE Trans. Anten. Propag.* **20**(1):106–107 (Nov. 1972).
39. K. Lee et al., A dual band phased array using interleaved waveguides and dipoles printed on high dielectric substrate, *Proc. IEEE Antennas and Propag. Society Int. Symp*. June 1984, Vol. 2, pp. 886–889.
40. R. Chu, K. Lee, and A. Wang, Multiband phased-array antenna with interleaved tapered-elements and waveguide radiators, *Proc. IEEE Antennas and Propagation Int. Symp.*, Dec. 1996, Vol. 3, pp. 21–26.
41. K. Lee, A. Wang, and R. Chu, Dual-band, dual-polarization, interleaved cross-dipole and cavity-backed disc elements phased array antenna, *Proc. IEEE Antennas and Propagation Int. Symp.*, July 1997, Vol. 2, pp. 694–697.
42. R. L. Haupt, Interleaved thinned linear arrays, *IEEE Trans. Anten. Propag.* **53**(9):2858–2864 (Sept. 2005).
43. A. Tennant, M. M. Dawoud, and A. P. Anderson, Array pattern nulling by element position perturbations using a genetic algorithm, *IEE Electron. Lett.* **30**(3):174–176 (Feb. 1994).
44. R. Haupt, Comparison between genetic and gradient-based optimization algorithms for solving electromagnetics problems, *IEEE Trans. Magn.* **31**(3):1932–1935 (May 1995).
45. R. L. Haupt, An introduction to genetic algorithms for electromagnetics, *IEEE Anten. Propag. Mag.* **37**(2):7–15 (April 1995).
46. M. G. Bray, D. H. Werner, D. W. Boeringer, and D. W. Machuga, Thinned aperiodic linear phased array optimization for reduced grating lobes during scanning with input impedance bounds, *Proc. 2001 IEEE Antennas and Propagation Society Int. Symp.*, July 2001, Vol. 3, pp. 688–691.
47. M. G. Bray, D. H. Werner, D. W. Boeringer, and D. W. Machuga, Matching network design using genetic algorithms for impedance constrained thinned arrays, *Proc. 2002 IEEE Antennas and Propagation Society Int. Symp.*, June 2002, Vol. 1, pp. 528–531.
48. M. G. Bray, D. H. Werner, D. W. Boeringer, and D. W. Machuga, Optimization of thinned aperiodic linear phased arrays using genetic algorithms to reduce grating lobes during scanning, *IEEE Trans. Anten. Propag.* **50**(12):1732–1742 (Dec. 2002).
49. Y. T. Lo and S. W. Lee, *Antenna Handbook: Theory, Applications, and Design*, Van Nostrand Reinhold, New York, 1988.
50. S. Mummareddy, D. H. Werner, and P. L. Werner, Genetic optimization of fractal dipole antenna arrays for compact size and improved impedance performance over scan angle, *Proc. 2002 IEEE Antennas and Propagation Society Int. Symp.*, June 2002, Vol. 4, pp. 98–101.
51. N. Toyama, Aperiodic array consisting of subarrays for use in small mobile earth stations, *IEEE Trans. Anten. Propag.* **53**(6):2004–2010 (June 2005).
52. D. H. Werner, W. Kuhirun, and P. L. Werner, The Peano-Gosper fractal array, *IEEE Trans. Anten. Propag.* **51**(8):2063–2072 (Aug. 2003).
53. D. H. Werner, W. Kuhirun, and P. L. Werner, Fractile arrays: A new class of tiled arrays with fractal boundaries, *IEEE Trans. Anten. Propag.* **52**(8):2008–2018 (Aug. 2004).

54. J. N. Bogard, D. H. Werner, and P. L. Werner, A comparison of the Peano-Gosper fractile array with the regular hexagonal array, *Microw. Opt. Technol. Lett.* **43**(6):524–526 (Dec. 2004).
55. J. N. Bogard, D. H. Werner, and P. L. Werner, Optimization of Peano-Gosper arrays for broadband performance using genetic algorithms to eliminate grating lobes during scanning, *Proc. 2005 IEEE Antennas and Propagation Society Int. Symp.*, July 2005, Vol. 1B, pp. 755–758.
56. J. N. Bogard and D. H. Werner, Optimization of Peano-Gosper fractile arrays using genetic algorithms to reduce grating lobes during scanning, *Proc. 2005 IEEE Int. Radar Conf.*, May 2005, pp. 905–909.
57. Y. Kim and D. L. Jaggard, The fractal random array, *Proc. IEEE* **74**(9):1278–1280 (1986).
58. J. S. Petko and D. H. Werner, The evolution of optimal linear polyfractal arrays using genetic algorithms, *IEEE Trans. Anten. Propag.* **53**(11):3604–3615 (Nov. 2005).
59. D. G. Kurup, M. Himdi, and A. Rydberg, Design of unequally spaced reflectarray, *IEEE Anten. Wireless Propag. Lett.* **2**:33–35 (2003).
60. F. Ares-Pena, Applications of genetic algorithms and simulated annealing to some antenna problems, in *Electromagnetic Optimization by Genetic Algorithms*, Y. Rahmat-Samii and E. Michielssen, eds., Wiley, New York, 1999, pp. 119–155.
61. R. J. Allard, D. H. Werner, and P. L. Werner, Radiation pattern synthesis for arrays of conformal antennas mounted on arbitrarily-shaped three-dimensional platforms using genetic algorithms, *IEEE Trans. Anten. Propag.* **51**(5):1054–1062 (May 2003).
62. W. L. Stutzman and G. A. Thiele, *Antenna Theory and Design*, Wiley, New York, 1981.
63. D. H. Werner and A. J. Ferraro, Cosine pattern synthesis for single and multiple main beam uniformly spaced linear arrays, *IEEE Trans. Anten. Propag.* **37**:1480–1484 (Nov. 1989).
64. E. M. Koper, W. D. Wood, and S. W. Schneider, Aircraft antenna coupling minimization using genetic algorithms and approximations, *IEEE Trans. Aerospace Electron. Syst.* **40**(2):742–751 (April 2004).
65. Y. C. Chung and R. Haupt, Low-sidelobe pattern synthesis of spherical arrays using a genetic algorithm, *Microw. Opt. Technol. Lett.* **36**(6):412–414 (March 2002).
66. M. Hoffman, Conventions for the analysis of spherical array, *IEEE Trans. Anten. Propag.* **AP-11**(4):390–393 (July 1963).
67. E. A. Wolf, *Antenna Analysis*, Artech House, Norwood, MA, 1988.
68. Y. C. Chung and R. L. Haupt, Adaptive nulling with spherical arrays using a genetic algorithm, *Proc. IEEE Antennas and Propagation Society Int. Symp.* Orlando, FL, July 1999, Vol. 3, pp. 2000–2003.
69. C. W. Brann and K. L. Virga, Generation of optimal distribution sets for single-ring cylindrical arc arrays, *Proc. IEEE Antennas and Propagation Society Int. Symp.*, Atlanta, GA, June 1998, Vol. 2, pp. 732–735.
70. A. Armogida, G. Manara, A. Monorchio, P. Nepa, and E. Pagana, Synthesis of point-to-multipoint patch antenna arrays by using genetic algorithms, *Proc. IEEE*

Antennas and Propagation Society Int. Symp. Salt Lake City, UT, July 2000, Vol. 2, pp. 1038–1041.

71. L. C. Kretly, A. S. Cerqueira Jr., and A. S. Tavora, A hexagonal antenna array prototype for adaptive system application, *Proc. 5th Int. Symp. Wireless Personal Multimedia Communications*, Oct. 2002, Vol. 2, pp. 757–761.
72. K. F. Sabet, D. P. Jones, J. C. Cheng, L. P. B. Katehi, K. Sarabandi, and J. F. Harvey, Efficient printed antenna array synthesis including coupling effects using evolutionary genetic algorithms, *Proc. IEEE Antennas and Propagation Society Int. Symp.* Orlando, FL, July 1999, Vol. 3, pp. 2084–2087.
73. J. G. Maloney, M. P. Kesler, L. M. Lust, L. N. Pringle, T. L. Fountain, P. H. Harms, and G. S. Smith, Switched fragmented aperture antennas, *Proc. IEEE Antennas and Propagation Int. Symp.*, Salt Lake City, UT, July 2000, Vol. 1, pp. 310–313.
74. L. N. Pringle, P. G. Friederich, S. P. Blalock, G. N. Kiesel, P. H. Harms, D. R. Denison, E. Kuster, T. L. Fountain, and G. S. Smith, GTRI reconfigurable aperture design, *Proc. IEEE Antennas and Propagation Int. Symp.*, San Antonio, TX, June 2002, Vol. 1, pp. 473–476.
75. L. N. Pringle, P. H. Harms, S. P. Blalock, G. N. Kiesel, E. J. Kuster, P. G. Friederich, R. J. Prado, J. M. Morris, and G. S. Smith, A reconfigurable aperture antenna based on switched links between electrically small metallic patches, *IEEE Trans. Anten. Propag.* **52**(6):1434–1445 (June 2004).
76. R. L. Li, V. F. Fusco, and R. Cahill, Pattern shaping using a reactively loaded wire loop antenna, *IEE Proc. Microw. Anten. Propag.* **148**(3):203–208 (June 2001).
77. W. H. Weedon, W. J. Payne, and G. M. Rebeiz, MEMS-switched reconfigurable antennas, *Proc. IEEE Antennas and Propagation Society Int. Symp.* July 2001, Vol. 3, pp. 654–657.
78. C. S. DeLuccia, D. H. Werner, P. L. Werner, M. F. Pantoja, and A. R. Bretones, A novel frequency agile beam scanning reconfigurable antenna, *Proc. 2004 IEEE Antennas and Propagation Int. Symp.*, Monterey, CA, June 21–26, 2004, Vol. II, pp. 1839–1842.
79. A. D. Chuprin, J. C. Batchelor, and E. A. Parker, Design of convoluted wire antennas using a genetic algorithm, *IEE Proc. Microw. Anten. Propag.* **148**(5):323–326 (Oct. 2001).
80. D. S. Linden, Optimizing signal strength in-situ using an evolvable antenna system, *Proc. NASA/DoD Conf. Evolvable Hardware*, July 2002, pp. 15–18.

5

Smart Antennas Using a GA

As the frequency spectrum becomes more crowded, wireless systems become more vulnerable to interference. When the mainbeam gain times the desired signal is less than the sidelobe gain times the interference signal, the desired signal is overwhelmed by the interference. A smart or adaptive antenna is one alternative for recovering desirable signals. A smart antenna adapts its receive and/or transmit pattern characteristics in order to improve the antenna's performance. It may place a null in the direction of an interference source or steer the mainbeam in the direction of a desired signal. At least two different antennas constitute a smart antenna, usually in the form of an antenna array. The amplitude and phase of the received signals are weighted and summed in such a way as to meet some desired performance expectation. MIMO (multiple input/multiple output) communications systems have adaptive transmit and receive antennas.

Using a GA as an adaptive antenna algorithm is a new approach to smart antennas. The original smart antenna was a sidelobe canceler [1]. It consists of a high-gain antenna that points at the desired signal. Interfering signals are assumed to enter the sidelobes. One or more low-gain antennas receive both the desired and interfering signals in its mainbeam (Fig. 5.1). The gain of the small antennas is approximately the same as the gain of the peak sidelobes of the high-gain antenna. Appropriately weighting the low-gain antenna signal and subtracting it from the high-gain antenna signal results in canceling the interference in the high gain antenna. One low-gain antenna is needed for

Genetic Algorithms in Electromagnetics, by Randy L. Haupt and Douglas H. Werner

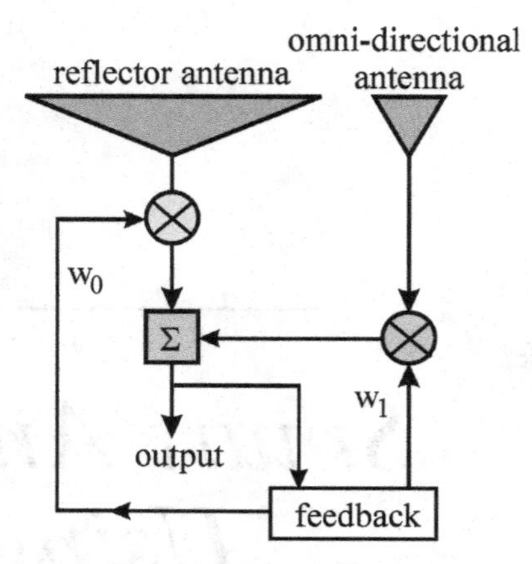

Figure 5.1. *Sidelobe cancellor.*

each interfering signal. The low-gain antenna minimally perturbs the main-beam of the high-gain antenna.

The sidelobe canceler works well for a high-gain dish antenna. With slight modification, the concept can be applied to the elements of a phased array [2]. The output of an adaptive array at a time sample n is given by

$$f(n) = \mathbf{w}^T(n)\mathbf{s}(n) \tag{5.1}$$

where $\mathbf{w}^T(n) = [w_1(n), w_2(n), \ldots, w_N(n)]$ = transpose of weight vector at time sample n

$s(n)$ = signal vector at time sample n

This output differs from the desired output at sample $n[d(n)]$ by

$$\varepsilon(n) = d(n) - \mathbf{w}^T(n)\mathbf{s}(n) \tag{5.2}$$

The mean-square error of (5.2) is

$$E[\varepsilon^2(n)] = E[d^2(n)] + \mathbf{w}^\dagger(n)\mathbf{R}(n)\mathbf{w}(n) - 2\mathbf{w}^\dagger(n)\mathbf{q}(n) \tag{5.3}$$

where † is the complex conjugate transpose of the vector and $E[\ \ ]$ is the expected value. The autocovariance matrix is defined to be

$$\mathbf{R} = E\begin{bmatrix} s_1^*(n)s_1(n) & s_1^*(n)s_2(n) & \cdots & s_1^*(n)s_N(n) \\ s_2^*(n)s_1(n) & s_2^*(n)s_2(n) & & \\ \vdots & & \ddots & \vdots \\ s_N^*(n)s_1(n) & & \cdots & s_N^*(n)s_N(n) \end{bmatrix} \tag{5.4}$$

and

$$\mathbf{q}(n) = E[d(n)\mathbf{s}(n)] \tag{5.5}$$

Taking the gradient of the mean square error with respect to the weights results in

$$\nabla E[\varepsilon^2(n)] = 2\mathbf{R}(n)\mathbf{w}(n) - 2\mathbf{q}(n) \tag{5.6}$$

The optimum weights make the gradient zero, so

$$2\mathbf{R}(n)\mathbf{w}_{\text{opt}}(n) - 2\mathbf{q}(n) = 0 \tag{5.7}$$

Solving for the optimum weights yields the optimal Wiener–Hopf solution [3]:

$$\mathbf{w}_{\text{opt}}(n) = \mathbf{R}^{-1}(n)\mathbf{q}(n) \tag{5.8}$$

Finding the solution requires two very important pieces of information:

1. The signal at each element
2. The desired signal

The signal at each element is found by placing a receiver at each element. If we know the desired signal, then why even have a receive antenna? Overcoming this problem is discussed in detail in Ref. 3.

Although the Wiener–Hopf solution is not a practical approach to adaptive antennas, it forms the mathematical basis of the adaptive least-mean-square algorithm (LMS) [4]. The LMS algorithm is very similar to the Howells–Applebaum adaptive algorithm but was developed independently. The LMS formula is given by

$$\mathbf{w}(n+1) = \mathbf{w}(n) + \mu\mathbf{s}(n)[d(n) - \mathbf{w}^{\dagger}\mathbf{s}(n)] \tag{5.9}$$

Note that the expected values are dropped because of the difficulty in implementing the expected value operator in real time. Instead, the expected values are replaced by the instantaneous values.

The LMS algorithm and variations thereof are canonical adaptive signal processing algorithms. They are based on the steepest-descent algorithm, which is easy to implement but can get stuck in a local minimum. A major problem with the LMS algorithm is the need to know the signals incident at each element in order to form the covariance matrix, resulting in impractical hardware requirements for the array:

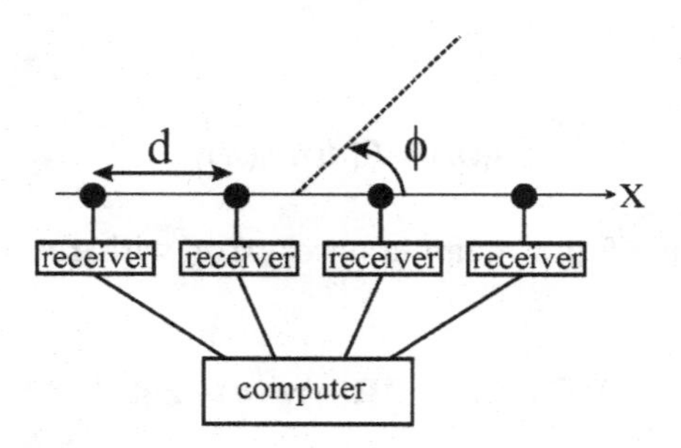

Figure 5.2. Diagram of an adaptive array.

1. The array needs a receiver at each element to detect the signals (see Fig. 5.2). Typically, an array has one receiver after all the signals are weighted and added together. Receivers are very expensive, so the cost of this type of an array is extremely high.
2. These receivers require calibration to ensure that the signals are weighted correctly [5].
3. The array needs variable amplitude and phase weights at each element. Usually, a phased array has only digital phase shifters to steer the beam. As you will see, the GA offers a powerful alternative to the LMS style of algorithm.

Another class of algorithms adjusts the phase shifter settings in order to reduce the total output power from the array [6–8]. These algorithms are cheap to implement because they use the existing array architecture without expensive additions, such as adjustable amplitude weights or more receivers. Their drawbacks include slow convergence and possibly high pattern distortions. One approach to phase only nulling is to use a random search algorithm [9]. Random search algorithms check a small number of all possible phase settings in search of the minimum output power. The search space for the current algorithm iteration can be narrowed around the regions of the best weights of the previous iteration. A second approach forms an approximate numerical gradient and uses a steepest-descent algorithm to find the minimum output power [10]. This approach has been implemented experimentally but is slow and gets trapped in local minima. As a result, the best phase settings to achieve appropriate nulls are seldom found. This chapter presents several applications of a GA for smart antennas. All are based on minimizing the total output power by making small adjustments that do not null the desired signal entering the mainbeam.

5.1 AMPLITUDE AND PHASE ADAPTIVE NULLING

The GA functions as an adaptive antenna algorithm by minimizing the total output power of the array [11]. This approach only works if the desired signal is not present or if the adaptive algorithm is constrained to making small amplitude and phase perturbations at each element. Although adapting while the desired signal is absent may work for a stationary antenna with relatively stationary interference sources, it would fail to provide reasonable protection for most communications and radar systems. Limiting the amplitude and phase perturbations at each element is actually quite easy to do, especially with a GA, since the GA inherently constrains its variables. Most phase shifters and attenuators are digital, so a binary GA naturally works with binary control signals.

In order to minimize total output power, which consists of the interference signal and possibly the desired signal, without sacrificing the desired signal requires the antenna to place nulls in the sidelobes and not the mainbeam. It's reasonable to assume that the desired signal enters the mainbeam and the interference enters the sidelobes. Large phase shifts are required to place a null in the mainbeam. Large reductions in the amplitude weights are required in order to reduce the mainbeam. Consequently, if only small amplitude and phase perturbations are allowed, then a null can't be placed in the mainbeam but can be placed in the sidelobes. Lower sidelobes require smaller perturbations to the weights in order to place the null. Using only a few least significant bits of the digital phase shifter and attenuator bits prevents the GA from placing nulls in the mainbeam. It's a natural mainbeam constraint. The amplitude and phase perturbations as a function of the number of least significant bits used out of a total of 6 bits are shown in Table 5.1.

Let's look at a 20-element, 20-dB, $\bar{n} = 3$ Taylor adaptive linear array with elements spaced half a wavelength apart. Assume the amplitude and phase weights have 6-bit quantization. If the only source enters the mainbeam, then the adaptive algorithm tries to reduce the mainbeam in order to reduce the total output power. Figure 5.3 shows the mainbeam reduction when 0–4 least significant bits (LSBs) in the amplitude weights are used for nulling. Zero bits corresponds to the quiescent pattern. A maximum reduction of 1 dB is possible using 4 bits of amplitude. In contrast, Figure 5.4 shows the mainbeam reduction when 0–4 LSBs least in the phase weights are used for nulling. Using 1–3 LSBs results in very little perturbation to the mainbeam. Unlike amplitude-only nulling, phase-only nulling causes beam squint. Also, 4 bits of phase had more

TABLE 5.1. Amplitude and Phase Values of Nulling Bits Assuming Amplitude and Phase Weights Totaling 6 Bits

Number of Nulling Bits	Amplitude with Maximum of 1	Phase (deg)
1	0, 0.015625	0, 5.625
2	0, 0.015625, . . . , 0.046875	0, 5.625, . . . , 16.875
3	0, 0.015625, . . . , 0.10938	0, 5.625, . . . , 39.375
4	0, 0.015625, . . . , 0.23438	0, 5.625, . . . , 84.375
5	0, 0.015625, . . . , 0.48438	0, 5.625, . . . , 174.38
6	0, 0.015625, . . . , 0.98438	0, 5.625, . . . , 354.38

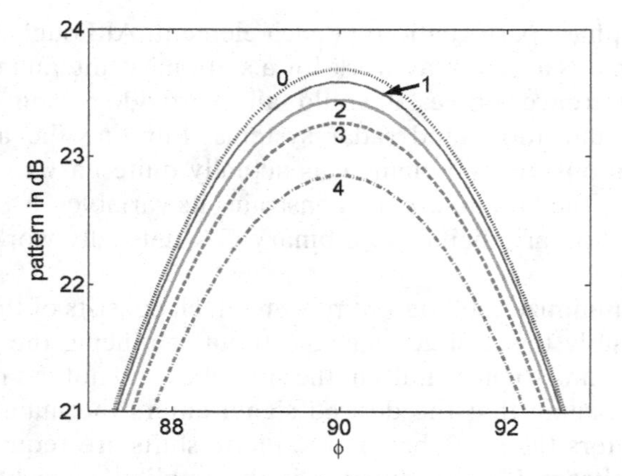

Figure 5.3. Maximum mainbeam reduction possible when 1–4 least significant bits out of 6 total bits in an amplitude weight are used to null a signal at $\phi = 90°$; 0 bits is the quiescent pattern.

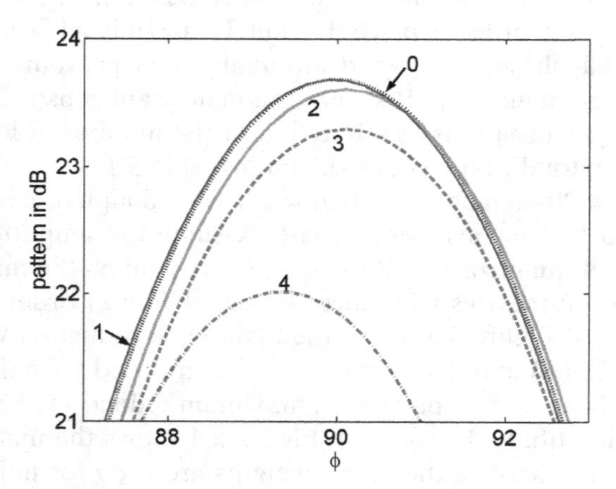

Figure 5.4. Maximum mainbeam reduction possible when 1–4 least significant bits out of 6 total bits in a phase weight are used to null a signal at $\phi = 90°$; 0 bits is the quiescent pattern.

effects on the mainbeam than did 4 bits of amplitude.[1] This example demonstrates that adaptive nulling with the LSBs would not cause significant degradation to the mainbeam.

A diagram of the adaptive array appears in Figure 5.5. The array has a standard corporate feed with variable weights at each element. The phase shifters in the weights are available for beamsteering as well as nulling. Since the phase shifters and attenuators are digital, their inputs are in binary format.

[1]The amplitude weights described here are based on a linear voltage scale. Commercial variable amplitude weights are based on a decibel power scale.

If both weights have 5-bit accuracy, then a subset of these bits is used for the nulling. Figure 5.6 demonstrates the process of taking the 3 LSBs out of the 5 bits for each element and placing them into a chromosome. Each chromosome forms a row in the population matrix. Parents are selected from the population and mate to form offspring as shown in Figure 5.7 (single-point crossover is

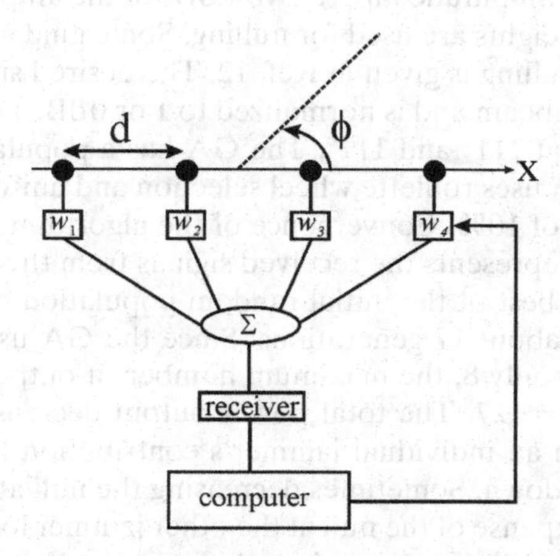

Figure 5.5. An adaptive antenna that minimizes the total output power.

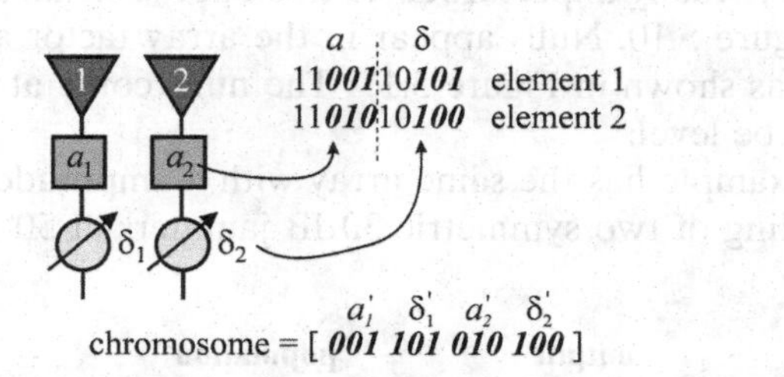

Figure 5.6. The least significant bits of the amplitude and phase weights are put in a chromosome.

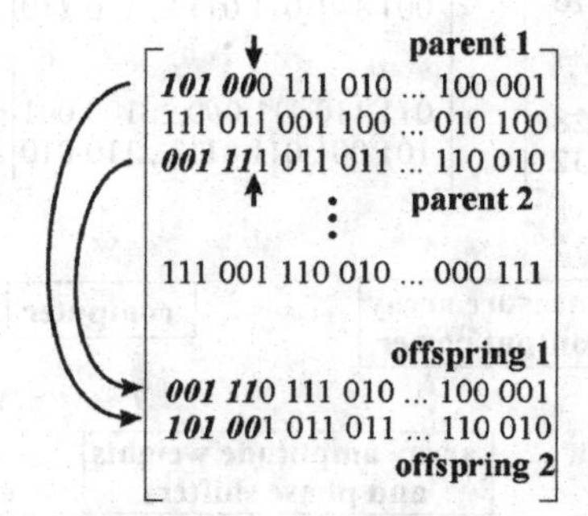

Figure 5.7. Two parents are selected from the mating pool. Two offspring are created using single-point crossover and placed into the population matrix to replace discarded chromosomes.

used here). Mutations occur inside the population (italicized digits in Fig. 5.8). Each chromosome in the population is then passed to the antenna to adjust the weights and the output power measured (Fig. 5.8).

As an example, consider a 20-element array of point sources spaced 0.5λ apart. The array has 6-bit amplitude and phase weights and a 20-dB, $\bar{n} = 3$ low-sidelobe Taylor amplitude taper. Two LSBs of the amplitude weights and three of the phase weights are used for nulling. Some guidance on the number of LSBs to use for nulling is given in Ref. 12. The desired signal is incident on the peak of the mainbeam and is normalized to 1 or 0 dB. Two 30-dB jammers enter the sidelobes at 111° and 117°. The GA has a population size of 8 and a 50% selection rate, uses roulette wheel selection and uniform crossover, and has a mutation rate of 10%. Convergence of the algorithm is shown in Figure 5.9. Generation −1 represents the received signals from the quiescent pattern. Generation 0 is the best of the initial random population of the GA. Performance levels off in about 17 generations. Since the GA uses elitism and has a population size of only 8, the maximum number of output power measurements is $8 + 17 \times 7 = 127$. The total power output decreases monotonically, but the reduction in an individual jammer's contribution to the total output power can go up or down. Sometimes decreasing the null at one jammer location is done at the expense of the null at the other jammer location. The output power due to the desired signal remains relatively constant, because the mainbeam remains virtually unperturbed. A track of the signal to interference ratio appears in Figure 5.10. Nulls appear in the array factor at the angles of the two jammers as shown in Figure 5.11. The nulls come at a cost of increased average sidelobe level.

The next example has the same array with 4 amplitude and 3 phase LSBs doing the nulling of two symmetric 30 dB jammers at 50° and 130°. Conver-

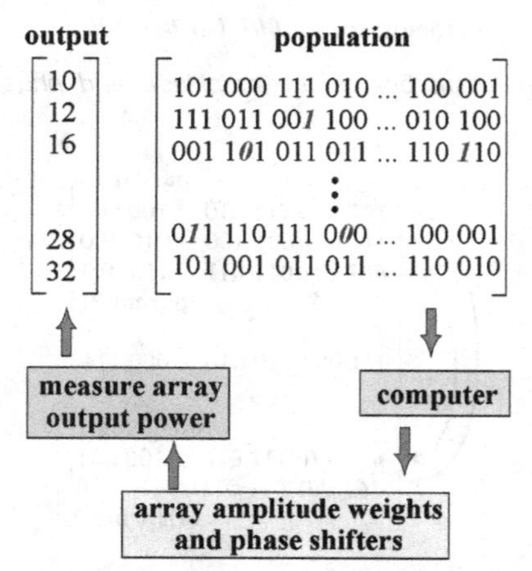

Figure 5.8. *Random bits in the population are mutated (italicized bits). The chromosomes are sent to the array one at a time and the total output power is measured. Each chromosome then has an associated output power.*

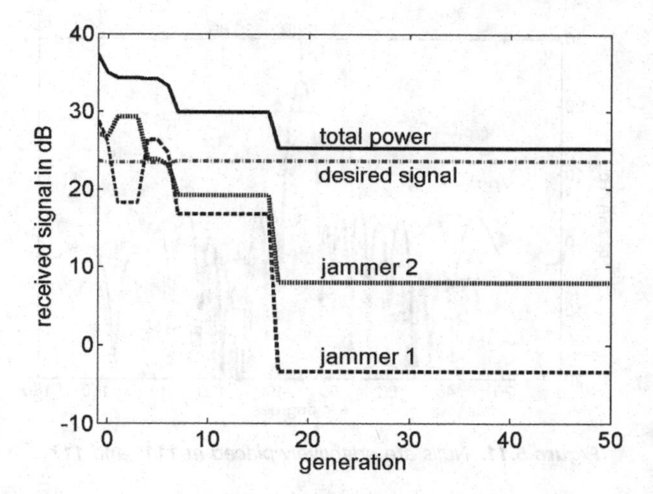

Figure 5.9. *The GA minimizes the total output power. In the process, it places nulls in the sidelobes and reduces the jammer power. Generation −1 is the quiescent pattern, and generation 0 is the best of the initial population.*

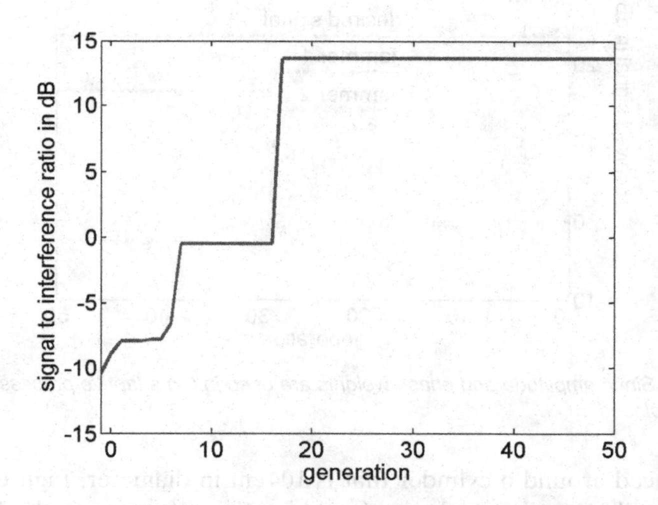

Figure 5.10. *Signal : interference ratio as a function of generation.*

gence is shown in Figure 5.12. The resulting array factor appears in Figure 5.13. More bits were needed to create adequate nulls at symmetric locations in the pattern. As a result, the sidelobes go up and the time needed to place the nulls is longer compared with nonsymmetric jammers.

This concept was experimentally demonstrated on a phased array antenna developed by the Air Force Research Laboratory (AFRL) at Hanscom AFB, MA [11]. The antenna has 128 vertical columns with 16 dipoles per column

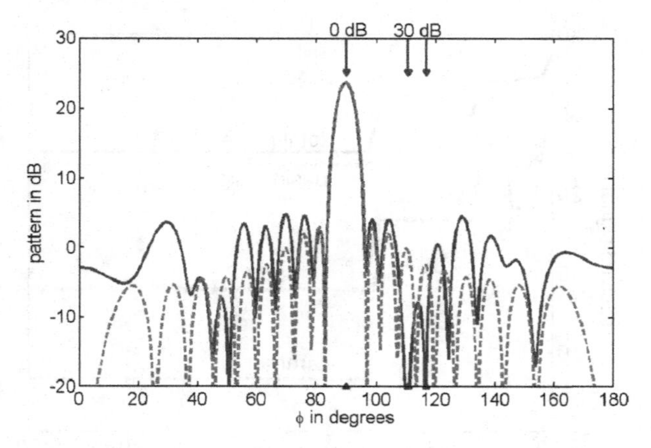

Figure 5.11. *Nulls are adaptively placed at 111° and 117°.*

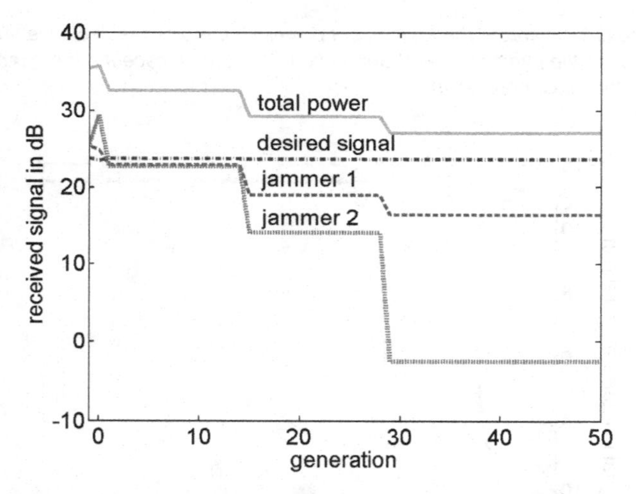

Figure 5.12. *Since amplitude and phase weights are used in the adaptive process, symmetric nulls can be placed.*

equally spaced around a cylinder that is 104 cm in diameter. Figure 5.14 is a cross-sectional view of the antenna. Summing the outputs from the 16 dipoles forms a fixed-elevation mainbeam pointing 3° above horizontal. Only eight columns of elements are active at a time. A consecutive eight elements form a 22.5° arc ($\frac{1}{16}$th of the cylinder), with the elements spaced 0.42λ apart at 5 GHz. Each element has an 8-bit phase shifter and 8-bit attenuator. The phase shifters have a LSB equal to 0.0078125π radians. The attenuators have an 80 dB range with the least significant bit equal to 0.3125 dB. The antenna has a quiescent pattern resulting from a 25 dB $\bar{n} = 3$ Taylor amplitude taper. Phase shifters compensate for the curvature of the array and unequal pathlengths through the feed network.

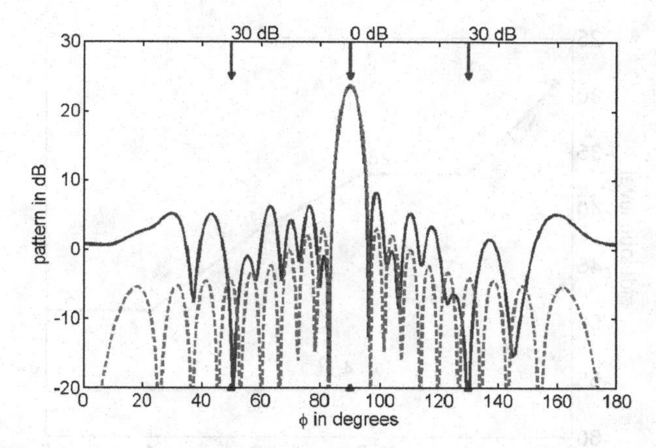

Figure 5.13. *Nulls are placed at symmetric locations in the antenna pattern.*

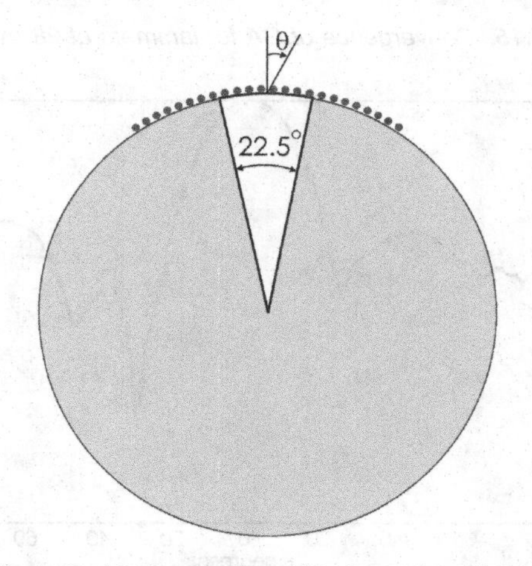

Figure 5.14. *The cylindrical array has 128 elements with 8 active at a time.*

A 5-GHz continuous-wave source served as the interference. Only the 4 LSBs of the phase shifters and attenuators were used. The GA had a population size of 16 chromosomes and used single-point crossover. Only one bit in the population was mutated every generation, resulting in a mutation rate of 0.1%. Nulling tended to be very fast with the algorithm placing a null down to the noise floor of the receiver in less than 30 power measurements. Two cases of placing a single null are presented here. The first example has the interference entering the sidelobe at 28° and the second example has the interference at 45°. Figure 5.15 plots the sidelobe level at 28° and 45° as a function of generation. The resulting far-field pattern measurements are shown in Figures 4.16 and 4.17 superimposed on the quiescent pattern. These examples demonstrate that the GA quickly places nulls in the sidelobes in the directions of the interfering signals by minimizing the total power output.

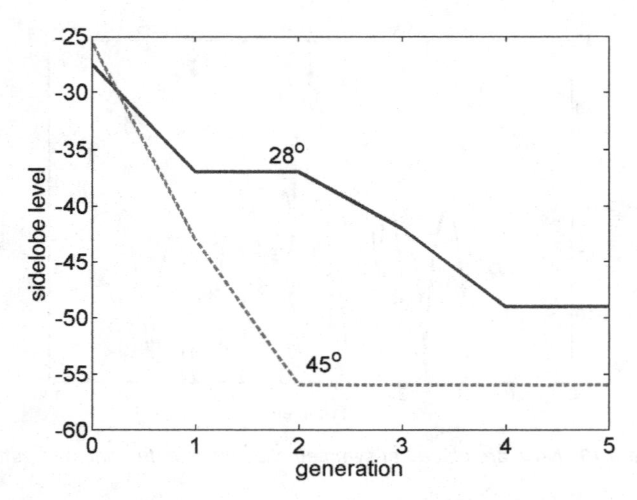

Figure 5.15. *Convergence of GA for jammers at 28° and 45°.*

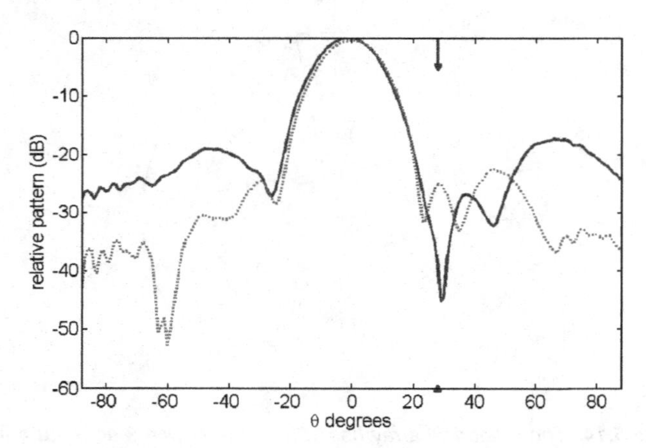

Figure 5.16. *Null placed in the far-field pattern at 28°.*

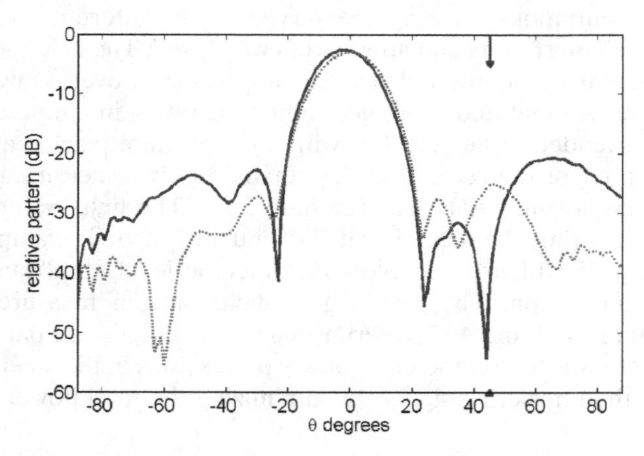

Figure 5.17. *Null placed in far-field pattern at 45°.*

5.2 PHASE-ONLY ADAPTIVE NULLING

Normally, a phased array does not have variable amplitude weights. Since a phased array does have phase shifters, nulling with only the phase shifters would significantly simplify the adaptive antenna design. Phase-only nulling differs from amplitude and phase nulling, because the weights in Figure 5.5 have amplitude of 1. The theory behind phase-only nulling is explained in Ref. 6. The authors present a beam-space algorithm derived for a low sidelobe array and assume that the phase shifts are small. When the direction of arrival for all the interfering sources is known, then cancellation beams are generated in the directions of the sources and subtracted from the original pattern. Adaptation consists of matching the peak of the cancellation beam with the culprit sidelobe and subtracting. A similar approach is also presented in Ref. 8. Other authors have suggested using gradient methods for phase-only adaptive nulling where the source locations do not have to be known [10,13].

Using a GA for adaptive phase-only nulling was first reported in 1997 [14]. Chromosomes only contain the phase nulling bits, so they are much shorter than the chromosomes in amplitude and phase nulling. Small population sizes help the algorithm converge reasonably rapidly. A GA can be real-time only if the population size is small. Otherwise, large population sizes require too many function evaluations (see Chapter 9).

For purposes of comparison, the two scenarios used in the amplitude–phase nulling examples are repeated here. The array has 20 elements spaced a half-wavelength apart and a 20-dB $\bar{n} = 3$ Taylor amplitude taper. Figures 5.18–5.20 show the results of phase-only nulling when the interference sources are at 111° and 117°. The GA successfully places the nulls in the sidelobes using only

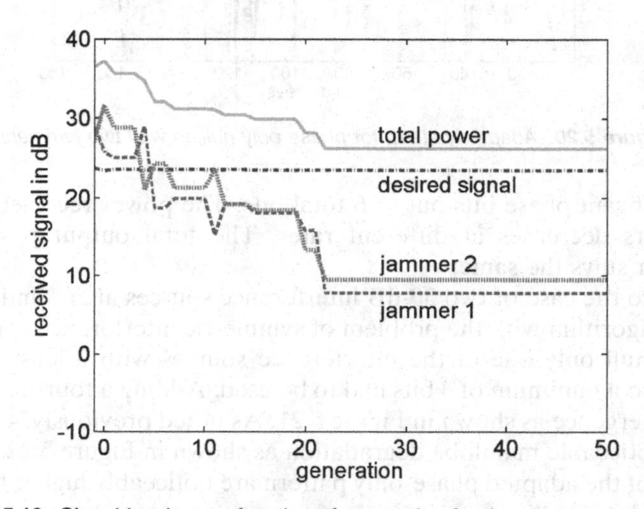

Figure 5.18. *Signal levels as a function of generation for the phase-only algorithm.*

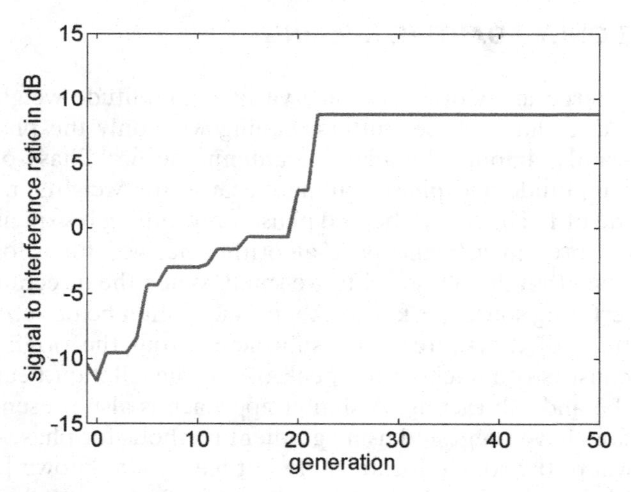

Figure 5.19. The signal:interference ratio for the phase-only adaptive algorithm with two jammers at 111° and 117°.

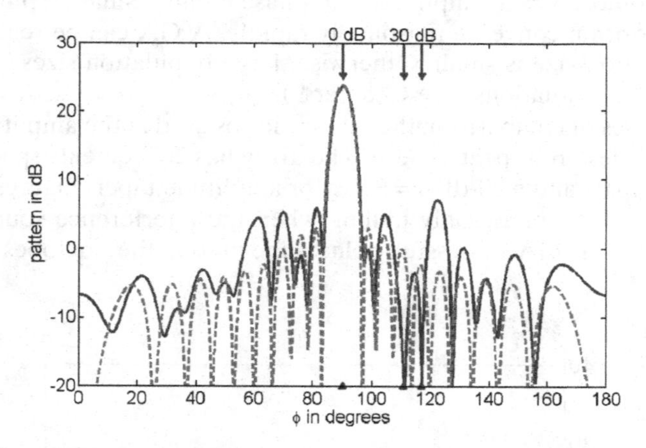

Figure 5.20. Adapted pattern for phase-only nulling with two jammers.

3 least significant phase bits out of 6 total bits. The power received from the two jammers decreases at different rates. The total output power either decreases or stays the same.

Moving to the case of two 30-dB interference sources at 50° and 130° confronts the algorithm with the problem of symmetric interference sources. The GA could null only one of the interference sources with 3 least significant phase bits, so a minimum of 4 bits had to be used. Adding a fourth bit resulted in nice convergence as shown in Figure 5.21. As noted previously, 4 phase bits results in noticeable mainlobe degradation as shown in Figure 5.22. The sidelobe levels of the adapted phase-only pattern are noticeably higher than in the amplitude-phase nulling case.

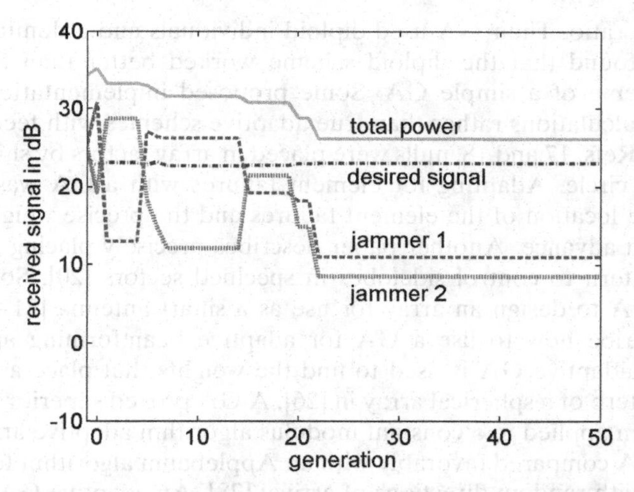

Figure 5.21. Convergence of the phase-only algorithm with symmetric jammers.

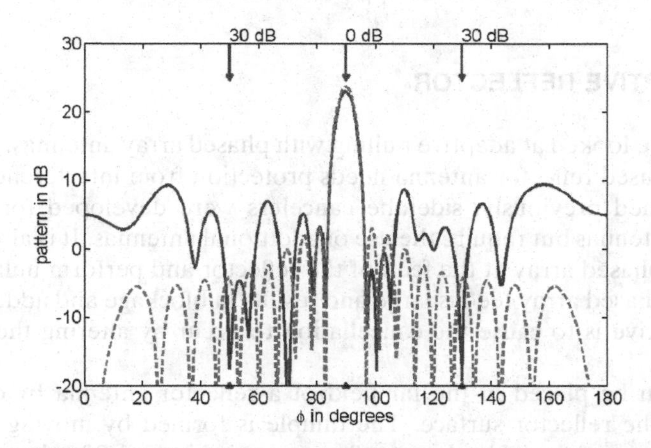

Figure 5.22. Adapted pattern for phase-only nulling with symmetric jammers.

Phase-only nulling has the advantage of simple implementation. The tradeoff is that more bits must be used to null interference signals that are at symmetric locations about the mainbeam. The additional nulling bits result in higher distortions in the mainbeam and sidelobes. Small phase shifts produce symmetric cancellation beams that are 180° out of phase. When they are added to the quiescent pattern to produce a null at one location, the symmetric sidelobe increases [8]. This problem can be overcome by either large phase shifts [15] or adding amplitude control.

A number of other adaptive nulling schemes with GAs have been tried. Weile and Michielssen [16] used a GA to optimize the signal-to-noise-plus-interference ratio. Their GA used diploid individuals and a dominance relation. They found that the diploid scheme worked better than the normal haploid scheme of a simple GA. Some proposed implementations involve open-loop calculations rather than true adaptive schemes with feedback. For instance, in Refs. 17 and 18, nulls were placed in array factors by shifting zeros on the unit circle. Adapting for element failures with a GA was shown in Ref. 19. The location of the element failures and the precise weighting must be known in advance. Another paper describes precisely placing nulls in an antenna pattern to control sidelobes in specified sectors [20]. Some papers applied a GA to design an array for use as a smart antenna [21–23]. Other papers describe how to use a GA for adaptive beamforming applications [24,25]. An adaptive GA is used to find the weights that place a null in the far-field pattern of a spherical array in [26]. A GA proved superior to steepest descent when applied to a constant modulus algorithm adaptive array [27]. A real-time GA compared favorably with an Applebaum algorithm for interfering signals with random directions of arrival [28]. An adaptive GA algorithm also place nulls in the far-field pattern of a hexagonal planar array [29]. The future looks bright for real-time applications of GAs.

5.3 ADAPTIVE REFLECTOR

So far, we've looked at adaptive nulling with phased array antennas. The more commonly used reflector antenna needs protection from interference as well. As mentioned previously, sidelobe cancelers were developed for use with reflector antennas but require the use of additional antennas. It is also possible to place a phased array at the feed of the reflector and perform nulling at the feed. The phased array feed is large and results in blockage and added weight. An alternative is to cause the cancellation to occur by altering the reflector surface.

Nulls can be placed in the far field of a reflector antenna by creating a dimple in the reflector surface. The dimple is formed by moving a flexible region of the surface. This idea was first proposed in Ref. 30, where an adaptive reflector surface was made using electrostatic actuators. The concept alters a portion of the reflector surface, as shown in Figure 5.23, in order to change the phase of the signal scattered from that region so that the scattering field cancels the signal entering a sidelobe. A very similar approach was implemented on the surface of an 85-cm-diameter offset wire mesh reflector at 10 GHz with 52 adjustable points that shape the reflector to produce a desirable pattern [31]. One of the adjustable points on the reflector surface was pulled back until a null formed in the desired direction. The authors stated that the adaptation should be done only in the absence of the desired signal. They said that two or more mesh control points are needed to place a single null and that nulls cannot be placed at certain angles. Experiments with altering the reflector surface were performed for the haystack main reflector [32]

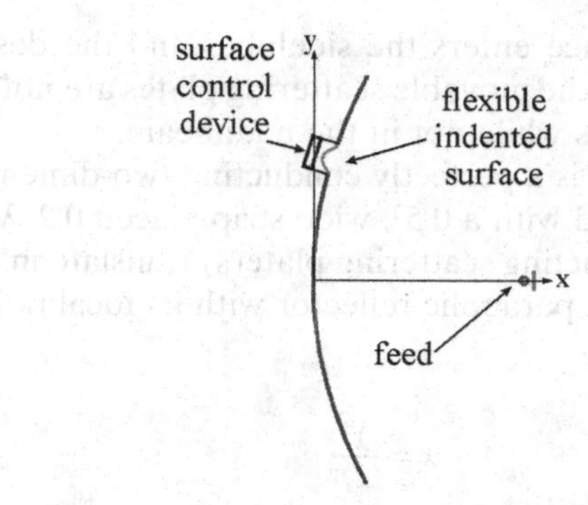

Figure 5.23. A small region of the reflector is flexible enough to move.

and subreflector [33] to reduce the root-mean-square (RMS) surface error. Piezoelectric actuators are another way to alter the shape of reflector antennas in order to steer the mainbeam or produce a desired pattern shape [34]. These actuators could be used for adaptive nulling.

Movable conducting plates parallel to the axis of the parabolic reflector can also place nulls in the sidelobes. The idea is that the scattered field from the plates will cancel the field generated by the reflector antenna in the desired direction of interference. Jacovanco reported results of an experiment in which he manually moved disks in and out from the surface of a parabolic reflector antenna [35,36]. A source was pointed at one of the receive sidelobes. He generated a null in the sidelobe by minimizing the measured signal received at the reflector feed. He reported that the disks must be large enough to scatter sufficient field amplitude to cancel a sidelobe but small enough to not perturb the mainbeam. Physical optics (PO) was used to model this concept, and some rules of thumb were developed to estimate disk size needed to cancel sidelobes [37]. The concept can be improved by making the disks from a metal grating rather than solid metal [38]. This approach permitted control over the polarization, consequently the amplitude of the scattered field when the disk is rotated. The authors reported only open loop nulling without any adaptation. Lam et al. [39] attached small reflectors to the interior or exterior of a main reflector surface. Using PO to model the concept, they rotated the auxiliary reflectors until the first two sidelobes of the antenna pattern were reduced by 10 dB. The diameters of the auxiliary reflectors were determined to be 0.3 times the size of the main reflector. When the auxiliary reflector was on the interior of the main reflector, the blockage caused a loss in gain and raised sidelobe levels in general. No mention was made of applying this concept to adaptive nulling or of the effect on the sidelobes beyond the first two.

The GA is a suitable candidate to adaptively control the Jacovanco nulling reflector [40,41]. As with the adaptive arrays presented in the previous sections, this algorithm works by minimizing the total output power of the reflector. If the interference enters the sidelobes and the desired signal the mainlobe, then, as long as the movable scattering plates are not too large, nulls are placed in the sidelobes while not in the mainbeam.

The parabolic reflector is a perfectly conducting two-dimensional reflector surface and has a line feed with a 0.5λ-wide strip placed 0.25λ behind it. The conformal perfectly conducting scattering plate(s) translate in the x direction. The equation describing a parabolic reflector with its focal point (F) on the x axis is

$$x = \frac{y^2}{4F} \tag{5.10}$$

A reflector having $F = 5\lambda$ and $D = 10\lambda$ is shown in Figure 5.24. The scattering plate extends a distance d in the x direction. If there is more than one scattering plate, then each one extends a distance d_n from the surface where $n = 1$ refers to the scattering plate with the smallest y coordinates. Increasing values of n refer to scattering plates with increasing y locations. Note that changing d_n has no effect on y coordinates. Plates do not overlap.

The surface currents on all perfectly conducting surfaces are found using the electric field integral equation formulation [40]

$$IH_0^{(2)}(k\rho_f) + \int_{C_0} J(\bar{\rho}')H_0^{(2)}(k|\bar{\rho} - \bar{\rho}'|)dc' + \sum_{n=1}^{N+1} \int_{C_n} J(\bar{\rho}')H_0^{(2)}(k|\bar{\rho} - \bar{\rho}''|)dc'' = 0 \tag{5.11}$$

where
$k = 2\pi/\lambda$
λ = wavelength
I = electric current
$H_0^{(2)}(\cdot)$ = zero-order Hankel function of the second kind

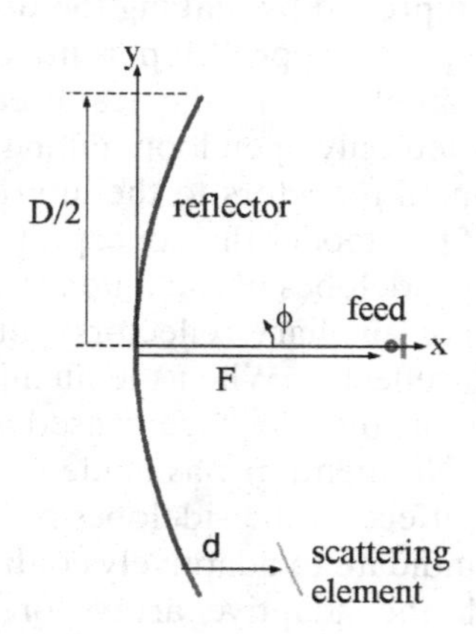

Figure 5.24. *Model of reflector with one scattering plate.*

J = surface current
C_0 = contour of main reflector surface
C_n = contour of scattering plates for $1 \leq n \leq N$
C_{N+1} = contour of subreflector
N = number of scattering plates
(x_f, y_f) = feed location
(x, y) = observation point
(x', y') = source point
$\bar{\rho}_f = x_f\hat{x} + y\hat{y}$ = feed vector
$\bar{\rho} = x\hat{x} + y\hat{y}$ = observation point vector
$\bar{\rho}' = x'\hat{x} + y'\hat{y}$ = source point reflector vector
$\bar{\rho}'' = (x' + \delta_n)\hat{x} + y' = \hat{y}$ source point on scattering plate vector

The currents are found on all the surfaces using the method of moments. Pulse basis functions and point matching reduce the integral equation to a matrix equation. The pulses are 0.05λ wide. Figure 5.25 shows the complete quiescent pattern of the parabolic reflector with no moving plates and a line source feed with a 0.5λ-wide strip placed 0.25λ behind it. This added strip behind the line source reduces spillover by increasing the feed directivity and adds a realistic aperture blockage (which accounts for the sidelobe variations in Fig. 5.25). Once the currents are found, then the far-field pattern is calculated via

$$FF(\theta, \varphi) = -e^{jk(x_f \sin\theta\cos\varphi + y_f \sin\theta\cos\varphi)} - \Delta\sum_{n=1}^{M} e^{jk(x_n \sin\theta\cos\varphi + y_n \sin\theta\cos\varphi)} J_n \tag{5.12}$$

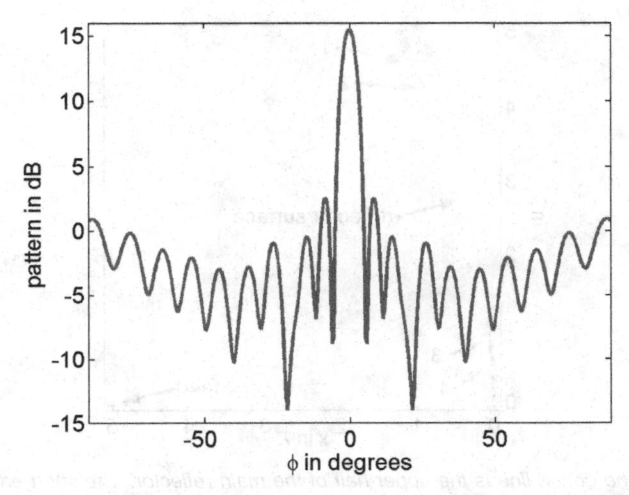

Figure 5.25. *Quiescent antenna pattern for the parabolic reflector. There are no scattering plates present.*

where M = total number of segments on all reflector parts
(θ, φ) = far-field direction
(x_n, y_n) = centerpoint of segment n
Δ = segment width
J_n = surface current on segment n

The number, size, and location of the scattering plates determine the reflector's ability to place nulls in its sidelobes. Figure 5.26 shows the upper half of the cylindrical parabolic reflector with scattering plates 1–4. Plates 1 and 2 are at the edge of the reflector, while 3 and 4 are at the center. Plates 1 and 4 are 0.5λ tall in the y direction, and plates 2 and 3 are 1.0λ tall. Only the fields due solely to these four possible scattering plates when $d = 0$ are shown in Figure 5.27 with the quiescent reflector pattern superimposed. The short plates have much wider 3-dB patterns but are approximately 6 dB below the peaks of the patterns of the long plates. This is logical, since the large plates are twice the size of the small plates. The center plates have higher peak scattering patterns than the edge plates, because the directional source at the feed has higher gain in the direction of the center plates. None of the plates reach the same level as the peak of any sidelobe, so total cancellation of a sidelobe is not possible. Increasing the size of the scattering plate increases its gain in the region near the mainbeam but decreases the gain outside that region. The long center plate would be much less effective at canceling the sidelobes in the quiescent pattern for $80^\circ \leq \varphi \leq 90^\circ$ because the pattern of the long center plate is about 30 dB below that of the short center plate. Consequently, increasing the size of the plate helps null sidelobes near the mainbeam but hinders null placement in sidelobes far from the mainbeam. The effect was noted but not explained

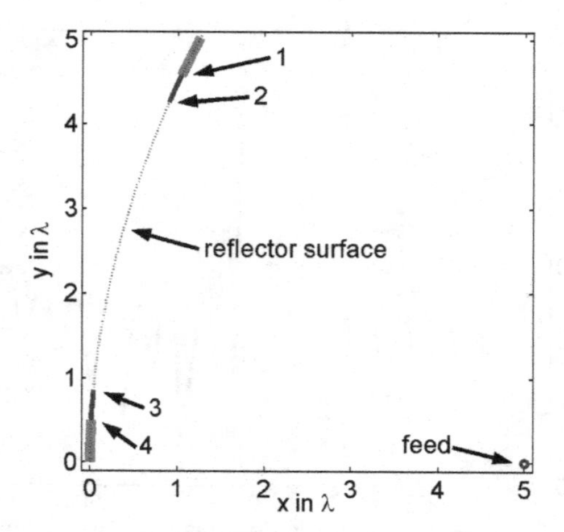

Figure 5.26. *The dotted line is the upper half of the main reflector. The short edge scattering plate (1) is 0.5λ long; the long edge scattering plate (2) is 1.0λ long; the short center scattering plate (3) is 0.5λ long; the long center scattering plate (4) is 1.0λ long.*

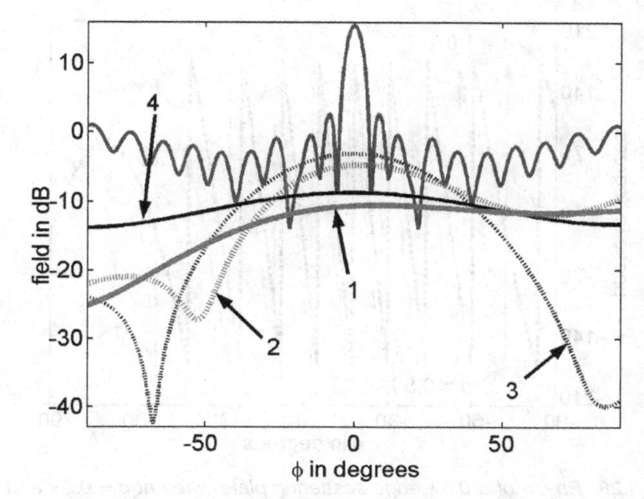

Figure 5.27. *These are the fields from the scattering plates in Figure 5.3 superimposed on the quiescent reflector pattern.*

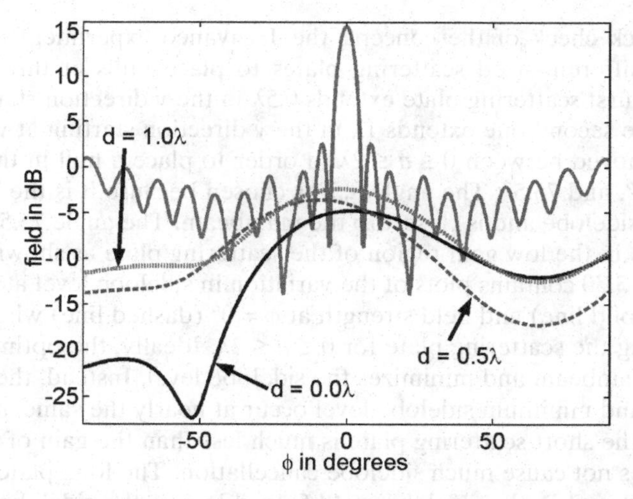

Figure 5.28. *The fields from the 0.5λ edge scattering plate when at d = 0.0λ, 0.5λ, and 1.0λ.*

in Ref. 31. Next, the 1.0λ long edge plate was moved. Figure 5.28 shows the scattered fields when the plate is at $d = 0$, 0.5λ, and 1.0λ. Moving the plate changes the scattered field amplitude near the mainbeam by a few decibels. The scattered field amplitude far from the mainbeam can vary as much as ~15 dB. A phase plot is shown in Figure 5.29 for $d = 0.5\lambda$ and $d = 1.0\lambda$ and $-90° \leq \varphi \leq 90°$. The phase is a linear function of φ until $\varphi > 45°$. At that point, the observation point moves behind the reflector, so the phase variation is no longer linear.

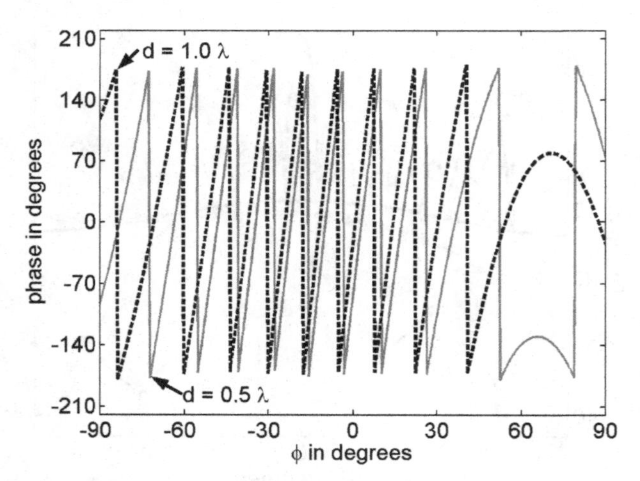

Figure 5.29. *Phase of a 0.5λ edge scattering plate when at d = 0.5λ and 1.0λ.*

As a quick check of the concept, the Jacavanco experiment is repeated using two different-sized scattering plates to place nulls at three different angles. The first scattering plate extends 0.5λ in the y direction starting at y = 4.5λ, and the second one extends 1λ in the y direction starting at y = 4λ. The plates are moved between $0 \leq d \leq 2\lambda$ in order to place a null in the sidelobe at 8.5°, 14.5°, and 75.5°. The angle 8.5° is chosen because it is the location of the highest sidelobe and is closest to the mainbeam. The angle 75.5° is chosen because it is in the low gain region of the scattering plate as shown in Figure 5.28. Figure 5.30 contains plots of the variation in sidelobe level at 8.5°, 14.5°, and 75.5° (solid line) and field strength at $\varphi = 0°$ (dashed line) when continuously moving the scattering plate for $0 \leq d \leq 2\lambda$. Ideally, the optimal d maximizes the mainbeam and minimizes the sidelobe level. Instead, the minimum mainbeam and minimum sidelobe level occur at nearly the same angle. Since the gain of the short scattering plate is much less than the gain of these sidelobes, it does not cause much sidelobe cancellation. The long plate, however, puts a 25-dB null in the sidelobe at 14.5° and lowers the sidelobe at 8.5° by about 3 dB. The long plate does not have much of an impact on the sidelobe level at 75.5° because of its very low gain in that direction. This example demonstrates the Jacavanco idea for placing a single null in the antenna pattern by manually adjusting the position of a perfectly conducting scattering plate. It also shows the dependence of the null depth on the size of the scattering plate and position and gain of the sidelobe.

A single scattering plate is adequate for placing a null in the antenna pattern at certain locations as was done by Jacavanco. An adaptive algorithm, however, is needed to make this approach credible in a realistic system. The results in Figure 5.30 show that there are several minima for a single movable

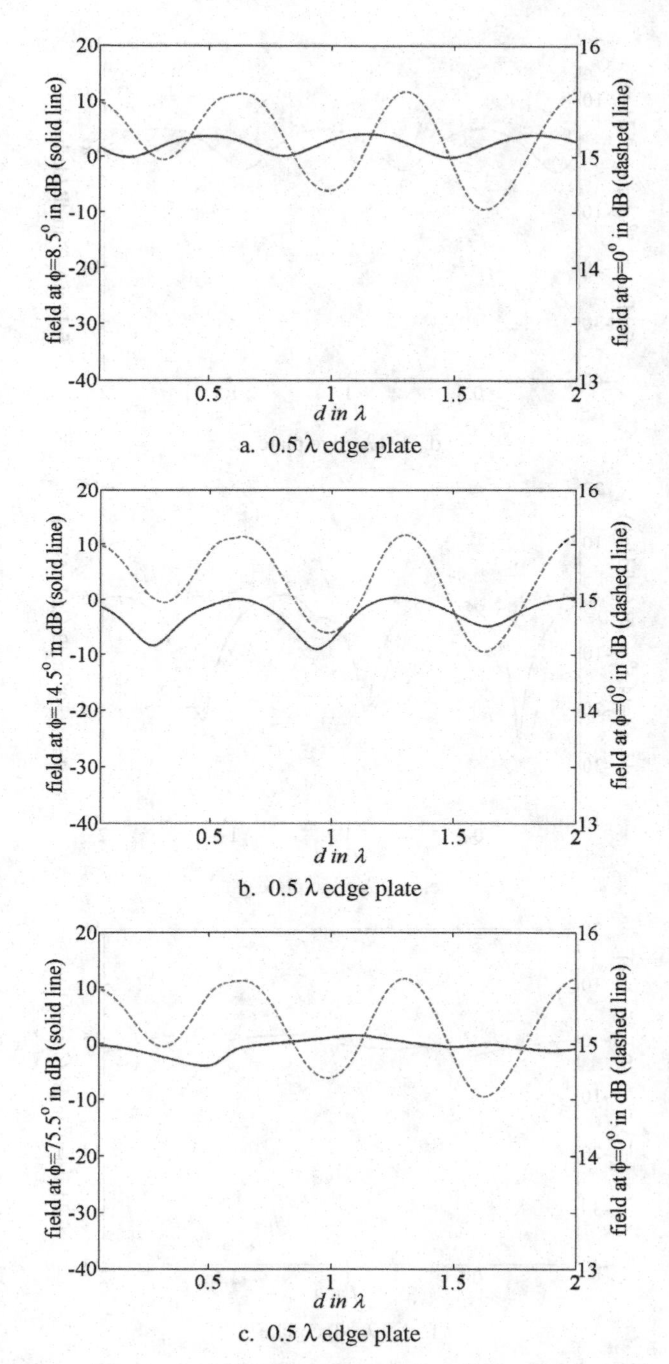

Figure 5.30. *The 0.5λ (a–c) and 1.0λ (d–f) edge scattering plates are continuously moved from $0 \leq d \leq 2\lambda$; these plots show the sidelobe level at three null locations and the mainbeam strength at $\varphi = 0°$.*

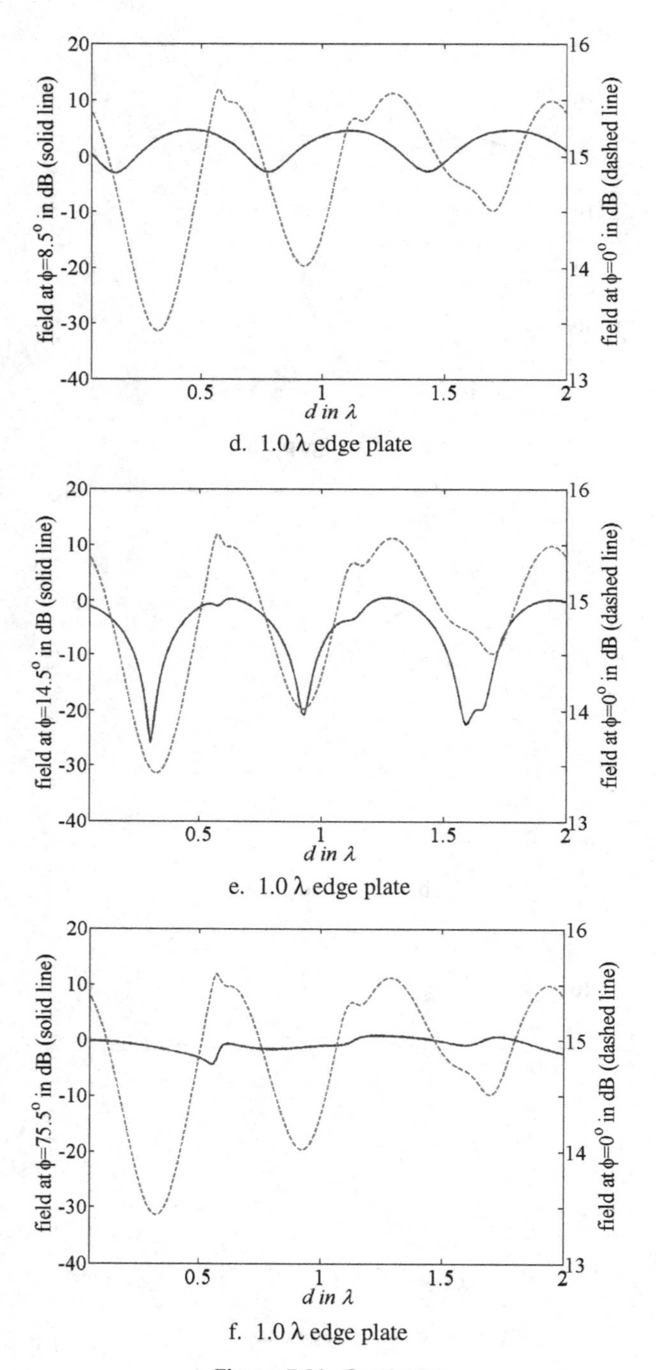

d. 1.0 λ edge plate

e. 1.0 λ edge plate

f. 1.0 λ edge plate

Figure 5.30. *Continued*

plate and one null. A local search algorithm might get caught in one of the local minima. The GA is a more appropriate choice in this case than traditional local minimization algorithms. In addition, placing multiple nulls using several scattering plates requires searching a very complex cost surface that would confuse local optimizers but not the GA.

The GA program for the results presented here has a population size of 8, a selection rate of 50%, and a mutation rate of 15%. These values are somewhat arbitrary but have worked well for other applications. The goal of the GA is to minimize the total output power by moving the scattering plates. A consequence of minimizing the total output power is that the algorithm also tries to null the desired signal. Keeping the scattering plates small, toward the edges, and less than 50% of the surface area of the reflector prevents the algorithm from nulling the mainbeam. The adaptive process described in Ref. 31 was capable of placing a null in the mainbeam. Consequently, the authors stated that the adaptive nulling must be done when the desired signal was not present.

The first example places a null at 14.5° using one 0.5λ-tall scattering plate at each edge of the reflector. Figure 5.31 is a graph of the GA convergence. The GA reaches a minimum in about 10 generations. Figure 5.32 shows the optimum positions of the plates. The resulting adapted pattern appears in Figure 5.33. There is a slight decrease in the mainbeam and a small increase in the sidelobe level. A relatively minor penalty was paid for placing a deep null in the direction of the interference source.

The next example places nulls at −35°, −14.5°, and 8.5° using two adjacent λ-tall scattering plates at each edge of the reflector. Figure 5.34 is a graph of the GA convergence. The GA reaches a minimum in 48 generations.

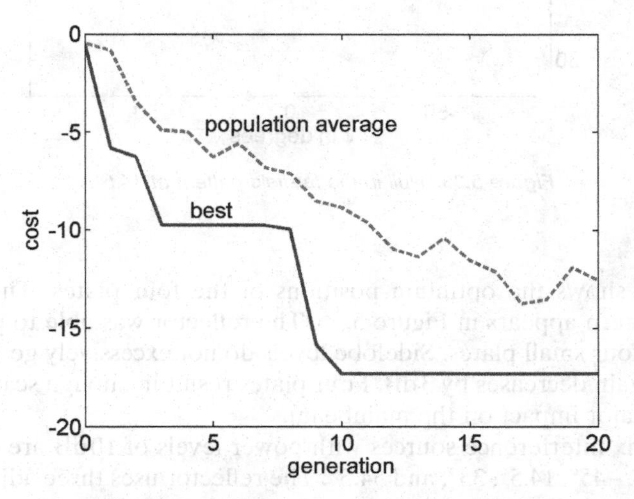

Figure 5.31. *The GA converges in ~10 generations.*

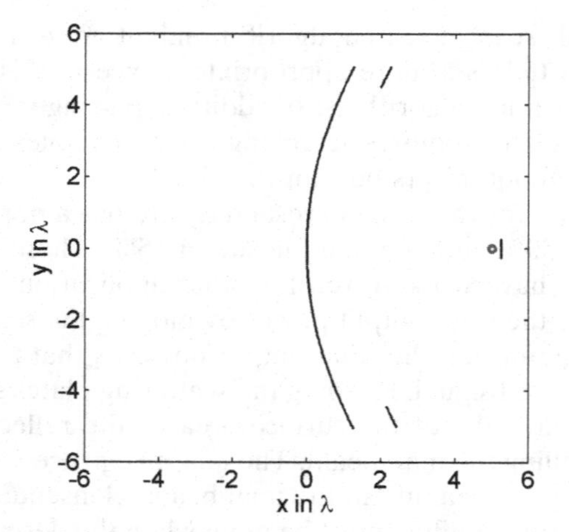

Figure 5.32. Placement of the scattering plates to put a null at 14.5°.

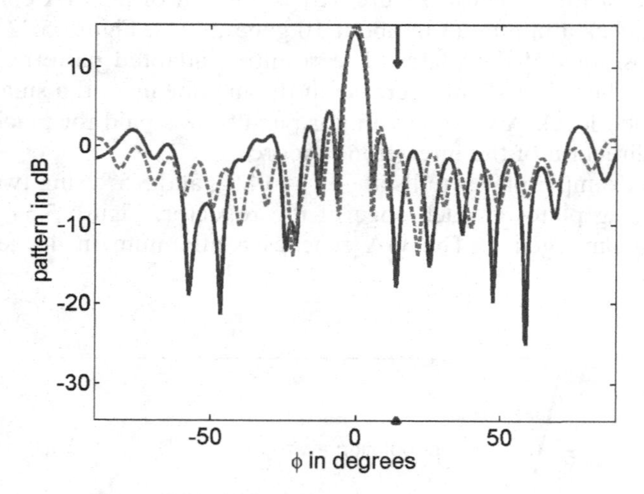

Figure 5.33. Null in the far-field pattern at 14.5°.

Figure 5.35 shows the optimum positions of the four plates. The resulting adapted pattern appears in Figure 5.36. The reflector was able to place three nulls using four small plates. Sidelobe levels do not excessively go up, but the mainbeam gain decreases by 3 dB. Four plates result in enough scattered field to have a major impact on the mainbeam.

Finally, six interference sources with power levels of 10 dB are incident at −75°, −64.5°, −45°, 14.5°, 35°, and 54.5°. The reflector uses three adjacent 0.5λ-

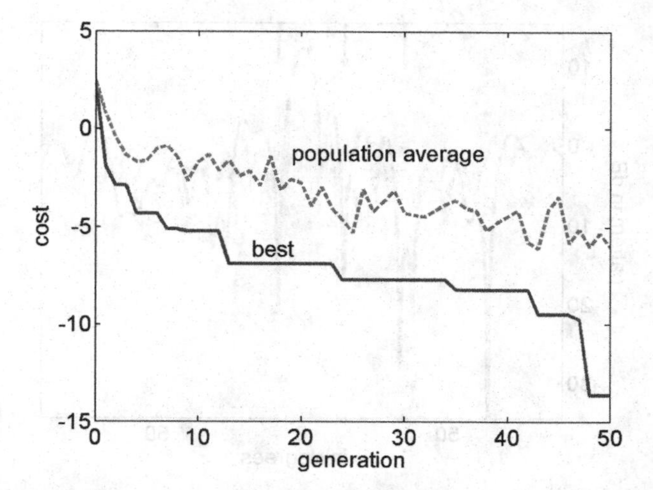

Figure 5.34. *Convergence of the GA when placing nulls in the far-field pattern at −35°, −14.5°, and 8.5°.*

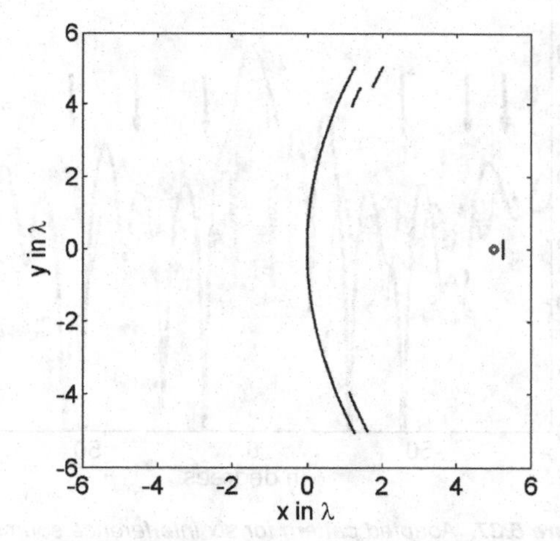

Figure 5.35. *Resulting locations of the four scattering plates for placing nulls at −35°, −14.5°, and 8.5°.*

tall scattering plates at each edge. Figure 5.37 is the resulting adapted antenna pattern. Not all the sources could be nulled, and the mainbeam underwent significant degradation. As with other adaptive schemes, this method needs enough degrees of freedom to deal with the number of jammers. Six plates in this small reflector were too many and resulted in a large decrease in antenna gain.

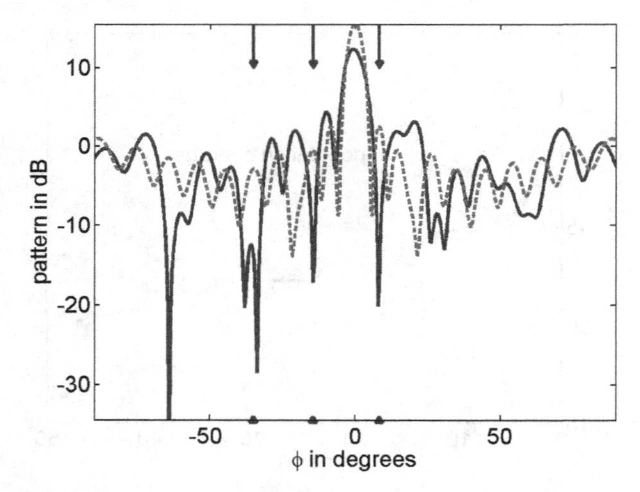

Figure 5.36. *Resulting nulls in the far-field pattern at −35°, −14.5°, and 8.5°.*

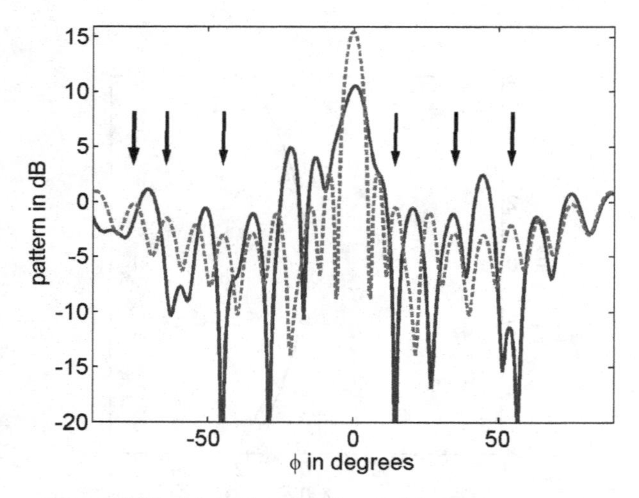

Figure 5.37. *Adapted pattern for six interference sources.*

5.4 ADAPTIVE CROSSED DIPOLES

The amount of power received by a moving antenna depends on the gains and polarization characteristics of the transmit and receive antennas. Antenna gain and polarization change as a function of angle. If the orientations of the transmit and receive antennas change as one or both antennas move, then the power received changes. The Friis transmission formula provides a way to calculate the received power in a communications system

$$P_r = \frac{P_t G_t(\theta_t, \phi_t) G_r(\theta_r, \phi_r) \lambda^2}{(4\pi r)^2} |\hat{\rho}_t(\theta_t, \phi_t) \cdot \hat{\rho}_r(\theta_r, \phi_r)|^2 \qquad (5.13)$$

where P_t = transmitter power

$G_t(\theta_t, \phi_t)$ = gain of transmit antenna

$G_r(\theta_r, \phi_r)$ = gain of receive antenna

r = distance between transmit and receive antennas

$\hat{\rho}_t(\theta_t, \phi_t)$, $\hat{\rho}_r(\theta_r, \phi_r)$ = polarization vectors of the transmit and receive antennas

The square of the magnitude of the dot product of the transmit and receive antenna polarization vectors is called the polarization loss factor (PLF). Assuming that the frequency, power transmitted, and separation distance remain the same, then the antenna gains and polarization vectors are the remaining ways to improve the link budget. Since these quantities are a function of angle, the antennas can be positioned to maximize the power transfer. When pointing the transmit and receive beams at each other is not possible, then adaptively altering the currents on the antennas to change the gains and polarization vectors is another way to improve the link budget.

The directivity and polarization of a crossed dipole antenna can be changed by varying the amplitude and phase of the currents fed to the individual dipoles. It has been experimentally shown that polarization diversity in the form of crossed dipoles improved a wireless link better than through spatial diversity (antenna separation) in a high-multipath environment [42]. Using three orthogonal crossed dipoles can also significantly increase the channel capacity of a wireless communication system inside a building [43]. Adaptive crossed dipoles change their polarization and gain to maximize the received power. A circularly polarized millimeter wave propagating through rain becomes elliptically polarized. The amount of depolarization is a function of the rainfall rate. In [44] an open-loop adaptive transmit antenna adjusted its polarization based on the amount of rainfall measured in the propagation path. Using an LMS (least-mean-square) algorithm to adapt the polarization and pattern of a two-element array of crossed dipoles improves the signal-to-interference-plus-noise ratio (SINR) [45]. The LMS algorithm was also used to find the complex weights of three orthogonal dipole antennas in order to improve the SINR. Rejection for interference signals at most angles of arrival and polarizations was possible [46]. A GA works well as an adaptive algorithm for a communications system that uses crossed dipoles [47].

In Figure 5.38 the transmit antenna is located at an angle of (θ_r, φ_r) from the receive antenna, and the receive antenna is located at an angle of (θ_t, φ_t) from the transmit antenna. In the nonadaptive mode, maximum power transfer occurs when $\theta_t = 0°$ and $\theta_r = 0°$. Maximum power transfer occurs when the mainbeams point at each other.

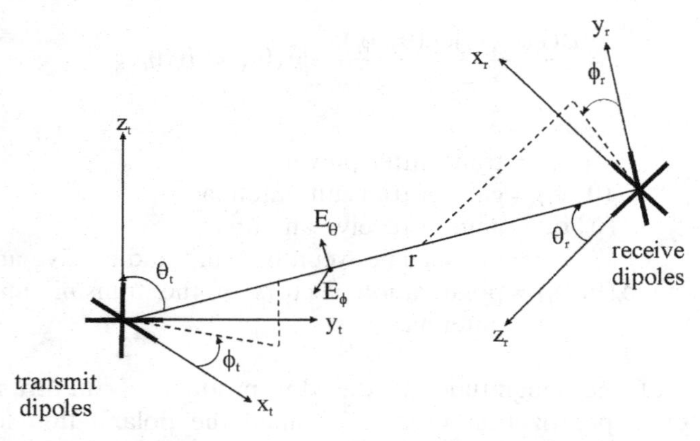

Figure 5.38. *Three orthogonal dipole antennas are used for the transmit and receive antennas.*

If the dipoles are short ($L \ll \lambda$), then the crossed dipole has a current that is the sum of the constant currents on each short dipole:

$$I(r') = I_x\hat{\boldsymbol{x}} + I_y\hat{\boldsymbol{y}} + I_z\hat{\boldsymbol{z}} \tag{5.14}$$

The magnetic vector potential associated with this current is

$$\boldsymbol{A} = \frac{\mu e^{-jkr}}{4\pi r}(I_x L_x\hat{\boldsymbol{x}} + I_y L_y\hat{\boldsymbol{y}} + I_z L_z\hat{\boldsymbol{z}}) \tag{5.15}$$

where r = distance from the origin to the field point at (x, y, z)
$L_{x,y,z}$ = dipole length in the x, y, and z directions
μ = permeability
$I_{x,y,z}$ = constant current in x, y, or z direction

The electric field far from the antenna is given by

$$\boldsymbol{E} = -j\omega\boldsymbol{A} \tag{5.16}$$

where ω = radial frequency. Substituting (5.15) into (5.16) leads to

$$\boldsymbol{E} = -j\frac{\omega\mu e^{-jkr}}{4\pi r}(I_x L_x\hat{\boldsymbol{x}} + I_y L_y\hat{\boldsymbol{y}} + I_z L_z\hat{\boldsymbol{z}})$$

Calculations are easier if the electric field is in spherical coordinates, so the rectangular components can be written in spherical coordinates as

$$E_\theta = -j\frac{\omega\mu e^{-jkr}}{4\pi r}(I_x L_x \cos\theta\cos\varphi + I_y L_y \cos\theta\sin\varphi - I_z L_z \sin\theta) \tag{5.17}$$

$$E_\varphi = -j\frac{\omega\mu e^{-jkr}}{4\pi r}(-I_x L_x \sin\varphi + I_y L_y \cos\varphi) \tag{5.18}$$

We are most interested in the directivity and polarization loss factor, which are given by

$$D(\theta,\varphi) = 4\pi\frac{|E_\theta(\theta,\varphi)|^2 + |E_\varphi(\theta,\varphi)|^2}{\int_0^{2\pi}\int_0^{\pi}\left[|E_\theta(\theta,\varphi)|^2 + |E_\varphi(\theta,\varphi)|^2\right]\sin\theta\, d\theta\, d\varphi} \tag{5.19}$$

$$|\hat{\rho}_t(\theta_t,\phi_t)\cdot\hat{\rho}_r(\theta_r,\phi_r)| = \frac{E_{\theta t}}{\sqrt{E_{\theta t}^2 + E_{\varphi t}^2}}\frac{E_{\theta r}}{\sqrt{E_{\theta r}^2 + E_{\varphi r}^2}} + \frac{E_{\varphi t}}{\sqrt{E_{\theta t}^2 + E_{\varphi t}^2}}\frac{E_{\varphi r}}{\sqrt{E_{\theta r}^2 + E_{\varphi r}^2}} \tag{5.20}$$

where $0 \le |\hat{\rho}_t(\theta_t,\phi_t)\cdot\hat{\rho}_r(\theta_r,\phi_r)| \le 1$. Transmit and receive quantities are represented by the t and r subscripts, respectively.

As an example, a satellite communications system has an earth station with a pair of orthogonal crossed dipoles in the x–y plane transmitting a circularly polarized field ($I_x = 1$, $I_y = j$, and $I_z = 0$) in the z direction. While θ increases from 0° to 90°, the polarization transitions from circular through elliptical and finally linear polarization at 90°. Figure 5.39 is a plot of the directivity and inverse axial ratio as a function of θ. Both quantities decrease as a function of θ. The axial ratio is the ratio of the major axis to the minor axis of the polarization ellipse. The inverse is often used to keep the quantity between zero and one.

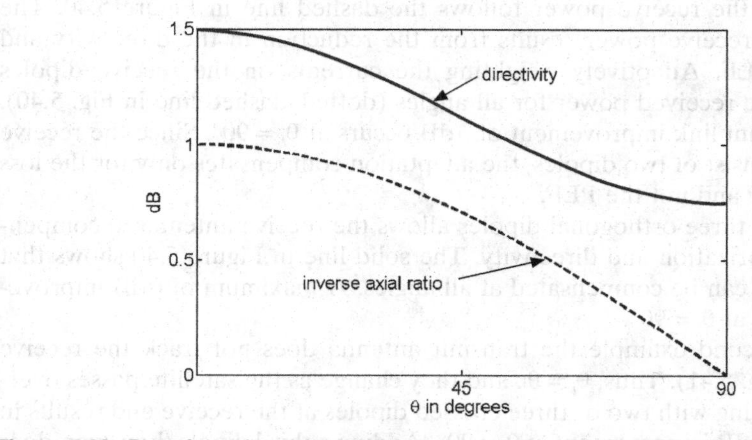

Figure 5.39. *Plot of the directivity and inverse axial ratio as a function of θ for two crossed dipoles at the transmit and receive ends.*

The directivities and polarizations of the transmit and receive antennas are the primary quantities in the Friis transmission formula that can be adaptively controlled by a GA. The fitness function uses the amplitude and phase weights for the received dipole and calculates the affected part of the Friis transmission formula

$$F(I_x, I_y, I_z) = D_t \times D_r \times |\hat{\rho}_t(\theta_t, \phi_t) \times \hat{\rho}_r(\theta_r, \phi_r)| \tag{5.21}$$

where D_t is the directivity of the transmit crossed dipoles in the direction of the receive crossed dipoles and D_r is the directivity of the receive crossed dipoles in the direction of the transmit crossed dipoles. The crossed dipoles have a maximum directivity close to 1, so the directivities are not normalized in the objective function. This fitness function has a maximum value of 2.25.

A GA optimized the continuous values of the amplitude and phase of the receive dipole currents in order to maximize (5.21). The GA has a population size of 8 and a mutation rate of 2%. It uses single-point crossover and has 50% replacement. The optimization process quickly improves the communications link but does not necessarily find the global minimum. A small population size and large mutation rate result in the best performance.

In all the examples presented here, the orientation of the ground and satellite antennas are assumed to change with time unless otherwise specified. Even though the distances between the antennas would also change, this variation is ignored. As the orientations of the antennas vary with time, so do their directivity and polarizations in the directions of each other.

The first example assumes that the transmit antenna tracks a satellite ($\theta_t = 0°$) and the receive antenna points at the ground (θ_r varies). The ground antenna transmits a circularly polarized signal with maximum directivity pointing at the moving receive antenna. If the transmit and receive antennas do not adapt, then the receive power follows the dashed line in Figure 5.40. The decrease in receive power results from the reduction in the directivity and increased PLF. Adaptively weighting the currents on the receive dipoles improves the received power for all angles (dotted–dashed line in Fig. 5.40). The maximum link improvement of 3 dB occurs at $\theta_r = 90°$. Since the receive antennas consist of two dipoles, the adaptation compensates only for the loss in directivity and not the PLF.

Adapting three orthogonal dipoles allows the receive antenna to compensate for polarization and directivity. The solid line in Figure 5.40 shows that the link loss can be compensated at all angles. A maximum of 6 dB improvement occurs at $\theta_t = 90°$.

In the second example the transmit antenna does not track the receive antenna (Fig. 5.41). Thus, $\theta_r = \theta_t$, and they change as the satellite passes overhead. Adapting with two or three crossed dipoles at the receive end results in as much as 3 dB improvement at $\theta_r = 90°$. Adding a third dipole, however, does

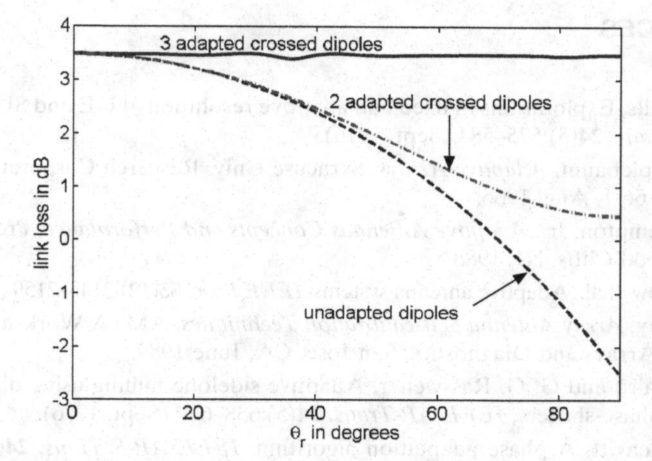

Figure 5.40. *The crossed dipole transmit antenna follows the receive antenna, so $\theta_t = 0°$ while the receive antenna moves.*

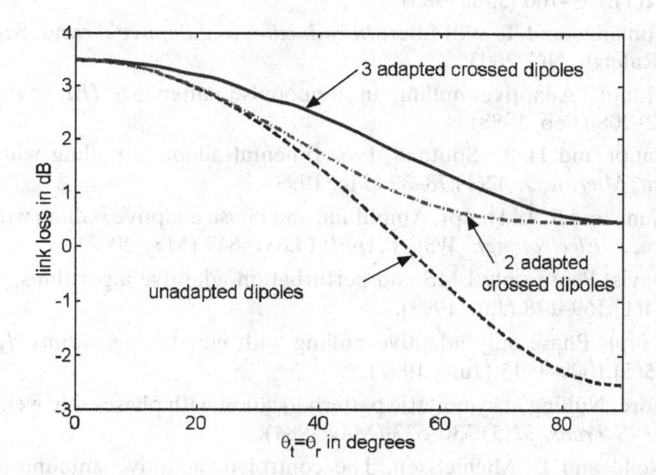

Figure 5.41. *The two antennas point at each other at $\theta_t = \theta_r = 0°$. The transmit antenna does not follow the receive antenna as the receive antenna moves.*

not provide significantly better improvement this time. At $\theta_t = 0°$, the transmit antenna transmits linear polarization, so the receive antenna can compensate only for the loss in directivity.

This technique works in the presence of noise and when the antennas move. In Ref. 47, it was found that a high mutation rate works best for a no-noise environment, and a low mutation rate works best in the presence of noise.

REFERENCES

1. P. Howells, Explorations in fixed and adaptive resolution at GE and SURC, *IEEE AP-S Trans.* **24**(5):575–584 (Sept. 1976).
2. S. P. Applebaum, *Adaptive Arrays*, Syracuse Univ. Research Corporation Report SPL TR 66-1, Aug. 1966.
3. R. T. Compton, Jr., *Adaptive Antennas Concepts and Performance*, Prentice-Hall, Englewood Cliffs, NJ, 1988.
4. B. Widrow et al., Adaptive antenna systems, *IEEE Proc.* **55**(12):2143–2159 (Dec. 1967).
5. R. Kinsey, *Array Antenna Self-calibration Techniques*, AMTA Workshop: Testing Phased Arrays and Diagnostics, San Jose, CA, June 1989.
6. C. A. Baird and G. G. Rassweiler, Adaptive sidelobe nulling using digitally controlled phase-shifters, *IEEE AP Trans.* **24**(5):638–649 (Sept. 1976).
7. M. K. Leavitt, A phase adaptation algorithm, *IEEE AP-S Trans.* **24**(5):754–756 (Sept. 1976).
8. H. Steyskal, Simple method for pattern nulling by phase perturbation, *IEEE AP-S Trans.* **31**(1):163–166 (Jan. 1983).
9. R. A. Monzingo and T. W. Miller, *Introduction to Adaptive Arrays*, SciTech Publishing, Raleigh, NC, 2003.
10. R. L. Haupt, Adaptive nulling in monopulse antennas, *IEEE AP-S Trans.* **36**(2):202–208 (Feb. 1988).
11. R. L. Haupt and H. L. Southall, Experimental adaptive nulling with a genetic algorithm, *Microw. J.* **42**(1):78–89 (Jan. 1999).
12. Y. C. Chung and R. L. Haupt, Amplitude and phase adaptive nulling with a genetic algorithm, *J. Electromagn. Waves Appl.* **14**:631–649 (May 2000).
13. R. M. Davis, Phase-only LMS and perturbation adaptive algorithms, *IEEE AES Trans.* **34**(1):169–178 (Jan. 1998).
14. R. L. Haupt, Phase-only adaptive nulling with genetic algorithms, *IEEE AP-S Trans.* **45**(5):1009–1015 (June 1997).
15. R. A. Shore, Nulling at symmetric pattern location with phase-only weight control, *IEEE AP-S Trans.* **32**(5):530–533 (May 1984).
16. D. S. Weile and E. Michielssen, The control of adaptive antenna arrays with genetic algorithms using dominance and diploidy, *IEEE AP-S Trans.* **49**(10):1424–1433 (Oct. 2001).
17. B. Chambers, A. P. Anderson, and R. J. Mitchell, Application of genetic algorithms to the optimization of adaptive antenna arrays and radar absorbers, *Proc. 1st Int. Conf. Genetic Algorithms in Engineering Systems: Innovations and Applications*, 94–99 (Sept. 1995).
18. R. J. Mitchell, B. Chambers, and A. P. Anderson, *Proc. 10th Int. Conf. Antennas and Propagation*, **1**:330–333 (April 1997).
19. B. K. Yeo, Array failure correction with a genetic algorithm, *IEEE AP-S Trans.* **47**(5):823–828 (May 1999).
20. Y. Lu and B. K. Yeo, Adaptive wide null steering for digital beamforming array with the complex coded genetic algorithm, *Proc. IEEE Phased Array Systems and Technology Symp.* 557–560 (May 2000).

21. G. Golino, A genetic algorithm for optimizing the segmentation in subarrays of planar array antenna radars with adaptive digital beamforming, *Proc. IEEE Phased Array Systems and Technology Symp.* 211–216 (Oct. 2003).
22. A. T. Bu et al., Design of the sector array antenna based on genetic algorithm for smart antenna system front end, *Proc. IEEE AP-S Symp.* **3**:686–689 (June 2003).
23. Y. Yashchyshyn and M. Piasecki, Improved model of smart antenna controlled by genetic algorithm, *Proc. CADSM*, 147–150 (Feb. 2001).
24. C. H. Hsu and T. M. Babij, Designing adaptive antenna uplink weighting vector to achieve optimization of radiation pattern, *Proc. IEEE SoutheastCon*, 245–249 (April 2002).
25. M. Vitale et al., Genetic algorithm assisted adaptive beamforming, *IEEE VTC Proc.* **1**:601–605 (Sept. 2002).
26. Y. C. Chung and R. L. Haupt, Adaptive nulling with spherical arrays using a genetic algorithm, *Proc. 1999 AP-S Symp.*, Orlando, FL, 2000–2003 (July 1999).
27. Y. Kimura and K. Hirasawa, A CMA adaptive array with digital phase shifters by a genetic algorithm and steepest descent method, *Proc. IEEE AP-S Symp.* **2**:914–917 (July 2000).
28. C. Sacchi et al., Adaptive antenna array control in the presence of interfering signals with stochastic arrivals: assessment of a GA-based procedure, *IEEE Wireless Commun. Trans.* **3**(4):1031–1036 (July 2004).
29. A. Massa et al., Planar antenna array control with genetic algorithms and adaptive array theory, *IEEE AP-S Trans.* **52**(11):2919–2924 (Nov. 2004).
30. R. L. Haupt, *Directional Antenna System Having Sidelobe Suppression*, US Patent 4,571,594 (Feb. 18, 1986).
31. A. D. Monk and P. J. B. Clarricoats, Adaptive null formation with a reconfigurable reflector antenna, *IEE Proc. Microw. Anten. Propag.* **142**(3):220–224 (June 1995).
32. M. S. Zarghamee, J. Antebi, and F. W. Kan, Optimal surface adjustment of haystack antenna, *IEEE AP-S Trans.* **43**(1):79–86 (Jan. 1995).
33. J. Antebi et al., A deformable subreflector for the haystack radio-telescope, *IEEE Anten. Propag. Mag.* **36**(3):19–28 (June 1994); corrections in **36**(4):34 (Aug. 1994).
34. H. Yoon, G. Washington, and W. H. Theunissen, Analysis and design of doubly curved piezoelectric strip-actuated aperture antennas, *IEEE AP-S Trans.* **48**(5):755–763 (May 2000).
35. D. Jacavanco, Controlled surface distortion effects, *Proc. 1984 Antenna Applications Symp.*, Robert Allerton Park, IL, Sept. 1984.
36. D. Jacavanco, *Reflector Antenna Having Sidelobe Suppression Elements*, U.S. Patent 4,631,547 (Dec. 23, 1986).
37. D. A. Trapp, *Physical Optics Model of Side Lobe Nulling by Discs on a Parabolic Reflector*, M.S. thesis, Air Force Institute of Technology, Dec. 1985.
38. J. L. Poirier, *Reflector Antenna Having Sidelobe Nulling Assembly with Metallic Gratings*, US Patent 4,725,847 (Feb. 16, 1988).
39. P. T. Lam et al., Sidelobe reduction of a parabolic reflector with auxiliary reflectors, *IEEE AP-S Trans.* **35**(12):1367–1374 (Dec. 1987).

40. R. Haupt, Adaptive nulling with a reflector antenna using movable scattering elements, *IEEE AP-S Trans.* **3**(2):887–890 (Feb. 2005).
41. N. Malik and R. Haupt, Near field nulling with a cylindrical reflector and moving plates, *Proc. IEEE AP-S Symp.* **3**:3019–3022 (June 2004).
42. A. Singer, Space vs. polarization diversity, *Wireless Rev.* 164–168 (Feb. 15, 1998).
43. M. R. Andrews, P. P. Mitra, and R. deCarvalho, Tripling the capacity of wireless communications using electromagnetic polarization, *Nature* **409**:316–318 (Jan. 18, 2001).
44. R. E. Marshall and C. W. Bostian, An adaptive polarization correction scheme using circular polarization, *Proc. IEEE Int. Antennas and Propagation Society Symp.*, Atlanta, GA, 395–397 (June 1974).
45. R. T. Compton, On the performance of a polarization sensitive adaptive array, *IEEE Trans. Anten. Propag.* **AP-29**:718–725 (Sept. 1981).
46. R. T. Compton, The tripole antenna: An adaptive array with full polarization flexibility, *IEEE Trans. Anten. Propag.* **AP-29**:944–952 (Nov. 1981).
47. R. L. Haupt, Adaptive crossed dipole antennas using a genetic algorithm, *IEEE Trans. Anten. Propag.* **AP-52**:1976–1982 (Aug. 2004).

6

Genetic Algorithm Optimization of Wire Antennas

6.1 INTRODUCTION

The advent of the GA has provided antenna engineers with a new and powerful design tool for wire antenna structures and systems. For example, the GA has been successfully applied to such complex design problems as wire antennas loaded with lumped elements to achieve broadband or multiband operation, the optimization of miniature two-dimensional meander-line and three-dimensional crooked-wire antennas, and the performance enhancement of Yagi–Uda arrays. These and other related applications of the GA to wire antenna design are discussed in the remainder of this chapter.

6.2 GA DESIGN OF ELECTRICALLY LOADED WIRE ANTENNAS

GAs have been demonstrated to be a powerful tool for the design of electrically loaded wire antennas. Several studies have shown the utility of GAs for solving this class of challenging design optimization problems, where a wire antenna is loaded with either stubs or lumped components such as *LC*, *RL*, or *RLC* networks. In some cases, the design of matching networks, transformers, and/or attenuators have also been included as part of the GA optimization process for loaded wire antennas.

A GA has been used [1] to design a monopole loaded with a modified folded dipole. The geometry has six design variables as shown in Figure 6.1.

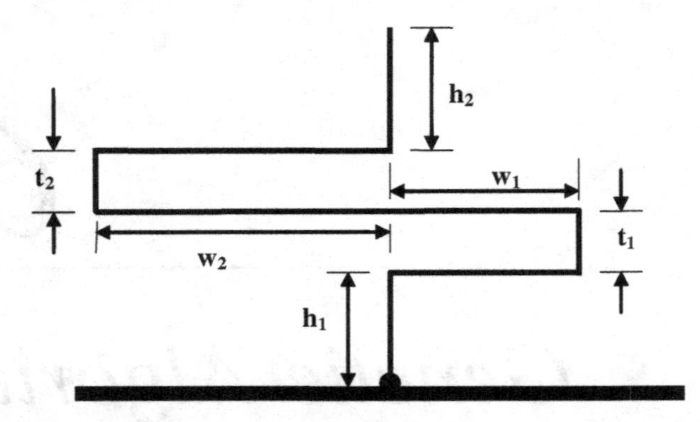

Figure 6.1. Geometry for a monopole loaded with a modified folded dipole.

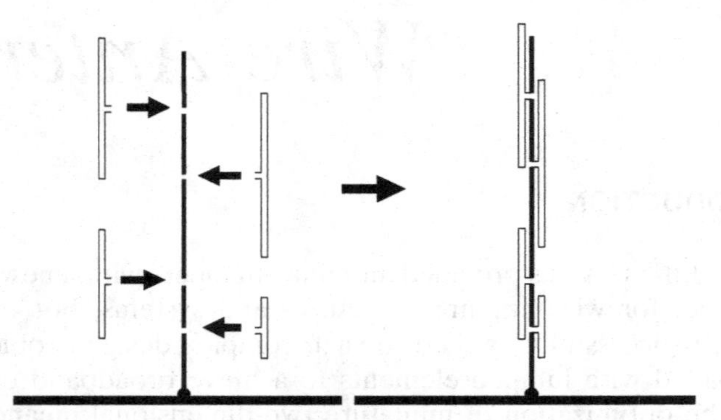

Figure 6.2. Example of a stub-loaded miniature multiband monopole antenna (P. L. Werner and D. H. Werner, © IEEE, 2005. [2]).

The objective is to find values for the variables that result in a radiation pattern with uniform hemispherical coverage. It was found that a GA can rapidly produce an antenna configuration that possessed a nearly uniform power pattern over the entire hemisphere.

A novel GA design methodology has been introduced for miniature multiband monopole and whip-type antennas [2]. The miniaturization and multiband response are simultaneously achieved by placing a fixed number of thin stubs at strategic locations along the antenna as illustrated in Figure 6.2. A GA is used to determine the optimal lengths and locations of the stubs required to achieve a specified percent reduction in monopole length. For example, a GA is used to design a miniature dual-band VHF monopole antenna capable of operating at 100 and 210 MHz. The maximum length of the monopole and stubs is fixed at 51 and 40 cm, respectively. The antenna was optimized over an infinite PEC ground plane, and the final structure evolved by the GA is shown in Figure 6.3. The lengths of the first and second stubs were selected by

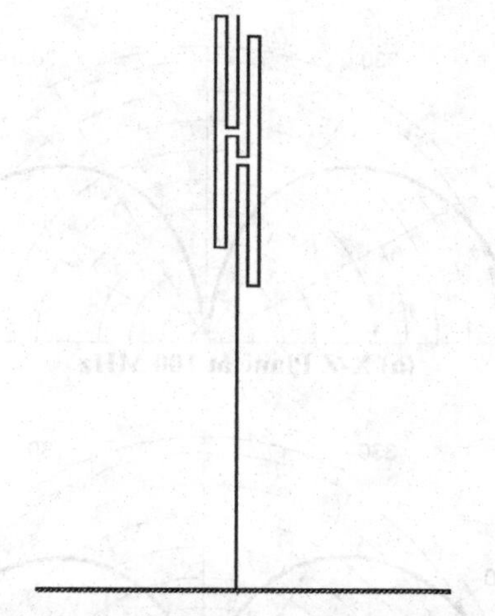

Figure 6.3. Dual-band stub-loaded VHF monopole antenna geometry evolved by the GA.

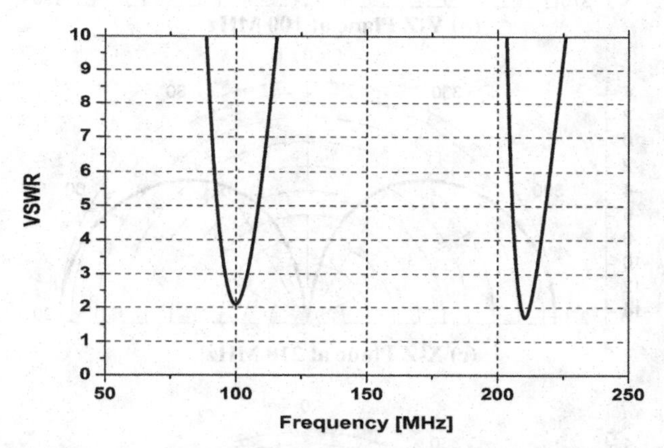

Figure 6.4. VSWR plot for the dual-band VHF monopole antenna shown in Figure 6.3 with respect to a 50Ω impedance.

the GA to be 20.53 and 22.14 cm, respectively, while the width of the stubs is 0.6 cm. The locations of the stubs chosen by the GA are 40.3 cm for the first stub and 38.3 cm for the second stub. The VSWR plot shown in Figure 6.4 clearly demonstrates the dual-band performance of the stub-loaded monopole. The radiation patterns for the two operating frequencies of the antenna appear in Figure 6.5, with a maximum gain of 4.95 dBi at 100 MHz and 4.84 dBi at 210 MHz. Finally, the total length of this antenna is 51 cm, which represents a 32% size reduction relative to a conventional quarter-wavelength monopole operating at 100 MHz.

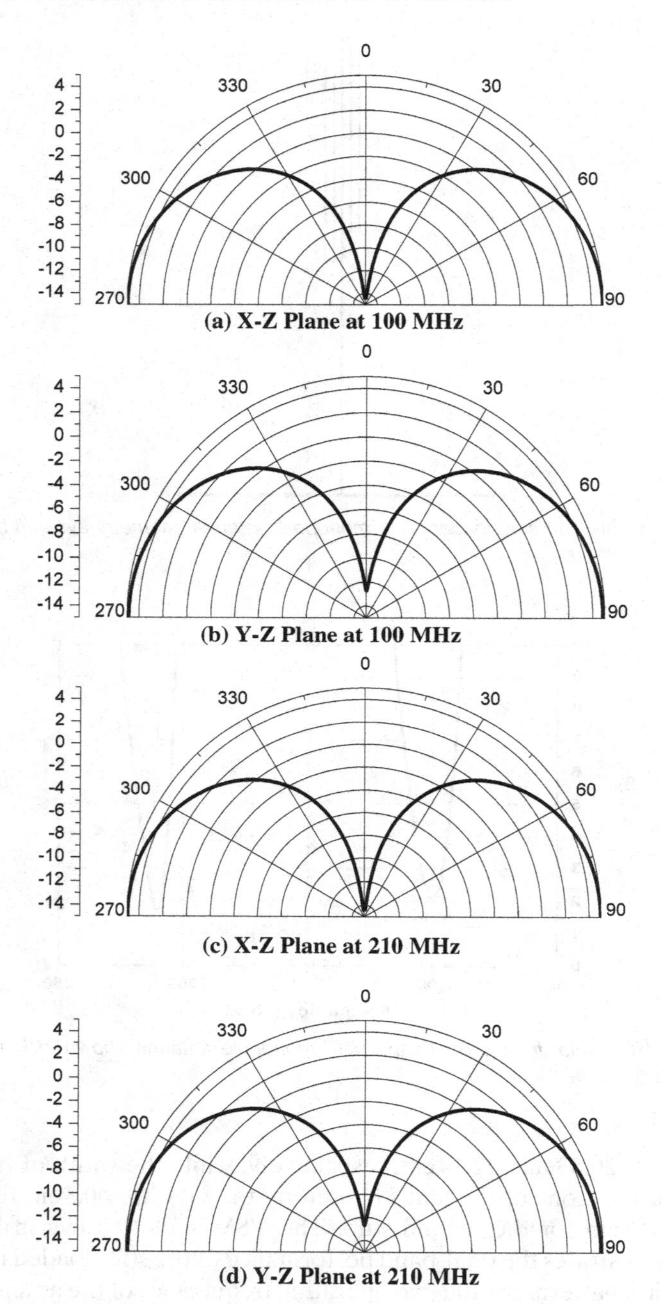

Figure 6.5. *Radiation patterns of antenna over an infinite ground plane for operating frequencies 100 and 210 MHz: (a) X–Z plane at 100 MHz; (b) Y–Z plane at 100 MHz; (c) X–Z plane at 210 MHz; (d) Y–Z plane at 210 MHz.*

Design approaches have been investigated [3–9] for the possibility of using a GA to synthesize wire antennas loaded with lumped components. For many of these designs, a GA is employed to simultaneously optimize the loading circuit variables, locations of the loads along the antenna, and the matching network variables. The objective is to evolve optimal loaded wire antenna designs that exhibit ultrawideband performance. Figure 6.6 illustrates the geometric configuration for a simple monopole antenna, which is loaded with parallel *RLC* resonant traps and a corresponding matching network. Loaded monopole antennas of this type have been considered [3,7]. For example, following the development outlined in Ref. 7, a micro-GA was used to optimize the values of three load circuits and one matching inductor in the monopole antenna/matching network system of Figure 6.6. The monopole is loaded with an inductor near its base and has two parallel inductor–resistor circuits near the middle and top. In order to simplify the problem for this example, the

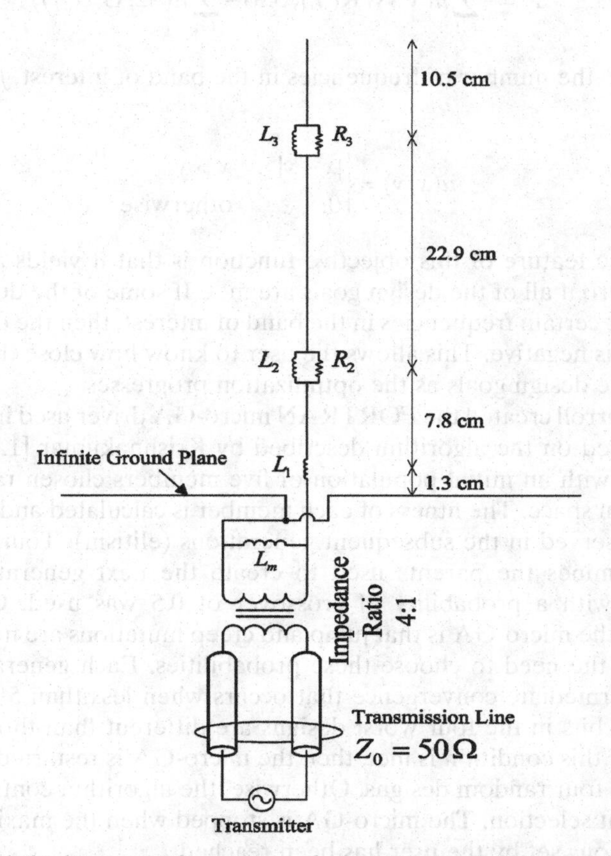

Figure 6.6. *Loaded wire monopole antenna of diameter d = 0.635 cm and total height h = 42.5 cm with matching network consisting of an inductor and 4 : 1 impedance ratio transformer.*

transformer impedance ratio, load positions, and load circuit topologies were chosen in advance of optimization. These variables were known beforehand to lead to good results for similar problems as described by Rogers [7,8]. The method of moments (MoM) was used to numerically analyze the basic antenna structure. Further details about the fast analysis, optimization, realization, and electrical measurements of loaded antennas similar to this example are available in Refs. 7 and 8.

The design goals for this antenna system are maximum voltage standing-wave ratio (VSWR) of 3.5 and minimum system gain at the horizon (G_{sys}) equal to −2 dBi over the frequency range 100–1500 MHz. The system gain is sometimes called the "realized gain" and is equivalent to the mismatch loss ($1 - |\Gamma|^2$) multiplied by the antenna gain. An objective function that describes these goals explicitly is

$$F = -\sum_{i=1}^{N^f} u(\mathrm{VSWR}(f_i), 3.5) - \sum_{i=1}^{N^f} u(-2, G_{sys}(f_i)) \tag{6.1}$$

where N^f is the number of frequencies in the band of interest, f_i is frequency, and

$$u(x, y) = \begin{cases} |x - y|^2, & x > y \\ 0, & \text{otherwise} \end{cases} \tag{6.2}$$

One nice feature of this objective function is that it yields a fitness value equal to zero if all of the design goals are met. If some of the design goals are not met for certain frequencies in the band of interest, then the objective function value is negative. This allows the user to know how close the system is to meeting the design goals as the optimization progresses.

D. L. Carroll created the FORTRAN micro-GA driver used in these studies [10,11] based on the algorithm described by Krishnakumar [12]. The micro-GA starts with an initial population of five members chosen randomly from the solution space. The fitness of each member is calculated and the best solution is preserved in the subsequent generations (elitism). Tournament selection determines the parents used to create the next generation. Uniform crossover with a probability of crossover of 0.5 was used. One desirable feature of the micro-GA is that jump and creep mutations are not used, which eliminates the need to choose these probabilities. Each generation is tested for an intermediate convergence that occurs when less than 5% of the total number of bits in the four worst designs are different than those of the best solution. If this condition is met, then the micro-GA is restarted with the best design and four random designs. Otherwise, the algorithm continues with the tournament selection. The micro-GA is stopped when the maximum number of generations set by the user has been reached.

In this example, the total number of bits in the chromosome string is 54, which leads to a total number of possibilities in the solution space equal to

1.8×10^{16}. The number of bits used to represent each variable and the corresponding variable resolutions are given in Table 6.1. An optimum solution that meets all of the design goals was found after 389 generations or 1945 function evaluations. One sees in Figure 6.7 that the VSWR is less than 3.5 over the band of 100–1500 MHz and in Figure 6.8 that the system gain is greater than −2 dBi over the 100–1500 MHz bandwidth. The overall convergence of the micro-GA for this example is graphed on a logarithmic scale in Figure 6.9. The micro-GA is successful in finding a solution that meets all design goals in a very efficient manner considering that only a small percentage of the designs in the whole solution space were evaluated.

Other more complex wire configurations loaded with lumped components have also been considered, including twin-whip antennas [3], folded monopoles [3], stacked dipoles [5,6], diamond and kite antennas [4], and even crooked-wire antennas [9]. To further demonstrate the versatility of this GA antenna design synthesis approach, we will next consider the bifolded monopole shown in Figure 6.10. In this example, a broadband antenna will be designed using a reactively loaded bifolded monopole [13], where a binary GA is employed to optimize the lengths of the monopole, location, and component

TABLE 6.1. Variable Ranges and Number of Possibilities for Micro-GA Optimization of Antenna and Matching Network of Figure 6.6[a]

	Minimum Value	Maximum Value	Number of Bits	Number of Possibilities	Resolution
L values	0.01 μH	1.1 μH	8	256	0.0043 μH
R values	100 Ω	2500 Ω	11	2048	1.17 Ω
L_m	0.01 μH	0.8 μH	8	256	0.0031 μH

[a] The three inductance values, two resistance values, and one L_m value make the total number of bits 54 and total number of possibilities 1.8×10^{16}.

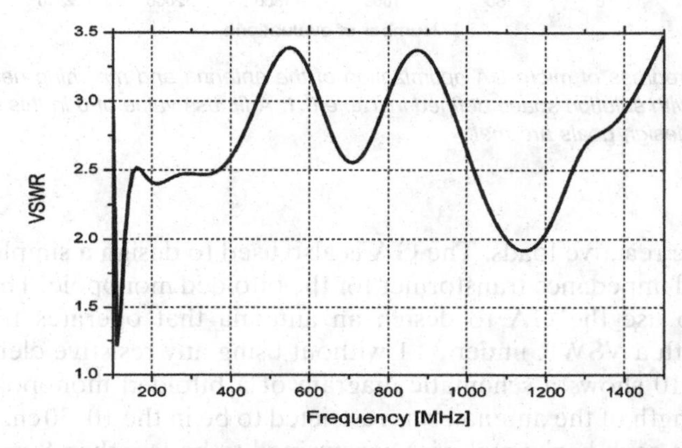

Figure 6.7. VSWR of antenna and matching network system of Figure 6.6 with L_1 = 0.01 μH, L_2 = 0.0442 μH, L_3 = 0.557 μH, L_m = 0.177 μH, R_2 = 1156 Ω, and R_3 = 529 Ω.

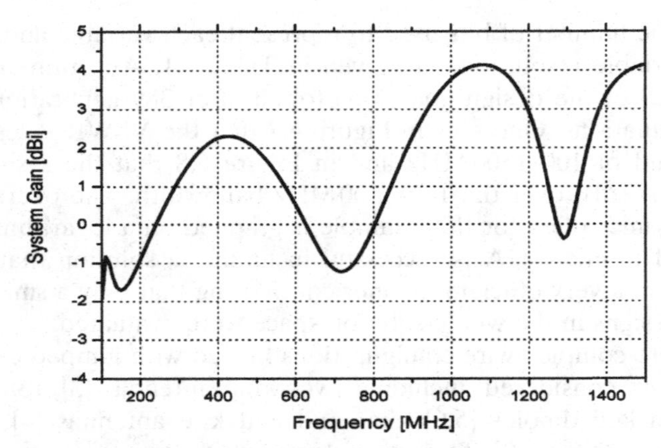

Figure 6.8. *System gain of antenna and matching network system of Figure 6.6 with* $L_1 = 0.01\,\mu H$, $L_2 = 0.0442\,\mu H$, $L_3 = 0.557\,\mu H$, $L_m = 0.177\,\mu H$, $R_2 = 1156\,\Omega$, *and* $R_3 = 529\,\Omega$.

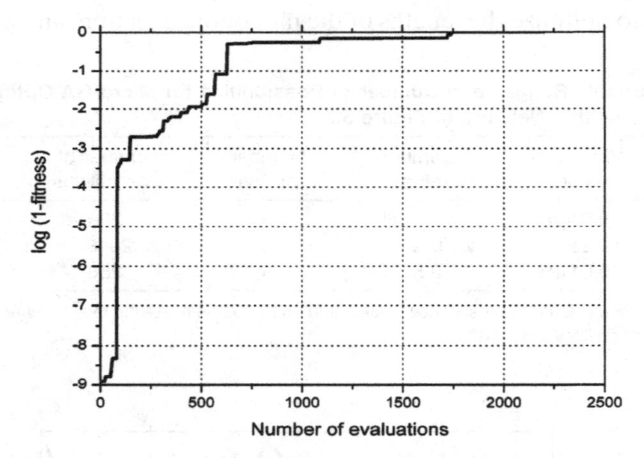

Figure 6.9. *Progress of micro-GA optimization of the antenna and matching network system of Figure 6.6 with solution space defined in Table 6.1. A fitness value of 0 in this example indicates that all design goals are met.*

values of the reactive loads. The GA is also used to design a simple matching network and impedance transformer for the bifolded monopole. The objective here was to use the GA to design an antenna that operates from 250 to 500 MHz with a VSWR under 3 : 1 without using any resistive elements.

Figure 6.10 shows a schematic diagram of a bifolded monopole antenna. The total length of the antenna was restricted to be in the 10–30 cm range with the length of each horizontal wire constrained to be less than 8 cm. The optimized width of this design is 8.875 cm and the length is 25.687 cm. In addition, the GA was given the task of placing four reactive LC loads anywhere on the

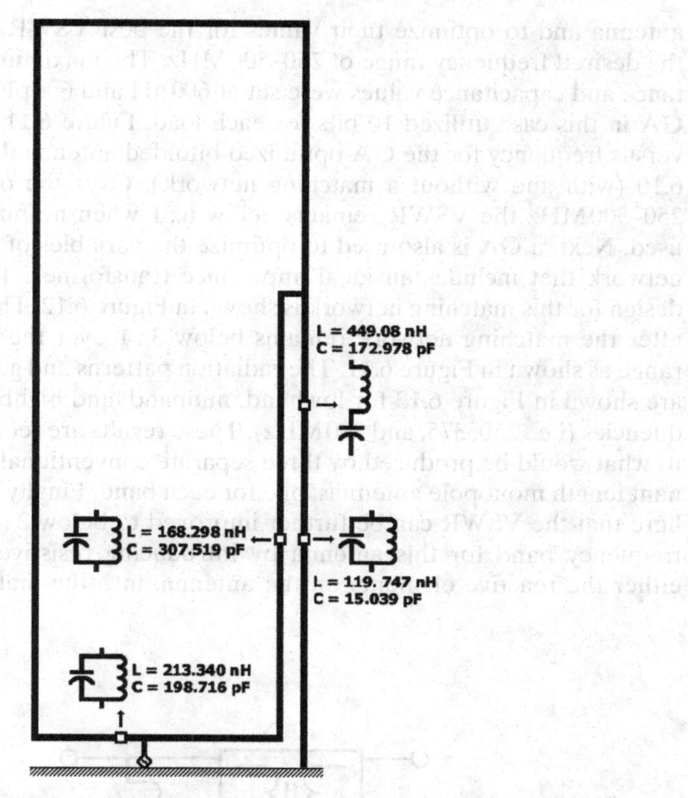

Figure 6.10. *Bifolded monopole with reactive series and parallel LC loads. The locations and component values of these loads were optimized using a binary GA.*

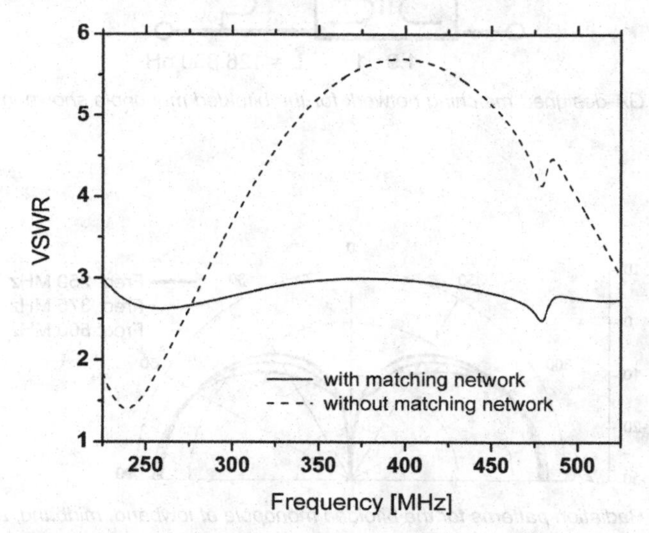

Figure 6.11. *VSWR versus frequency for the bifolded monopole with and without the matching network.*

antenna and to optimize their values for the best VSWR performance over the desired frequency range of 250–500 MHz. The maximum allowable inductance and capacitance values were set at 600 nH and 600 pF, respectively. The GA in this case utilized 10 bits for each load. Figure 6.11 shows the VSWR versus frequency for the GA optimized bifolded antenna illustrated in Figure 6.10 (with and without a matching network). Over the optimized range of 250–500 MHz the VSWR remains below 6 : 1 when no matching network is used. Next, a GA is also used to optimize the variables of a simple matching network that includes an ideal impedance transformer. The GA optimized design for this matching network is shown in Figure 6.12. The resulting VSWR after the matching network remains below 3 : 1 over the desired frequency range as shown in Figure 6.11. The radiation patterns and gain for this antenna are shown in Figure 6.13 for lowband, midband, and highband operating frequencies (i.e., 250, 375, and 500 MHz). These results are seen to be very similar to what would be produced by three separate conventional narrowband resonant length monopole antennas: one for each band. Finally, it should be noted here that the VSWR can be further improved to below 2 : 1 over the desired frequency band for this antenna by introducing resistive components into either the reactive elements on the antenna, into the matching network, or

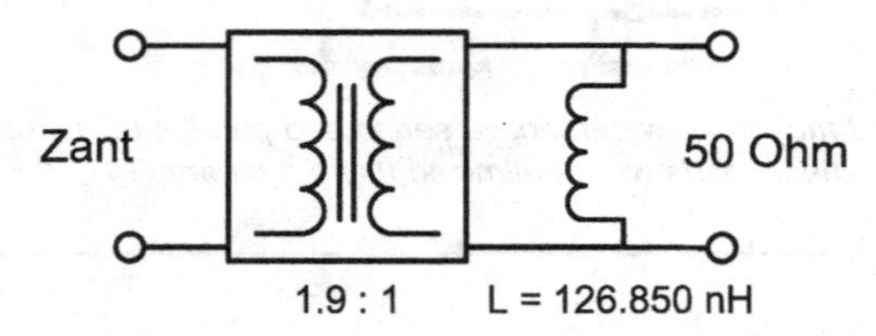

Figure 6.12. GA-designed matching network for the bifolded monopole shown in Figure 6.10.

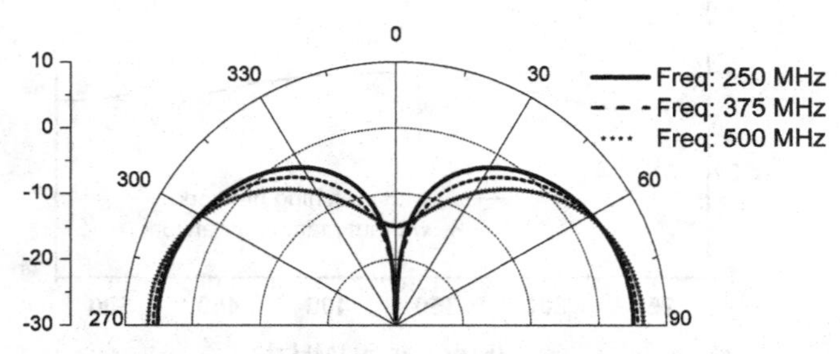

Figure 6.13. Radiation patterns for the bifolded monopole at lowband, midband, and highband (θ plot with $\varphi = 90°$).

both. However, this will also have the adverse affect of simultaneously reducing the overall system gain.

The use of dielectric bead loading has been investigated [14] as an alternative to lumped elements for controlling the currents on wire antennas. In this case a 4λ-long wire antenna is considered, and a GA is used to determine the optimal locations of the dielectric beads to provide performance comparable to a half-wave dipole. It has been shown [14] that this approach can yield results very similar to those obtained using resonant *RLC* traps or chokes, but with the advantage of being less bulky.

The majority of the work on the design synthesis of electrically loaded antennas has been done by combining a GA with full-wave analysis techniques that are based on a thin-wire method of moments formulation in the frequency domain (MoM-FD) [1–9,14]. However, some more recent work has considered a direct GA-based optimization of resistively loaded wires in the time domain [15,16]. In this case, a thin-wire method of moments technique is formulated directly in the time domain (MoM-TD) and used in conjunction with a GA. This approach has been used successfully to design optimal broadband thin-wire antennas for various applications including ground-penetrating radar (GPR).

6.3 GA DESIGN OF THREE-DIMENSIONAL CROOKED-WIRE ANTENNAS

The genetic antenna (also known as the "crooked-wire antenna") and the process of creating it was invented in early 1995 by Dr. Edward E. Altshuler and Dr. Derek S. Linden, and patented by the U.S. Air Force [17]. The process was first applied to design an antenna with hemispherical coverage and right-hand circular polarization using a single feedpoint. Research regarding this antenna has been published in many journal articles, conference papers, and other works [18–36]. This section will explain the unique features of crooked-wire antennas and present an example of an application effectively solved by such an antenna.

A crooked-wire genetic antenna is an antenna that neither has a preconceived method of operation nor uses a preexisting design, beyond very basic operational constraints (such as a monopole-like means of excitation and the presence of a ground plane). Its method of operation is determined by an optimization algorithm (which is not necessarily a GA). It is not constrained to be a predefined antenna design (Yagi, helix, spiral, etc.), but essentially forms its own method of operation during optimization.

Crooked-wire antennas are generally random-looking in shape (although they are far from random), and many of those who see them for the first time remark that they "look like paper clips." Because these antennas are so under-constrained, they tend to appear very different from each other, even when they are optimized to solve the same problem and have similar performance characteristics. As a result, they are almost impossible to classify by their specific shape. Instead, they have been given the name "genetic antennas" regardless of their resulting shape to indicate that their defining characteristic is their genetic code, which is manipulated by an optimizer to define its electromagnetic characteristics and hence its final shape. Note again that the genetic antenna received its name from the underlying and defining genetic code, and not because it is optimized by a GA.

The genetic antenna usually is subject to very basic constraints such as maximum size, number of wires and loads, and the number of feedpoints and their corresponding locations. For an antenna to be classified as a genetic antenna, these constraints cannot be so strict that they define a specific method of operation, but they can be set to help push the optimization toward a method of operation that solves the problem effectively, as will be demonstrated by the example considered here.

The first application of a genetic antenna was to solve a satellite communications problem, which has been published in several references [19,20,23,25,28]. As it turned out, this was a very good application for the genetic antenna, producing excellent results that were very counterintuitive and paving the way for many additional designs.

Ground-based transceivers for earth-to-satellite communication links require antennas that are circularly polarized and have near-hemispherical coverage. Circular polarization is necessary for systems operating at frequencies below 3 GHz, since the Faraday rotation produced by the ionosphere can cause a linearly polarized wave to rotate out of alignment with the receiving antenna. In a worst-case scenario, the incoming wave becomes cross-polarized so that no signal is received. A circularly polarized signal eliminates this problem.

Near-hemispherical coverage is also desirable since the earth-based antenna is often required to receive a signal from a satellite anywhere above the horizon, except at low elevation angles (at low angles, signals have multipath components that can disrupt system performance). Usually, helical or patch antennas are used for this application, but these antennas are generally narrowband and require a phasing network, which increases their complexity and cost. In addition, these antennas are usually somewhat directional, meaning that they need to be pointed more precisely toward the satellite to be used effectively—an inconvenience for mobile systems.

At this point we consider a specific design example for the genetic crooked-wire antenna. The antenna design requires a flat gain above 10° elevation (i.e., near hemispherical coverage) for a right-hand circularly polarized signal, and an operating frequency centered around 1600 MHz. It was decided to make the antenna vehicle-mounted, so it was designed over an infinite ground plane. Also, for simplicity in construction and lower cost, only a single feedpoint was specified. Since near-hemispherical coverage was desired for this antenna, it was expected to be confined to a cube 0.5λ on a side. This design space is shown in Figure 6.14.

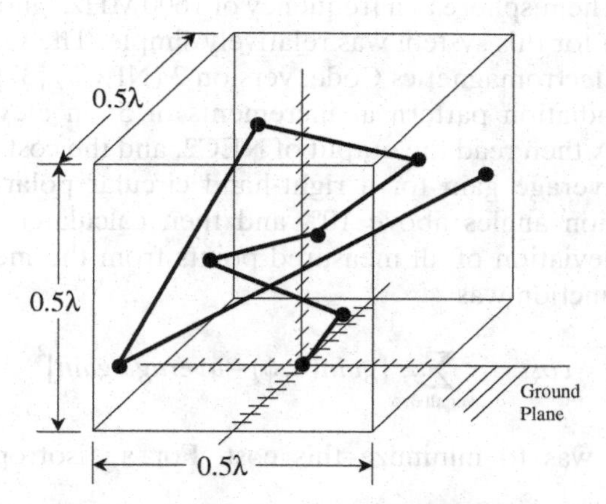

Figure 6.14. *Search space for crooked-wire genetic antenna (E. Altshuler and D. Linden, © IEEE, 1997, [23]).*

Both real and binary GAs have been used to design this antenna. When this design was first optimized using the binary GA, 5 bits were allowed for each component of the (X, Y, Z) coordinate for the beginning and end of each wire. In other words, each axis of the design space had 32 levels that could be selected from one side of the search space to the other, and there were, therefore, 32^3 possible vertices at which the wires could be connected. Five bits were chosen because that allowed the GA to work in units of $\frac{1}{64}$th of a wavelength, which was close to the limit of fabrication tolerances at the chosen frequency of 1600 MHz. For the real GA, one real gene specified each variable.

Next, the number of wires and the connection scheme that a design could use had to be specified. Initially, antennas consisting of 5, 6, 7, and 8 connected wire segments were investigated. Preliminary results showed the 7-wire genetic antenna to perform slightly better than the 5-, 6-, and 8-wire antennas, so the 7-wire genetic antenna was chosen to investigate in detail. In addition, the decision was made to connect all wires in series for simplicity. This constraint seems to allow for quite a bit of flexibility while maintaining ease of construction.

With 5 bits for each axis coordinate, 3 axis coordinates per point, and 7 points to be designated (1 point per wire, since each wire starts at the previous wire's endpoint, and the first wire begins at the origin), each 7-wire genetic antenna required a binary chromosome with $5 \times 3 \times 7 = 105$ bits. Each antenna with a real chromosome required 21 real genes. The bits for each coordinate were placed next to each other in the chromosome, as follows:

$$[X_1, Y_1, Z_1, X_2, Y_2, Z_2, X_3, Y_3, Z_3, \ldots, X_7, Y_7, Z_7]$$

The goal was to obtain right-hand circular polarization 10° above the horizon over the hemisphere at a frequency of 1600 MHz, ignoring impedance, the cost function for this system was relatively simple. The GA program used the Numerical Electromagnetics Code, version 2 (NEC2) [37] to compute the hemispherical radiation pattern at increments of 5° in elevation and 5° in azimuth. The GA then read the output of NEC2, and the cost function routine calculated the average gain for a right-hand circular polarization (RHCP) wave for elevation angles above 10°, and then calculated the sum of the squares of the deviation of all measured points from the mean. In equation form, the cost function was

$$cost = \sum_{\text{over all } \theta, \phi} [gain(\theta, \phi) - average\ gain]^2 \tag{6.3}$$

The GA's goal was to minimize this cost. For an isotropic gain pattern $cost = 0$.

The antenna design evolved by the GA has an unusual shape, as seen by the photograph in Figure 6.15. The coordinates for its vertices are listed in Table 6.2. The computed radiation patterns of the antenna over an infinite ground plane are shown in Figure 6.16 for elevation cuts corresponding to azimuth angles of 0°, 45°, 90°, and 135° at a frequency of 1600 MHz. Note that the response to a circularly polarized wave varies by less than 4 dB for angles above 10° over the horizon.

The measured normalized radiation patterns for the crooked-wire genetic antenna over a ground plane are shown in Figure 6.17 for the same elevation cuts that were previously computed at a frequency of 1600 MHz. There is

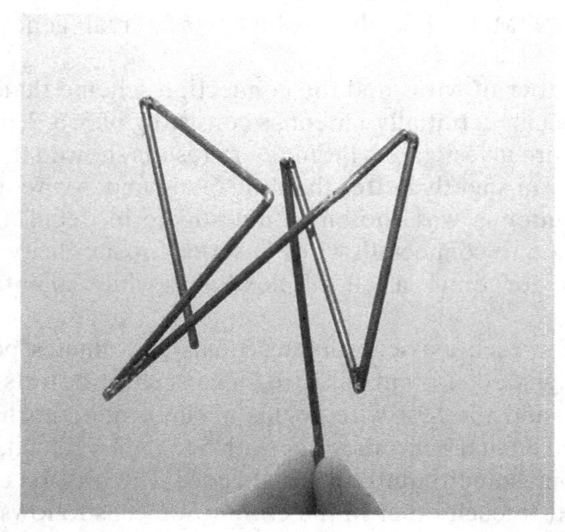

Figure 6.15. *Photograph of the seven-wire genetic antenna; height of antenna is 8.7 cm (photo provided by Derek S. Linden).*

TABLE 6.2. Coordinates of 7-Wire Genetic Antenna

7-Wire Genetic Antenna with Ground Plane					
Startpoint (Coordinates in Meters)			Endpoint (Coordinates in Meters)		
X	*Y*	*Z*	*X*	*Y*	*Z*
0.0000	0.0000	0.0000	−0.0166	0.0045	0.0714
−0.0166	0.0045	0.0714	−0.0318	−0.0166	0.0170
−0.0318	−0.0166	0.0170	−0.0318	−0.0287	0.0775
−0.0318	−0.0287	0.0775	−0.0318	0.0439	0.0140
−0.0318	0.0439	0.0140	−0.0318	0.0045	0.0624
−0.0318	0.0045	0.0624	−0.0106	0.0378	0.0866
−0.0106	0.0378	0.0866	−0.0106	0.0257	0.0230

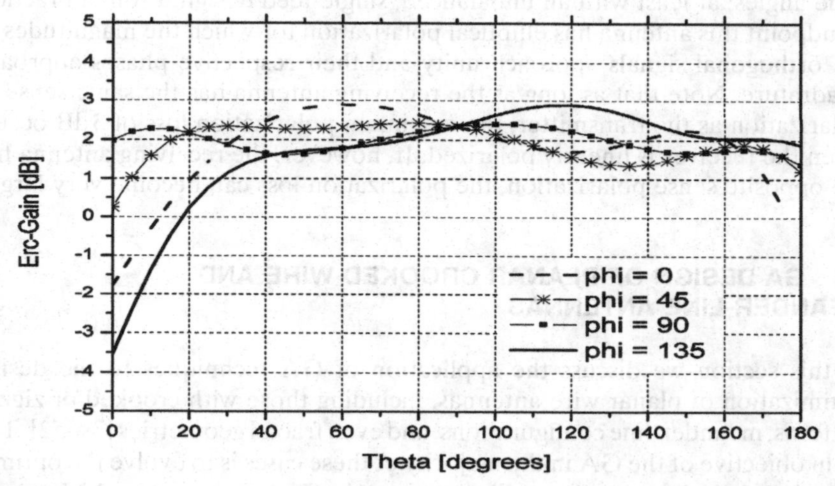

***Figure 6.16.** Computed ϕ dependence of 1600-MHz seven-wire genetic antenna with infinite ground plane (calculated using WIPL-D [38]).*

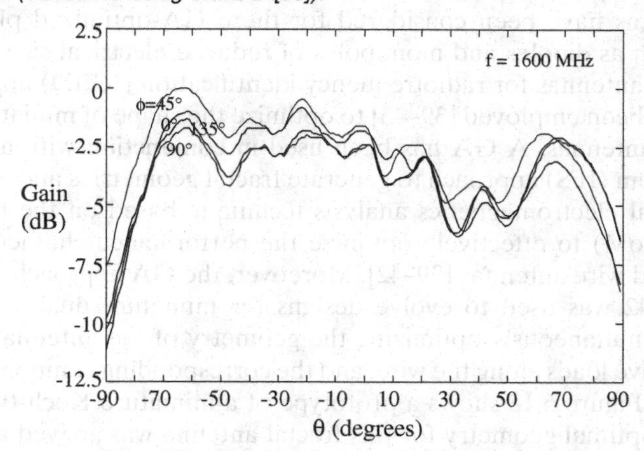

***Figure 6.17.** Measured ϕ dependence of seven-wire genetic antenna with finite ground plane (E. Altshuler and D. Linden, © IEEE, 1997, [23]).*

approximately a 6 dB variation in the field above an elevation angle of 10° as compared to the computed variation of ~4 dB. This discrepancy can for the most part be attributed to the fact that the measurements were made over a 1.2 × 1.2-m ground plane, whereas the computations for the GA were done for an infinite ground plane. The ripples in the pattern arise from reflections from the edges of the ground plane. The rolloff at the edges is also due to the finite size of the ground. Note in particular that the four azimuth angles produce very similar results, demonstrating the uniformity of the design. These angles were produced by rotating the antenna only—the ground plane was not rotated—showing more clearly that the variation is due to the ground plane and not the antenna.

It should be mentioned that true circular polarization is not achievable over wide angles, at least with an unbalanced, single-feed design. From a practical standpoint this antenna has elliptical polarization for which the magnitudes of the orthogonal signals approach unity and their respective phases approach quadrature. Note that as long as the receiving antenna has the same sense of polarization as the transmitter, the maximum polarization loss of 3 dB occurs when the receiver is linearly polarized. If, however, the receiving antenna has the opposite sense polarization, the polarization loss can become very large.

6.4 GA DESIGN OF PLANAR CROOKED-WIRE AND MEANDER-LINE ANTENNAS

In this section we discuss the application of GA techniques to the design optimization of planar wire antennas, including those with crooked or zigzag patterns, meander-line configurations, and even fractal geometries [39–52]. The main objective of the GA in the majority of these cases is to evolve the optimal shape of the wire antenna, subject to certain size constraints, which would provide the best possible performance in terms of bandwidth and/or efficiency. Several designs have been considered for these GA-optimized planar wire antennas such as dipoles and monopoles of reduced electrical size as well as miniaturized antennas for radiofrequency identification (RFID) applications.

GAs have been employed [39–43] to optimize the shape of miniature Koch-type fractal antennas. A GA has been used in conjunction with an iterated function system (IFS) approach to generate fractal geometries and a full-wave computational electromagnetics analysis technique based on the method of moments (MoM) to effectively optimize the performance characteristics of fractal-shaped wire antennas [39–42]. Moreover, the GA approach developed in Refs. 39–42 was used to evolve designs for miniature dual-band fractal dipoles by simultaneously optimizing the geometry of the antenna, the locations of reactive loads along the wire, and the corresponding component values of the loads. Figure 6.18 shows a prototype of a miniature Koch-type fractal dipole. The optimal geometry for this fractal antenna was arrived at by combining the robustness of a GA together with the versatility and efficiency of

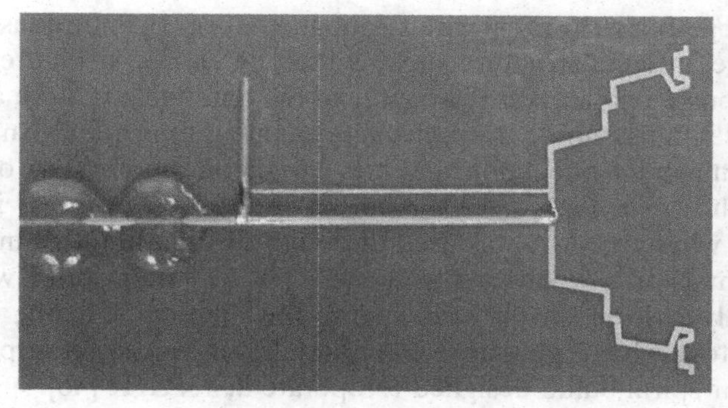

Figure 6.18. Photograph of a miniature GA optimized Koch fractal dipole antenna with balun (photo provided by Michael J. Wilhelm).

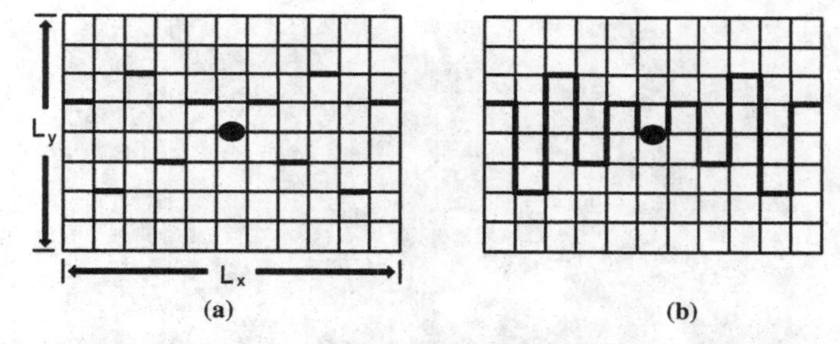

Figure 6.19. Procedure for creating a miniature linear meander-line dipole antenna. Steps 1 and 2 are illustrated in panels (a) and (b), respectively.

an IFS. Another GA design technique for miniature Koch-like fractal monopole antennas is considered in Ref. 43, with comparisons to the performance characteristics of other optimized nonfractal designs such as zigzag and meander-line antennas. A multiobjective GA was applied in these cases to the design of the wire antennas with the goal of optimizing their bandwidth and efficiency while reducing their resonance frequency. Finally, the performance of other types of GA optimized fractal and nonfractal small wire monopole antenna geometries were compared in Ref. 44.

GA optimization techniques used to evolve miniature meander-line antenna elements (also referred to as *stochastic antennas*) with linear and circular polarization are described in the literature [45–47]. To initiate the process of generating linear meander-line or stochastic elements, the overall maximum projected lengths L_X and widths L_Y of a candidate antenna are predetermined by the size of a fixed grid as shown in Figure 6.19a. The number of horizontal wire segments in each of the vertical columns is preset at 2^N, where N is a positive integer. The GA will then select one of the 2^N horizontal elements in each column of the grid. These selected horizontal elements are then connected together by vertical wire segments to form a continuous conducting meander-line antenna structure such as the one shown in Figure 6.19b. The excitation source is placed at the center of the antenna structure as indicated in Figure 6.19 and assumed fixed throughout the optimization. An example of a miniature meander-line dipole antenna designed using this GA optimization process is shown in Figure 6.20. Another interesting example is shown in Figure 6.21, where the GA was used to design a miniature 50Ω meander-line dipole antenna for a credit-card-size 916MHz AM transmitter with components directly written on glass, including the antenna. Finally, Figure 6.22 shows a photograph of a miniature GA optimized meander-line dipole antenna with a planar spiral balun designed to operate at 1.4GHz [46].

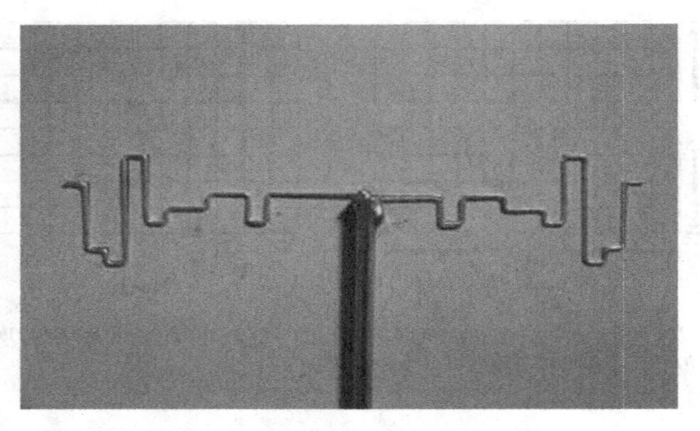

Figure 6.20. *Photograph of a miniature GA-optimized meander-line (stochastic) dipole antenna (photo provided by Michael J. Wilhelm).*

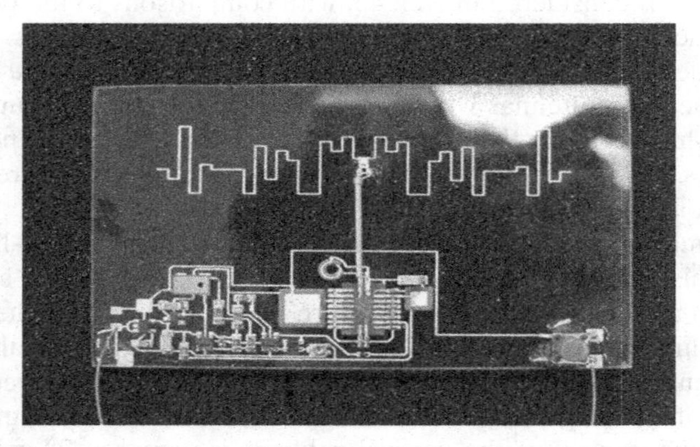

Figure 6.21. *Miniature credit-card-size 916MHz AM transmitter with components directly written on glass including a miniature 50Ω meander-line dipole antenna (photo provided by Michael J. Wilhelm).*

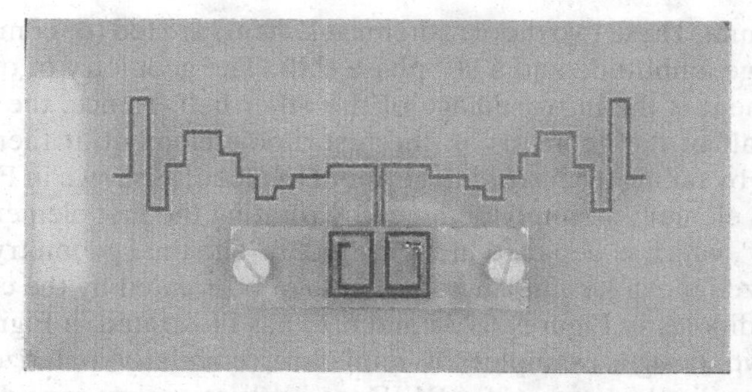

Figure 6.22. *Photograph of a miniature GA-optimized meander-line dipole antenna with a planar spiral balun designed to operate at 1.4 GHz (K. O'Connor, R. Libonati, J. Culver, D. H. Werner, and P. L. Werner, © IEEE, 2003, [46]).*

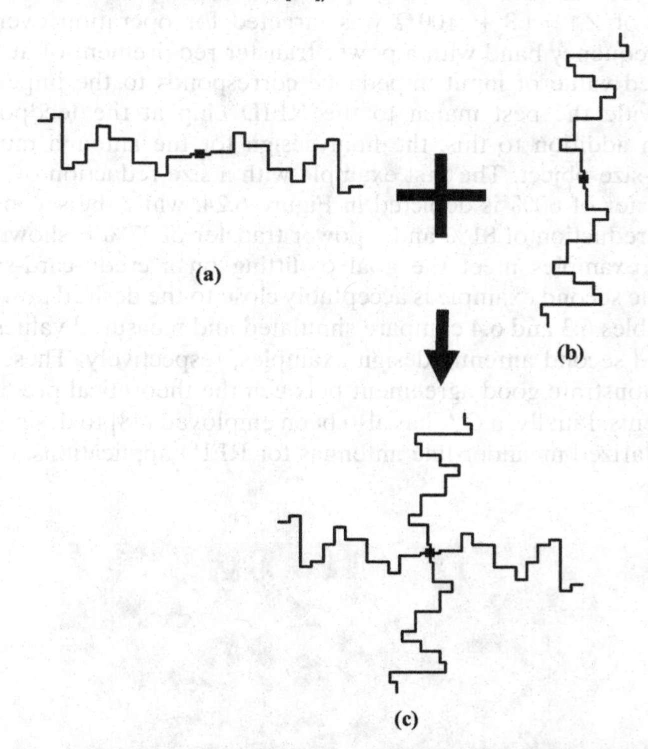

Figure 6.23. *Procedure for creating a miniature crossed meander-line dipole antenna.*

The GA optimization techniques introduced by the Werners and colleagues [45,46] for creating small planar linearly polarized meander-line dipoles were modified and extended [47] to synthesize small antennas that are circularly polarized. The steps in the algorithmic procedure developed for generating these miniature circularly polarized antennas are graphically illustrated in Figure 6.23. Two meander-line elements are required in forming a crossed dipole antenna. These two dipole antenna elements are fed (or combined) with equal voltage amplitude and a 90° phase shift. The geometry of one half of a dipole element is the mirror image of the other half. Hence, the GA selects only one half of the geometry in the first dipole element; it then forms the second half by taking its mirror image about the feed, as shown in Figure 6.23a. The second element is simply obtained by rotating the first element in Figure 6.23a by 90°, which is depicted in Figure 6.23b. The final geometry of a circularly polarized meander-line antenna can then be created by the combination of the two dipoles in Figures 6.23a and 6.23b as illustrated in Figure 6.23c.

Next, two design examples of miniature, circularly polarized, crossed meander-line dipole antennas for RFID tag applications are considered. These designs were synthesized using the GA optimization procedure introduced in Ref. 45. To demonstrate the design potential of this process, a desired input impedance of $Z_{in} = 68 + j100\,\Omega$ was targeted for operation over the 902–928 MHz frequency band with a power transfer requirement of at least 80%. The targeted value of input impedance corresponds to the impedance that would provide the best match to the RFID chip at the feedpoint of the antenna. In addition to this, the final design for the antenna must fit on a credit-card-size object. The first example with a size reduction of 86% and a power transfer of 60% is depicted in Figure 6.24, while the second example with a size reduction of 81% and a power transfer of 77% is shown in Figure 6.25. Both examples meet the goal of fitting on a credit-card-size object; however, the second example is acceptably close to the desired power transfer of 80%. Tables 6.3 and 6.4 compare simulated and measured values of Z_{in} for the first and second antenna design examples, respectively. These tabulated results demonstrate good agreement between the theoretical predictions and measurements. Finally, a GA has also been employed [48] to design miniature linearly polarized meander-line antennas for RFID applications.

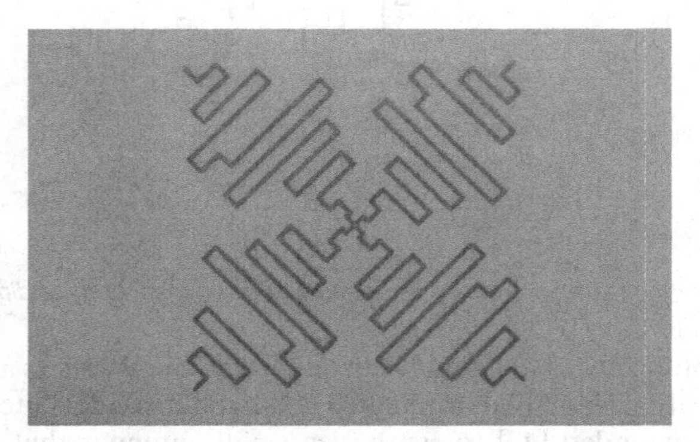

Figure 6.24. *Photograph of a 915 MHz miniature credit-card-size GA optimized meander-line crossed-dipole RFID tag antenna (P. L. Werner, M. Wilhelm, R. Salisbury, L. Swann, and D. H. Werner, © IEEE, 2003, [47]). The antenna is on a substrate composed of 0.060 in. (1.5 mm) RO4003 dielectric.*

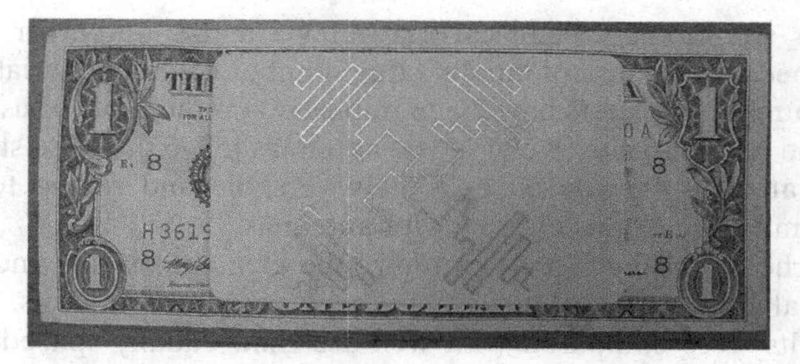

Figure 6.25. *Photograph of a 915 MHz miniature credit-card-size GA-optimized meander-line crossed-dipole RFID tag antenna (P. L. Werner, M. Wilhelm, R. Salisbury, L. Swann, and D. H. Werner, © IEEE, 2003, [47]). The antenna is on a substrate composed of 0.060 in. (1.5 mm) RO4003 dielectric.*

TABLE 6.3. Simulated and Measured Input Impedance Values for the Meander-Line Crossed Dipole RFID Tag Antenna Shown in Figure 6.24

Frequency (MH$_z$)	Simulated Z_{in} (Ω)	Measured Z_{in} (Ω)
902	14.0 + j89.4	13.8 + j89
915	14.8 + j108	15.9 + j107
928	15.7 + j128	14.9 + j125

TABLE 6.4. Simulated and Measured Input Impedance Values for the Meander-Line Crossed Dipole RFID Tag Antenna Shown in Figure 6.25

Frequency (MH$_z$)	Simulated Z_{in} (Ω)	Measured Z_{in} (Ω)
902	23.5 + j87.9	23.6 + j95.7
915	25.0 + j107	25.4 + j115
928	26.9 + j128	26.2 + j136

While generating these planar meander-line antenna designs, the GA selects the optimal geometry by evolving until it finds the configuration that best meets the desired design objectives. A full-wave MoM analysis technique is used in conjunction with the GA to evaluate the radiation characteristics of individual population members (i.e., candidate antenna designs). Typically, a population of >100 candidate antennas is employed in the optimization process.

6.5 GA DESIGN OF YAGI–UDA ANTENNAS

The design of Yagi–Uda antennas typically involves adjusting the number of wires and the corresponding wire lengths, the spacing between the wires, the radius of the wires, and even in some cases the shape of the wires. This design process has traditionally been carried out via a trial-and-error procedure. However, because of the potentially large number of design variables, a considerable number of papers have been published on the topic of using GAs to optimize the performance of Yagi–Uda antennas [53–65]. These studies demonstrate that GAs provide an extremely versatile and robust tool for the design optimization of this class of wire antennas.

Approaches for using GAs to optimize the element spacing and lengths of conventional Yagi–Uda antennas have been presented in Refs. 53 and 54. These results were compared with well-designed equally spaced Yagi–Uda arrays and shown to provide superior performance characteristics. The use of GAs for the optimization of gain, impedance, and bandwidth in the design of Yagi–Uda antennas was considered in Ref. 55. Multiobjective GA techniques have been investigated [56–58] as an effective means of evolving Pareto optimal solutions, which enable the selection of variables in accordance with the Yagi–Uda antenna design requirements.

GAs have also been used to develop unconventional designs for Yagi–Uda antennas, such as those considered in [54,59–62]. For example, Linden and Altshuler [54] developed a design approach using a GA to determine the optimal way to rotate the individual wires in a Yagi–Uda array in order to achieve a desired polarization. This technique was demonstrated by evolving an optimal design for a rotated Yagi with the goal of achieving a circularly polarized gain pattern. Another technique [59] uses a GA to optimize the configuration of a V-shape Yagi–Uda array. In this case the design variables are the spacing, lengths, and angles of the wire elements. GAs have also been used to optimize the performance of Yagi fractal arrays in Ref. 60, where it has been suggested that a certain degree of miniaturization relative to the design of conventional Yagi–Uda antennas could be achieved by using fractal-shape wires. Moreover, the objective for these designs was to achieve a maximum gain of 10–13 dB with a real part of the input impedance as close as possible to 50 Ω.

An effective technique based on particle swarm optimization (PSO) for designing miniature Yagi–Uda arrays with meander-line elements has been proposed [61,62]. To accomplish miniaturization of such arrays, a simple but effective grid searching technique was introduced for use in conjunction with the PSO. Designs utilizing both parallel and planar grids have been considered. It was found that the overall array length and width could be reduced by as much as 70% and 44%, respectively, compared to conventional Yagis. Finally, a comparison of the miniature stochastic Yagis designed using PSO was made to similar designs obtained via GA.

To illustrate this novel GA design technique for miniature Yagi–Uda antennas, two examples of a three-element array will be presented here: one having planar elements and the other having parallel elements. In the first configuration, the grids are assumed to be coplanar as shown in Figure 6.26. This will be referred to as the *planar element configuration*. In the second configuration, the grids are rotated 90° so they are parallel to each other as shown in

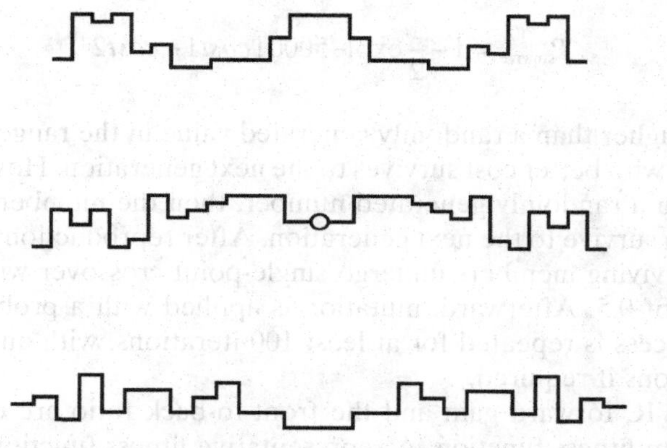

Figure 6.26. *Planar stochastic meander-line three-element Yagi–Uda antenna designed using a binary GA.*

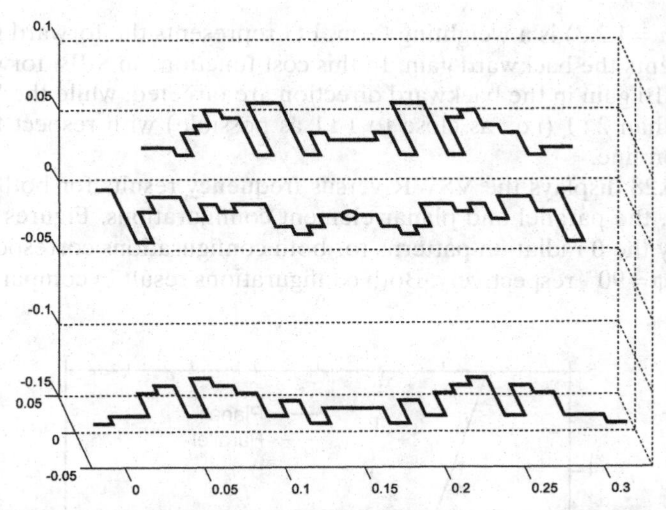

Figure 6.27. *Parallel stochastic meander-line three-element Yagi–Uda antenna designed using a binary GA. The units of the horizontal and vertical axes (abscissa and ordinate) are in meters.*

Figure 6.27. This will be referred to as the *parallel element configuration*. The advantage of this latter configuration is that the elements can be placed closer together consequently allowing for increased mutual coupling.

A binary-valued GA with a population of 100 members is used where the wires on each row are represented by 3 bits and each element distance by 10. In each generation, the cost of every population member is compared to the cost of a randomly selected member. If a new member has a better cost (*cost*1) than the other randomly selected member (*cost*2), it is assigned a probability of survival given by

$$P_{\text{survive}} = 1 - \frac{1}{2}\exp\{-5000\,|\,cost1 - cost2\,|\} \tag{6.4}$$

If P_{survive} is higher than a randomly generated value in the range of [0, 1], then the member with better cost survives to the next generation. However, if P_{survive} is lower than a randomly generated number, then the member with less cost is allowed to survive to the next generation. After reproduction of the population, the surviving members undergo single-point crossover with a crossover probability of 0.5. Afterward, mutation is applied with a probability of 0.05 and this process is repeated for at least 100 iterations, with an option to run more iterations if required.

The VSWR, forward gain and the front-to-back ratio are optimized in a weighted sum fitness function. A representative fitness function is defined as

$$cost = [k_1\,(\text{VSWR} - 1)^2] + [k_2(\text{FG} - 8)^2] + [k_3(\text{BG} + 10)^2] \tag{6.5}$$

where k_n (n = 1,2,3) is a weighting term, FG represents the forward gain, and BG represents the backward gain. In this cost function, an 8dBi forward gain and a −10dBi gain in the backward direction are targeted, while the VSWR is to be less than 2 : 1 (i.e., as close to 1 : 1 as possible) with respect to a 50Ω transmission line.

Figure 6.28 displays the VSWR versus frequency results for both designs, specifically, the parallel and planar element configurations. Figures 6.29 and 6.30 display the θ radiation patterns for both configurations corresponding to $\varphi = 0°$ and $\varphi = 90°$, respectively. Both configurations result in comparable per-

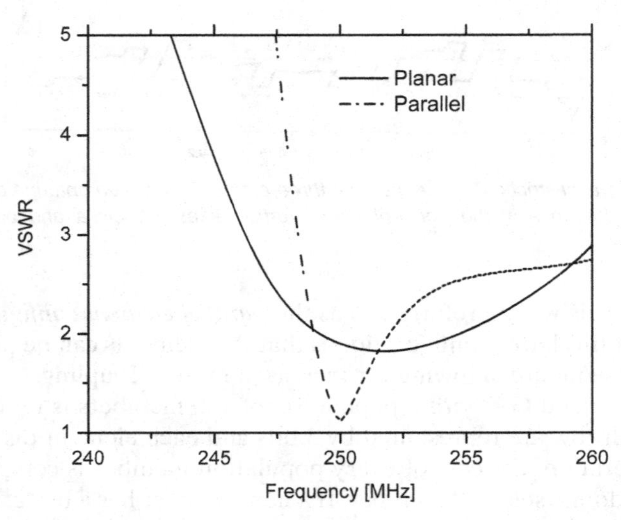

Figure 6.28. VSWR versus frequency for both planar and parallel three-element Yagi–Uda designs.

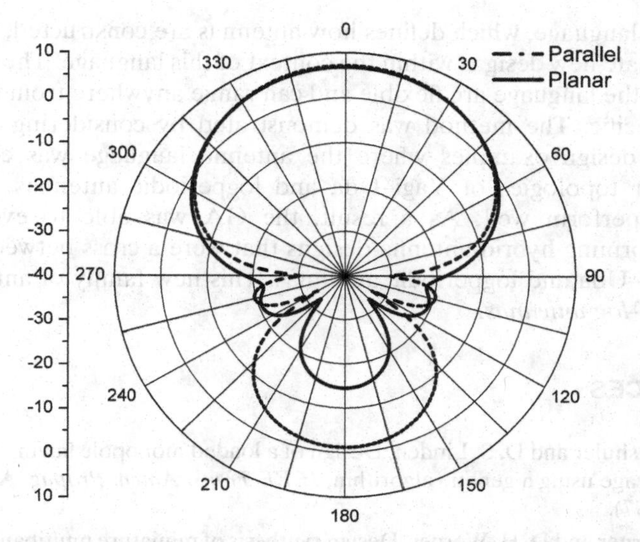

Figure 6.29. *θ radiation patterns (in dBi) with φ = 0° for both planar and parallel three-element miniature Yagi–Uda designs.*

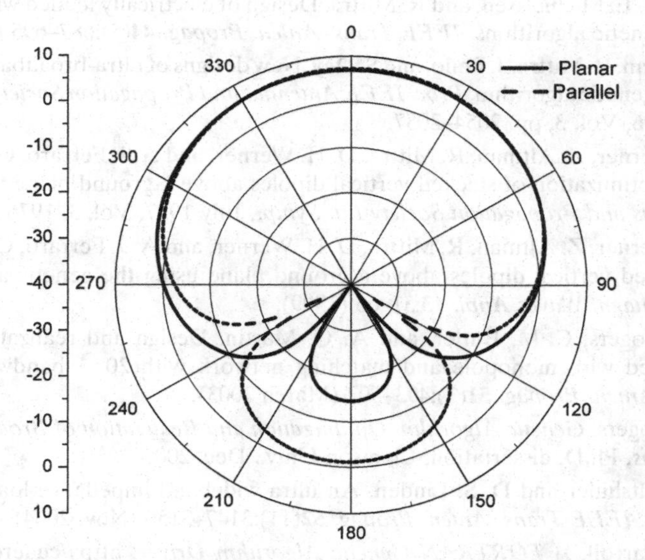

Figure 6.30. *θ radiation patterns (in dBi) with φ = 90° for both planar and parallel three-element miniature Yagi–Uda designs.*

formance for a 50% projected length reduction and a 68% boom length reduction in the case of the parallel element configuration.

Jones and Joines introduced a novel methodology for optimizing both the topology and the variables of an antenna design [63]. The approach combines an antenna language, which defines how antennas are constructed, with a GA that will create new designs within the context of this language. The grammatical rules of the language are flexible and can range anywhere from very vague to very specific. The method was demonstrated by considering some very interesting design examples where the antenna language was confined to the familiar topologies of Yagi–Uda and logperiodic antennas, which are known to perform well. As a result, the GA was able to evolve some better-performing hybrid antenna designs that were a cross between conventional Yagi–Uda and logperiodic antennas. This new family of antennas was called *Yagi–log antennas*.

REFERENCES

1. E. E. Altshuler and D. S. Linden, Design of a loaded monopole having hemispherical coverage using a genetic algorithm, *IEEE Trans. Anten. Propag.* **AP-45**(1):1–4 (Jan. 1997).
2. P. L. Werner and D. H. Werner, Design synthesis of miniature multiband monopole antennas with application to ground-based and vehicular communication systems, *Anten. Wireless Propag. Lett.* **4**:104–106 (2005).
3. A. Boag, E. Michielssen, and R. Mittra, Design of electrically loaded wire antennas using genetic algorithms, *IEEE Trans. Anten. Propag.* **44**(5):687–695 (May 1996).
4. Z. Altman, R. Mittra, J. Philo, and S. Dey, New designs of ultra-broadband antennas using a genetic algorithm, *Proc. IEEE Antennas and Propagation Society Int. Symp.*, July 1996, Vol. 3, pp. 2054–2057.
5. P. L. Werner, Z. Altman, R. Mittra, D. H. Werner, and A. J. Ferraro, Genetic algorithm optimization of stacked vertical dipoles above a ground plane, *Proc. IEEE Antennas and Propagation Society Int. Symp.*, July 1997, Vol. 3, 1976–1979.
6. P. L. Werner, Z. Altman, R. Mittra, D. H. Werner, and A. J. Ferraro, Optimization of stacked vertical dipoles above a ground plane using the genetic algorithm, *J. Electromagn. Waves Appl.* **13**:51–66 (1999).
7. S. D. Rogers, C. M. Butler, and A. Q. Martin, Design and realization of GA-optimized wire monopole and matching network with 20 : 1 bandwidth, *IEEE Trans. Anten. Propag.* **51**(3):493–502 (March 2003).
8. S. D. Rogers, *Genetic Algorithm Optimization and Realization of Broadband Wire Antennas*, Ph.D. dissertation, Clemson Univ., Dec. 2000.
9. E. E. Altshuler and D. S. Linden, An ultra wideband impedance-loaded genetic antenna, *IEEE Trans. Anten. Propag.* **52**(11):3147–3150 (Nov. 2004).
10. D. L. Carroll, *A FORTRAN Genetic Algorithm Driver*, http://cuaerospace.com/carroll/ga.html.
11. D. L. Carroll, Chemical laser modeling with genetic algorithms, *AIAA J.* **34**(2):338–346 (Feb. 1996).
12. K. Krishnakumar, Micro-genetic algorithms for stationary and non-stationary function optimization, *SPIE: Intelligent Control and Adaptive Systems*, Philadelphia, 1989, Vol. 1196, pp. 289–296.

13. L. Mattioni and G. Marrocco, Design of a broadband HF antenna for multimode naval communications, *Anten. Wireless Propag. Lett.* **4**:179–182 (2005).
14. T. F. Kennedy, S. A. Long, and J. T. Williams, Dielectric bead loading for control of currents on electrically long dipole antennas, *Proc. IEEE Antennas and Propagation Society Int. Symp.*, June 2004, Vol. 4, pp. 4420–4423.
15. M. Fernandez-Pantoja, A. Monorchio, A. Rubio-Bretones, and R. Gomez-Martin, Direct GA-based optimisation of resistively loaded wire antennas in the time domain, *Electron. Lett.* **36**(24):1988–1990 (Nov. 2000).
16. A. Bretones, C. Moreno de Jong van Coevorden, M. Pantoja, F. Garcia Ruiz, S. Garcia, and R. Gomez Martin, GA design of broadband thin-wire antennas for GPR applications, *Proc. 3rd Int. Workshop on Advanced Ground Penetrating Radar*, May 2005, pp. 143–146.
17. E. E. Altshuler and D. S. Linden, *A Process for the Design of Antennas Using Genetic Algorithms*, LIS Patent 5,719,794 (Feb. 17, 1998).
18. D. S. Linden and E. E. Altshuler, Automating wire antenna design using genetic algorithms, *Proc. 1995 Antenna Applications Symp.*, Allerton Park, Monticello, IL, Sept. 1995.
19. D. S. Linden and E. E. Altshuler, Automating wire antenna design using genetic algorithms, *Microwave J.* **39**:74–86 (March 1996).
20. D. Linden, Using a real chromosome in a genetic algorithm for wire antenna optimization, *Proc. IEEE Antennas and Propagation Society Int. Symp.*, July 1997, Vol. 3, pp. 1704–1707.
21. E. Altshuler and D. Linden, Design of a vehicular antenna for GPS/Iridium using a genetic algorithm, *Proc. IEEE Antennas and Propagation Society Int. Symp.*, July 1997, Vol. 3, pp. 1680–1683.
22. E. E. Altshuler and D. S. Linden, Design of wire antennas using a genetic algorithm, *J. Electron. Defense* 50–52 (July 1997).
23. E. Altshuler and D. Linden, Wire-antenna designs using genetic algorithms, *IEEE Anten. Propag. Mag.* **39**(2):33–43 (April 1997).
24. D. Van Veldhuizen, B. Sandlin, E Marmelstein, G. Lamont, and A. Terzuoli, Finding improved wire-antenna geometries with genetic algorithms, *Proc. IEEE Int. Conf. Computational Intelligence*, May 1998, pp. 102–107.
25. D. Linden and E. Altshuler, Wiring like mother nature [antenna design], *IEEE Potentials*. **18**(2):9–12 (April–May 1999).
26. D. Linden and E. Altshuler, Evolving wire antennas using genetic algorithms: A review, *Proc. 1st NASA/DoD Workshop on Evolvable Hardware*, July 1999, pp. 225–232.
27. E. E. Altshuler, Design of a vehicular antenna for GPS/Iridium using genetic algorithms, *IEEE Trans. Anten. Propag.* **48**(6):968–972 (June 2000).
28. D. Linden, Wire antennas optimized in the presence of satellite structures using genetic algorithms, *IEEE Aerospace Conf. Proc.*, March 2000, Vol. 5, pp. 91–99.
29. D. S. Linden and R. T. MacMillan, Increasing genetic algorithm efficiency for wire antenna design using clustering, *ACES Special J. Genet. Algorithms.* **15**(2):75–86 (July 2000).

30. E. Altshuler, Electrically small self-resonant wire antennas optimized using a genetic algorithm, *IEEE Trans. Anten. Propag.* **50**(3):297–300 (March 2002).
31. S. Rengarajan and Y. Rahmat-Samii, On the cross-polarization characteristics of crooked wire antennas designed by genetic algorithms, *Proc. IEEE Antennas and Propagation Society Int. Symp.*, June 2002, Vol. 1, pp. 706–709.
32. H. Choo, R. Rogers, and H. Ling, Design of electrically small wire antennas using genetic algorithm taking into consideration of bandwidth and efficiency, *Proc. IEEE Antennas and Propagation Society Int. Symp.*, June 2002, Vol. 1, pp. 330–333.
33. E. E. Altshuler, Electrically small genetic antennas immersed in a dielectric, *Proc. IEEE Antennas and Propagation Society Int. Symp.*, June 2004, Vol. 3, pp. 2317–2320.
34. E. E. Altshuler and D. S. Linden, An ultra wideband impedance-loaded genetic antenna, *IEEE Trans. Anten. Propag.* **52**(11):3147–3151 (Nov. 2004).
35. H. Choo, R. Rogers, and H. Ling, Design of electrically small wire antennas using a Pareto genetic algorithm, *IEEE Trans. Anten. Propag.* **53**(3):1038–1046 (March 2005).
36. E. Altshuler, A method for matching an antenna having a small radiation resistance to a 50-ohm coaxial line, *IEEE Trans. Anten. Propag.* **53**(9):3086–3089 (Sept. 2005).
37. G. J. Burke and A. J. Poggio, *Numerical Electromagnetics Code (NEC)—Method of Moments*, Report UCID18834, Lawrence Livermore Laboratory, Jan. 1981.
38. B. M. Kolundzija, J. S. Ognjanovic, T. K. Sarkar, and R. F. Harrington. WIPL-program for analysis of metallic antennas and scatterers, *Proc. 9th Int. Conf. Antennas and Propagation* (ICAP '95), Conf. Publication 407, April 4–7, 1995.
39. D. H. Werner and P. L. Werner, Genetically engineered dual-band fractal antennas, *Proc. IEEE Antennas and Propagation Society Int. Symp.*, July 2001, Vol. 3, pp. 628–631.
40. D. H. Werner, P. L. Werner, and K. Church, Genetically engineered multiband fractal antennas, *Electron. Lett.* **37**(19):1150–1151 (Sept. 2001).
41. D. H. Werner, P. L. Werner, J. Culver, S. Eason, and R. Libonati, Load sensitivity analysis for genetically engineered miniature multiband fractal dipole antennas, *Proc. IEEE Antennas and Propagation Society Int. Symp.*, June 2002, Vol. 4, pp. 86–89.
42. D. H. Werner and S. Ganguly, An overview of fractal antenna engineering research, *IEEE Anten. Propag. Mag.* **45**(1):38–57 (Feb. 2003).
43. M. Fernandez Pantoja, F. Garcia Ruiz, A. Rubio Bretones, R. Gomez Martin, J. Gonzalez-Arbesu, J. Romeu, and J. Rius, GA design of wire pre-fractal antennas and comparison with other Euclidean geometries, *Anten. Wireless Propag. Lett.* **2**(1):238–241 (2003).
44. M. Pantoja, F. Ruiz, A. Bretones, R. Martin, S. Garcia, J. Gonzalez-Arbesu, J. Romeu, J. Rius, D. H. Werner, and P. L. Werner, GA design of small wire antennas," *Proc. IEEE Antennas and Propagation Society Int. Symp.*, June 2004, Vol. 4, pp. 4412–4415.

45. P. L. Werner and D. H. Werner, A design optimization methodology for multiband stochastic antennas, *Proc. IEEE Antennas and Propagation Society Int. Symp.*, June 2002, Vol. 2, pp. 354–357.

46. K. O'Connor, R. Libonati, J. Culver, D. H. Werner, and P. L. Werner, A planar spiral balun applied to a miniature stochastic dipole antenna, *Proc. 2003 IEEE Int. Symp. Antennas and Propagation and URSI North American Radio Science Meeting*, Columbus, OH, June 2003, Vol. 3, pp. 938–941.

47. P. L. Werner, M. Wilhelm, R. Salisbury, L. Swann, and D. H. Werner, Novel design techniques for miniature circularly-polarized antennas using genetic algorithms, *Proc. IEEE Antennas and Propagation Society Int. Symp.*, June 2003, Vol. 1, pp. 145–148.

48. G. Marrocco, A. Fonte, and F. Bardati, Evolutionary design of miniaturized meander-line antennas for RFID applications, *Proc. IEEE Antennas and Propagation Society Int. Symp.*, June 2002, Vol. 2, pp. 362–365.

49. A. Chuprin, J. Batchelor, and E. Parker, Design of convoluted wire antennas using a genetic algorithm, *IEE Proc. Microw. Anten. Propag.* **148**(5):323–326 (Oct. 2001).

50. A. Kerkhoff and H. Ling, The design and analysis of miniaturized planar monopoles, *Proc. IEEE Antennas and Propagation Society Int. Symp.*, June 2002, Vol. 4, pp. 30–33.

51. H. Choo and H. Ling, Design of planar, electrically small antennas with inductively coupled feed using a genetic algorithm, *Proc. IEEE Antennas and Propagation Society Int. Symp.*, June 2003, Vol. 1, pp. 300–303.

52. D. H. Werner and P. L. Werner, The design optimization of miniature low profile antennas placed in close proximity to high-impedance surfaces, *Proc. IEEE Antennas and Propagation Society Int. Symp.*, June 2003, Vol. 1, pp. 157–160.

53. E. Jones and W. Joines, Design of Yagi-Uda antennas using genetic algorithms, *IEEE Trans. Anten. Propag.* **45**(9):1386–1392 (Sept. 1997).

54. D. Linden and E. Altshuler, Evolving wire antennas using genetic algorithms: A review, *Proc. 1st NASA/DoD Workshop on Evolvable Hardware*, July 1999, pp. 225–232.

55. D. Correia, A. Soares, and M. Terada, Optimization of gain, impedance and bandwidth in Yagi-Uda antennas using genetic algorithm, *Proc. SBMO/IEEE MTT-S, APS and LEOS—IMOC '99, Int. Microwave and Optoelectronics Conf.*, Aug. 1999, Vol. 1, pp. 41–44.

56. N. Venkatarayalu and T. Ray, Single and multi-objective design of Yagi-Uda antennas using computational intelligence, *Proc. 2003 Congress on Evolutionary Computation*, Dec. 2003, Vol. 2, pp. 1237–1242.

57. Y. Kuwahara, Multiobjective optimization design of Yagi-Uda antenna, *IEEE Trans. Anten. Propag.* **53**(6):1984–1992 (June 2005).

58. P. K. Varlamos, P. J. Papakanellos, S. C. Panagiotou, and C. N. Capsalis, Multi-objective genetic optimization of Yagi-Uda arrays with additional parasitic elements, *IEEE Anten. Propag. Mag.* **47**(4):92–97 (Aug. 2005).

59. B. Austin and W. Liu, An optimized shaped Yagi-Uda array using the genetic algorithm, *Proc. IEE Natl. Conf. Antennas and Propagation*, Aug. 1999, pp. 245–248.

60. S. Zainud-Deen, D. Mohrram, and H. Sharshar, Optimum design of Yagi fractal arrays using genetic algorithms, *Proc. 21st Natl. Radio Science Conf.*, March 2004, Vol. B19, pp. 1–10.
61. Z. Bayraktar, P. L. Werner, and D. H. Werner, Miniature three-element stochastic Yagi-Uda array optimization via particle swarm intelligence, *Proc. 2005 IEEE Antennas and Propagation Society Int. Symp. and USNC/URSI Natl. Radio Science Meeting*, Washington, DC, July 3–8, 2005, Vol. 2B, pp. 263–266.
62. Z. Bayraktar, P. L. Werner, and D. H. Werner, The design of miniature three-element stochastic Yagi-Uda arrays using particle swarm optimization, *IEEE Anten. Wireless Propag. Lett.* **5**(1):22–26 (Dec. 2005).
63. E. A. Jones and W. T. Joines, Genetic design of linear antenna arrays, *IEEE Anten. Propag. Mag.* **42**(3):92–100 (June 2000).
64. H. Aliakbarian and F. Arazm, Sensitivity analysis in genetic design of wire antennas, *Proc. 3rd Int. Conf. Computational Electromagnetics and Its Applications*, Nov. 2004, pp. 149–152.
65. X. Chen, K. Huang, and X. Xu, Automated design of a three-dimensional fishbone antenna using parallel genetic algorithm and NEC, *IEEE Anten. Wireless Propag. Lett.* **4**:425–428 (2005).

7

Optimization of Aperture Antennas

This chapter presents GA optimization of reflector, horn, and microstrip antennas. These antennas have very complicated cost functions compared to wire antenna and arrays of isotropic point sources. Since the cost functions take a long time to calculate a single solution, they tend to have few input variables. GAs work well on aperture antennas, because the cost surfaces are often multimodal and variable constraints are easy to implement.

7.1 REFLECTOR ANTENNAS

Scattering from the edges of a reflector antenna determines the sidelobe level and beamwidth of the reflector. Attempts at controlling this scattering have been successful. One approach is to place an absorber inside a cylinder that fits around the reflector edge [1]. This "shroud" creates a "tunnel antenna" [2] that has a reduced front-to-back ratio and sidelobe levels compared to the parabolic reflector alone. Other edge treatments include serrated edges and rolled edges [3] for compact range reflectors. These treatments create a large quiet zone in a compact range. Another approach is to taper the resistivity of the reflector edge to lower sidelobe levels [5].

As of the writing of this book, there has been little application of the GA to the design of reflector antennas. Cost functions for reflector antennas tend to take much longer to compute than, for example, the cost function associated with a linear array of point sources. Since the GA must make many cost

---

Genetic Algorithms in Electromagnetics, by Randy L. Haupt and Douglas H. Werner

function evaluations, reflector antennas have not been the focus of GA applications. Vall-llossera et al. [6] used a GA to shape the beam of an array-fed mesh reflector antenna. A GA outperformed conjugate gradient in finding an optimal beamshape to cover Japan from a satellite reflector antenna [7]. The reflector surface had several control points that were moved in order to change the locations of nulls and maxima in the cost function. The far-field patterns were found using physical optics. The GA has also been applied to minimize the sidelobe level of a corner reflector fed by a Yagi–Uda antenna [8]. The cost function employed the Numerical Electromagentics Code (NEC) called by MATLAB and accepted various design parameters from the Yagi–Uda antenna while keeping the wire reflector antenna constant. A GA was used to optimize the gain and cross-polarization of the pattern of an offset dual Cassegrain reflector [9]. The axis tilt angle and eccentricity were the input variables to the cost function. Another beam shaping reflector synthesis approach using a GA was reported [10] in which the shape of an offset reflector was varied in order to produce a pattern with maximum gain over the country of Brazil. Physical optics was also used in this analysis. Several papers have appeared on the GA optimization of microstrip reflectarrays. These arrays consist of microstrip patches arranged so that they produce a desired scattering pattern due to an incident source, such as a horn antenna. Both the sidelobe level [11,12] and bandwidth performance have been optimized [13].

As an example, consider a cylindrical parabolic reflector antenna with a line source feed at the focus. The reflector diameter is 13.8λ and the focus is 5λ from the vertex. A GA shapes the last 2λ of each edge of the reflector in order to minimize the maximum sidelobe level. By assuming that the shaping is symmetric, only one edge needs to be optimized by the GA. The same method-of-moments (MoM) model used in Chapter 5 for the parabolic cylinder antenna is employed here. The reflector surface is approximated by 0.05λ segments. The last 40 segments have adjustable orientations but remain contiguous. The first adjustable segment has a pivot point on the last reflector segment (Fig. 7.1). Its endpoint is movable. When its endpoint is set by the optimization, then that endpoint serves as the pivot point for the next segment. Thus, the chromosome contains 40 continuous elements with values between 0 and 1. These values translate to relative angles through the formula

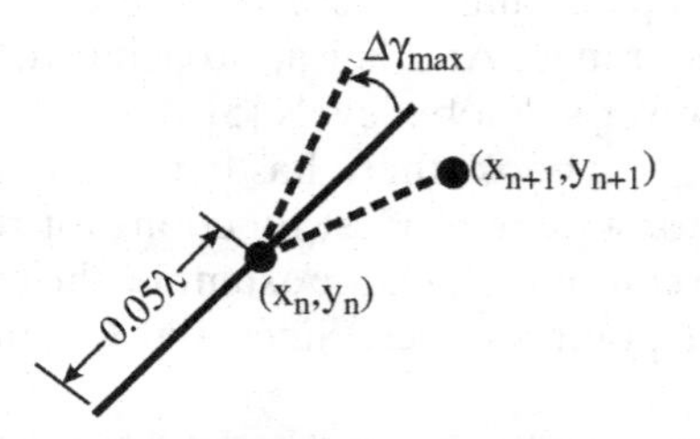

Figure 7.1. *Appending a new strip to the last one to form the reflector surface.*

$$\gamma_n = \tan^{-1}\left(\frac{y_n - y_{n-1}}{x_n - x_{n-1}}\right) \tag{7.1}$$

where (x_n, y_n) are the endpoints of a 0.05λ segment. The last segment on the reflector is defined by a parabola. The next segment can vary $\pm\Delta\gamma_{\max}$ from the angle γ_n formed by the last segment as shown in Fig. 7.1. The parameter $\Delta\gamma_n$ is given as

$$\Delta\gamma_n = (2c_n - 1)\Delta\gamma_{\max} \tag{7.2}$$

when a chromosome is represented by

$$c = [c_1, c_2, \ldots, c_n, \ldots, c_N] \tag{7.3}$$

and $0 \le c_n \le 1$. Once the relative angle between the old and new segments is determined, then the endpoint of the new segment is given by

$$\begin{aligned} x_{n+1} &= x_n + 0.5\cos(\gamma_n + \Delta\gamma_n) \\ y_{n+1} &= y_n + 0.5\sin(\gamma_n + \Delta\gamma_n) \end{aligned} \tag{7.4}$$

After the new reflector surface is formed, the surface currents are found using the method of moments and the far field determined through integration of the currents.

Prior to optimization, the feed is a line source with a $\lambda/2$-wide reflector placed $\lambda/4$ behind the feed. The antenna has a maximum sidelobe level of −14.48 dB relative to the mainbeam. Two 2λ appendages are added to each edge. These edges are bent by the GA to reduce the maximum sidelobe level of the reflector. After running a GA using continuous values for the chromosome for 1000 iterations, the algorithm found a reflector configuration that yielded a maximum sidelobe level of −16.10 dB relative to the mainbeam. Figure 7.2 shows the convergence of the GA as a function of the average and

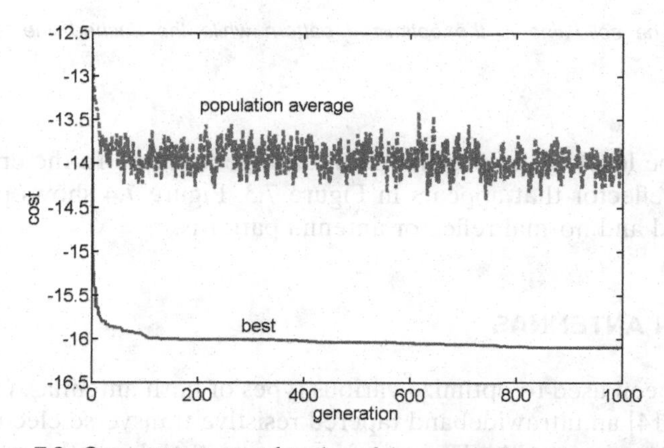

Figure 7.2. Convergence as a function of the population average and best.

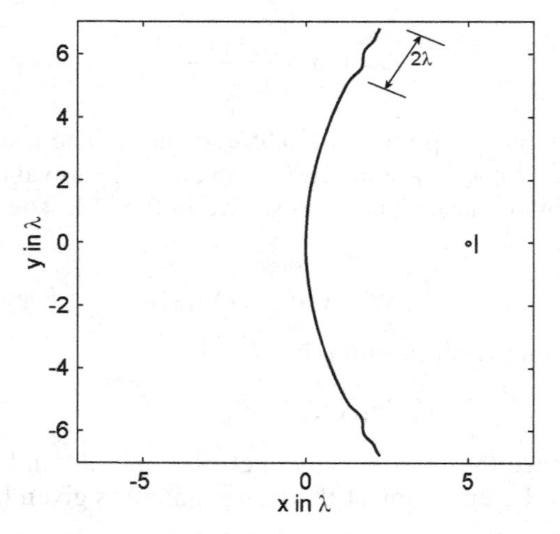

Figure 7.3. The shape of the surface 2λ from each edge is optimally shaped.

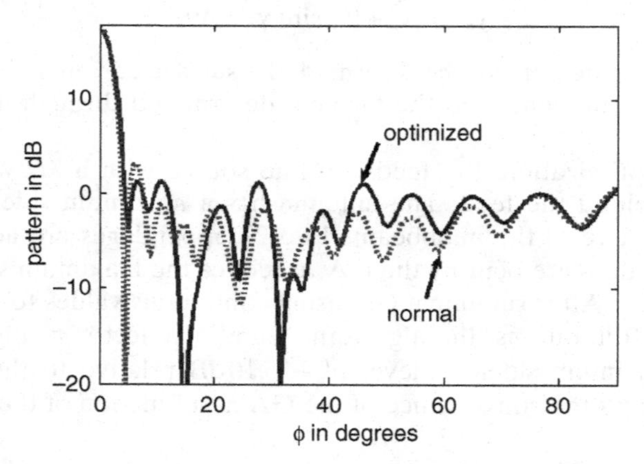

Figure 7.4. The solid line is the optimized pattern while the dashed line is the normal pattern.

best sidelobe level of the population at each generation. In the end, the GA found the reflector that appears in Figure 7.3. Figure 7.4 shows plots of the unperturbed and normal reflector antenna patterns.

7.2 HORN ANTENNAS

GAs have been used to optimize various types of horn antennas. Chung et al. optimized [14] an ultrawideband tapered resistive transverse electromagnetic (TEM) horn using a GA. The cost function tried to keep the antenna 3 dB beamwidth at 47° over the frequency range of 2–18 GHz. The input parameters to the cost function were the flare angle, the length of the perfect electric conductor, the major and minor axes of the ellipse that models the curved edge, and the angle of the ellipse. The resulting horn had a 49° beamwidth over the desired bandwidth. Results were confirmed by experimental measurements. A GA was also applied to finding the optimal choke distribution for a horn at X band [15]. A wideband corrugated multisectional conical horn was designed using a GA [16]. The cost function input was the length and number of sections, flare angles, and depth of the first slot in each horn section. The objectives were the input admittance, the VSWR, and the desired copolarization and cross-polarization patterns. The resulting horn had a VSWR <1.6 ranging from 11 to 18 GHz. Other authors also reported success at optimizing corrugated horns using a GA [17–19].

The phase center of a horn antenna is the point that is the center of a sphere of constant phase radiated from the horn. Analytical methods have been used to find the phase center of rectangular horns [20]. Generally, a horn antenna does not have a single phase center [21]. Consequently, placing the horn at the focal point of a reflector becomes difficult. The horn can be placed so that the halfway point between the two phase centers is located at the focal point of the reflector. Another way is to place the horn antenna at the point that results in the maximum reflector gain. Finally, a horn can be designed in such a way that the E-plane and H-plane phase centers are collocated [22].

Improved modeling of horn antennas allows a better calculation of the phase center. For instance, a MoM solution of a pyramidal horn provides the currents necessary to calculate the radiated far field. The phase should be constant over a sphere that has a phase center somewhere inside the horn. Assume that the horn is centered on and points along the x axis as shown in Figure 7.5. The phase center is found by first calculating the electric field far from the horn over a small angular range in the E and H planes. Since the phase is calculated relative to the origin of the coordinate system, translating the coordinate system along the x axis until the field phase is constant, will put the origin at the phase center of the horn. Calculating the mean square error

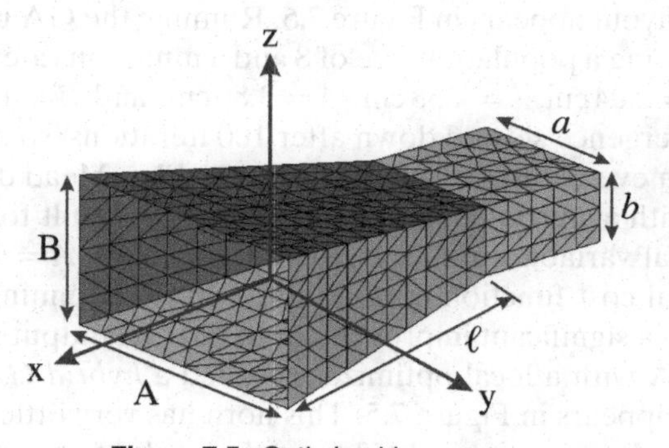

Figure 7.5. Optimized horn antenna.

of the phase provides a good measure of the phase variations over two arcs of a sphere that has its center at the approximate phase center of the horn. Since the horn is polarized in the z direction, only the θ component of the electric field will be calculated

$$\sigma_\theta = \sqrt{\frac{1}{N}\left[\angle E_\theta(\theta_n, \phi = 0^\circ) - \overline{\angle E_\theta(\theta_n, \phi = 0^\circ)}\right]^2} \tag{7.5}$$

$$\sigma_\phi = \sqrt{\frac{1}{N}\left[\angle E_\theta(\theta = 90^\circ, \phi_n) - \overline{\angle E_\theta(\theta = 90^\circ, \phi_n)}\right]^2} \tag{7.6}$$

where N = number of angles
$\angle E$ = phase of E
$\bar{E}$ = mean of E

The mean-square error is calculated in the E and H planes for $N = 11$ points given by

$$\theta_n, \phi_n = 0^\circ, 5^\circ, 10^\circ, \ldots, 45^\circ \tag{7.7}$$

When $\sigma_\theta = 0$, the phase center in the E plane is at the origin; when $\sigma_\phi = 0$, the phase center in the H plane is at the origin.

The goal is to design a pyramidal horn at 10.34 GHz that has σ_θ and σ_ϕ as small as possible. Assume the waveguide feeding the horn has dimensions a = 2.29 cm and b = 1.02 cm, and the coaxial feed is $\lambda/4$ from the back wall. The cost function for the GA is given by

$$\sigma_{\text{ph}}(p_c, \ell, A, B) = \max\{\sigma_\theta, \sigma_\phi\} \tag{7.8}$$

where p_c = distance of phase center from horn aperture, $0 \le p_c \le \ell + 2\,\text{cm}$
ℓ = distance from waveguide/horn junction and horn aperture, $0.5\lambda \le \ell \le 4.5\lambda$
A = horn dimension in z direction, $2.29\,\text{cm} \le A \le 11.43\,\text{cm}$
B = horn dimension in y direction, $1.02\,\text{cm} \le B \le 7.11\,\text{cm}$

The basic horn layout appears in Figure 7.5. Running the GA using continuous variables and having a population size of 8 and a mutation rate of 20% resulted in values of p_c = 0.04 cm, ℓ = 4.38 cm, A = 3.80 cm, and B = 1.65 cm after 100 iterations. Convergence slowed down after 100 iterations, so a local optimizer is used to find an even better result. Letting a Nelder–Mead downhill simplex algorithm run with a starting point given by the best result found using a GA produced optimal variables of p_c = 0.28 cm, ℓ = 4.31 cm, A = 4.22 cm, and B = 2.19 cm. The final cost function output is $\sigma_{\text{ph}} = 0.57^\circ$. Running the local optimizer resulted in a significant improvement to the GA output with little effort. Combining a GA with a local optimizer is called a *hybrid GA*. The resulting optimized horn appears in Figure 7.5. This horn has very little phase variation in the E- and H-plane cuts over $\pm 45^\circ$ from the peak of the mainbeam.

7.3 MICROSTRIP ANTENNAS

Next to wire antennas and arrays, microstrip antennas have received the most attention in terms of using the GA as a robust and powerful design optimization tool. GAs have proved to be particularly useful in the design of microstrip patch antennas since there are typically several design parameters that require simultaneous optimization as well as multiple and sometimes competing performance goals. For example, the design of stacked patch antennas can involve adjusting a relatively large number of parameters, including the length and width of each metallic patch, the thickness and material properties of each dielectric layer, and usually one or more additional parameters associated with the design of a feed for the antenna (e.g., coaxial probe, microstrip line, recessed microstrip line, or aperture coupled feed). The performance goals for microstrip stacked patch antennas can include broadband or multiband operation, maximum gain, no surfaces waves, high radiation efficiency, linear or circular polarization, and low cross-polarized radiation.

A GA approach has been applied [23] to aid in the design of a coaxially fed circularly polarized rectangular microstrip antenna. To this end, the GA is used to optimize the size (i.e., length and width) and feeding point of the antenna based on a relatively complex cost function derived from the cavity model. The cost function takes into account the input impedance, effective loss tangent, and the axial ratio when evaluating the fitness of a particular design. A GA technique has been introduced and shown [24] to be an effective optimization tool for the design of dual-frequency probe-fed microstrip antennas. The dual-band performance is achieved by using the GA to optimize the positions of multiple slots in the patch or multiple short-circuiting strips between the patch and the ground plane. A design procedure for a circularly polarized microstrip antenna with a matched dual-feed network is presented in Ref. 25. In order to achieve the desired performance goals, a GA was employed to optimize eight critical design parameters subject to practical constraints of the feed network configuration. Finally, a GA has been applied [26] to improve the bandwidth of microstrip antennas by optimizing the design of the feed network.

GAs have also been used effectively to optimize microstrip antenna designs by evolving new shapes for resonant metallic patch structures [27–35]. In these cases the GA is used as an optimizer for the microstrip patch antenna shape in order to achieve a broadband or multiband operation. The patch antenna geometry consists of an $N \times N$-pixel grid of binary values indicating the presence (1) or absence (0) of metal on a given pixel. By pixelizing the antenna geometry in this way, it may be readily incorporated into a binary chromosome representation for the GA. Other design parameters such as the substrate thickness and dielectric constant can also be included as additional genes in the chromosome. The fitness of each candidate microstrip antenna design is measured against a desired broadband or multiband response. The majority of the approaches considered in the literature use a subdomain MoM formulation with Rao–Wilton–Glisson (RWG) basis functions to evaluate the fitness of each candidate antenna design. Alternatively, a GA optimization methodology has been proposed [34,35] for microstrip patch antennas using a finite-difference time-domain (FDTD) technique to perform the required full-wave electromagnetic analysis.

Microstrip patch antennas are in widespread use because of their simple construction, good performance, low profile, and low cost. A typical single-layer microstrip patch antenna has modest broadside gain and a narrow impedance bandwidth. It is possible to increase the bandwidth of these antennas by adding a superstrate and parasitic patch to the single-layer antenna. By carefully adjusting of the parameters of these stacked patch antennas, it is possible to obtain wideband and dual-band VSWR responses.

GAs have been successfully applied to optimize designs of probe-fed stacked patch antennas [34–39]. The design objective in all cases was to achieve broadband operation. Designs having circular polarization, in addition to a broad impedance bandwidth, have been considered [37]. In order to achieve this goal, the cost function for the GA used in Ref. 37 consisted of two parts: one that takes into account the axial ratio and another that depends on the input impedance matching. The specific form of the cost function used in that study [37] is given by

$$cost = \sum_{i=1}^{N_f} A_i + \begin{cases} 0, & \text{if} \quad |\Gamma_i| < \Gamma_{\max} \\ 2|\Gamma_i|, & \text{if} \quad |\Gamma_i| \geq \Gamma_{\max} \end{cases} \tag{7.9}$$

where A_i denotes the axial ratio in decibels and Γ_i is the reflection coefficient of the antenna at a specified frequency. $\Gamma_{\max}$ is the maximum tolerable value for the input reflection coefficient, and N_f represents the total number of frequency sampling points. Furthermore, in Ref. 39, the practical case is considered where a GA is used to optimize the performance of a truncated stacked patch antenna (i.e., a stacked patch antenna with finite dielectric layers and a finite ground plane).

The geometry of a probe-fed microstrip stacked patch antenna and its nine basic design parameters are shown in Figure 7.6. The design parameters include

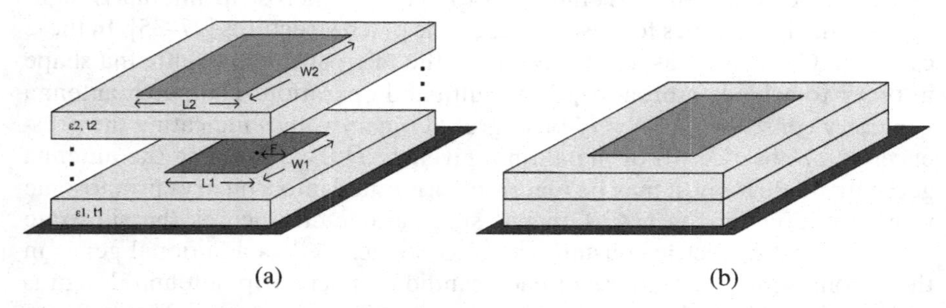

Figure 7.6. (a) Exploded view of a stacked patch antenna with its design parameters; (b) standard view of a stacked patch antenna.

the length (L_1) and width (W_1) of the lower patch, the length (L_2) and width (W_2) of the upper patch, the dielectric constant (ε_1) and thickness (t_1) of the lower substrate, the dielectric constant (ε_2) and thickness (t_2) of the upper substrate, and the feed location from the edge of the lower patch (F). By adjusting these parameters it is possible to tailor the bandwidth and gain of the antenna. In the design of these antennas, one could use a procedure that is based on a combination of physics-based reasoning and trial-and-error. While this procedure can work, it often proves to be quite tedious and time-consuming. Therefore, a design procedure that is based on a GA optimization is much more favorable.

Designs are presented for stacked patch antennas that are optimized via a GA for a broadband response and a dual-band response. For both designs the GA is used to optimize the nine parameters of the antenna to achieve a desired impedance response and broadside gain. The analysis of the antennas is carried out using method-of-moments (MoM) software that is linked with the GA. The MoM stacked patch model assumes infinite dielectric layers and an infinite ground plane. The feed wire radius for both designs was fixed at 0.25 mm.

The GA that is used in the optimizations is binary-coded with each of the nine genes of the chromosome encoded by 16 bits. A population size of 100 is used, and the optimization is carried out through 100 generations or until an acceptable design is obtained. Mating is performed using tournament selection and single-point crossover with a crossover rate of 50%. In order to introduce diversity into the population, a creep mutation rate of 4% and a jump mutation rate of 4% are applied to the chromosomes after the crossovers are performed. Elitism is enforced to carry the best population members over to the next generation.

The goal of the first stacked patch optimization is to achieve a design that has a broadband response centered at 8.5 GHz. Additionally, the antenna should have a broadside gain that is greater than 5 dB over the desired bandwidth. The performance of the antenna is evaluated at three sampling frequencies: 7.5, 8.5, and 9.5 GHz. The following cost function is used to judge the performance of the antennas:

$$cost = \sum_{i=1}^{N} P_i \tag{7.10}$$

$$P_i = \begin{cases} a_i \text{VSWR}_i + 10 & \text{if} \quad \text{gain}_i(\text{dB}) < 5 \\ a_i \text{VSWR}_i & \text{if} \quad \text{gain}_i(\text{dB}) \geq 5 \end{cases} \tag{7.11}$$

where N is the number of sampling frequencies that are used to evaluate the response of the antenna, VSWR_i is an element of the sampled VSWR data, a_i is a weighting factor, and $gain_i$ is the sampled broadside gain.

A crest in the VSWR response at the center of the band often results when a stacked patch antenna is designed for a wideband response. In order to achieve a more balanced response, the weighting factors in (7.11) can be adjusted to place more emphasis on minimizing the VSWR at the center frequency of the band. For this particular design, the weighting factor at the center of the band is made 1.5 times larger than the other weighting factors.

The optimized broadband design has a bandwidth (VSWR ≤ 2) of 27% that is centered around 8.5 GHz. A plot of the VSWR versus frequency is shown in Figure 7.7. A plot of the broadside gain versus frequency is shown in Figure 7.8. The parameters of the optimized design are listed in Table 7.1.

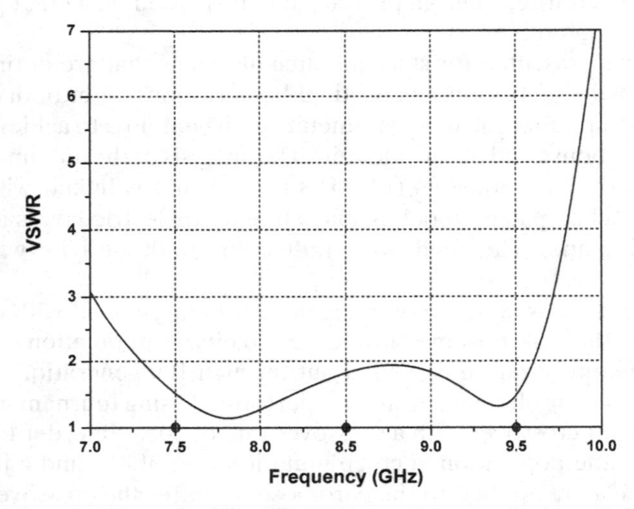

Figure 7.7. Voltage standing-wave ratio (VSWR) versus frequency for the optimized broadband stacked patch antenna. The markers indicate the sampling frequencies used by the GA.

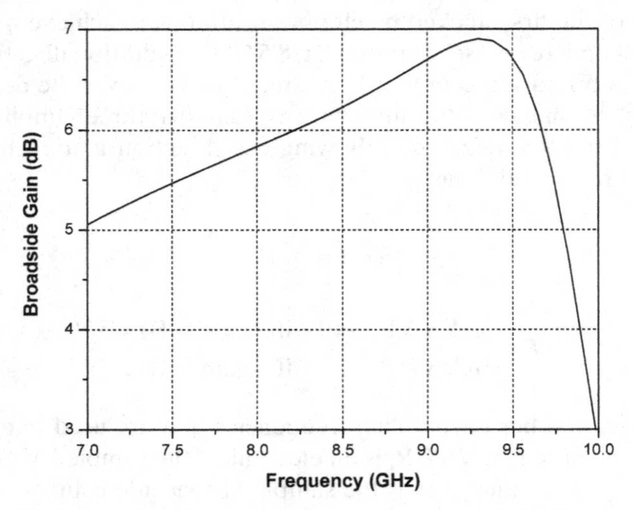

Figure 7.8. Broadside gain versus frequency for the optimized broadband stacked patch antenna.

TABLE 7.1. Optimized Parameters of Broadband and Dual-Band Stacked Patch Designs

	Broadband Design	Dual-Band Design
L_1 (mm)	11.85	13.75
W_1 (mm)	7.24	6.63
ε_1	2.49	2.22
t_1 (mm)	2.18	2.77
L_2 (mm)	11.96	12.97
W_2 (mm)	13.21	12.76
ε_2	1.27	1.09
t_2 (mm)	1.47	0.55
F (mm)	1.46	2.06

The goal of the second stacked patch antenna optimization is to obtain a design that exhibits a dual-band response and modest broadside gain (≥5 dB) at 7.5 and 9.75 GHz. Similar to the method used in broadband optimization, three sampling frequencies are chosen to evaluate the performance of the antenna. Two of the frequencies correspond to the centers of the desired operation bands. The other sampling frequency is located between the bands at 8.75 GHz. The fitness function that is used to evaluate the antennas is similar to the one used in the broadband stacked patch optimization (7.10). For this particular design, all of the weighting factors are set to unity magnitude. At the two operating bands the sign of the weighting factors is positive, and at 8.75 GHz the sign is negative. The aim of this weighting scheme is to minimize the VSWR at the two operating bands and maximize the VSWR in between the bands.

The optimized design has bands at the desired frequencies of 7.5 and 9.75 GHz. At the center of these bands the VSWR values are 1.27 and 1.2, respectively. A plot of VSWR versus frequency is shown in Figure 7.9. The design has a broadside gain of 7.5 dB at 7.5 GHz and 8.4 dB at 9.75 GHz. A plot of the broadside gain versus frequency is shown in Figure 7.10. The parameters of the optimized design are listed in Table 7.1.

When an antenna is placed within an array environment, its performance is altered by mutual coupling between the antenna and other elements of the array. Therefore, it seldom suffices to design a standalone antenna that is to be placed within an array environment. It is desirable to optimize the antenna while taking into account its interaction with the other elements of the array.

A design is presented for an infinite periodic array of stacked patch antennas that is optimized by a GA to have a broadband response [40]. Full-wave simulations are used to analyze the performance of the antenna in an array environment while accounting for mutual coupling effects. The full-wave simulations are carried out using the periodic finite-element–boundary integral (PFEBI) method [41,42]. A GA is used to optimize the nine parameters of the stacked patch antenna (Figure 7.6) to obtain the desired broadband response.

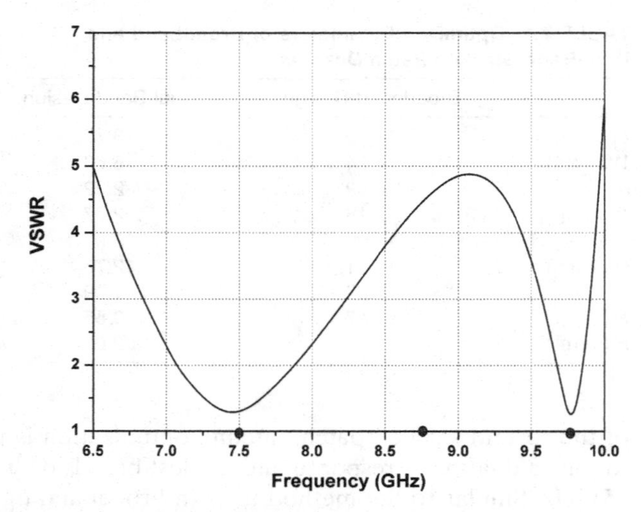

Figure 7.9. *VSWR versus frequency for the optimized dual-band stacked patch antenna. The markers indicate the sampling frequencies used by the GA.*

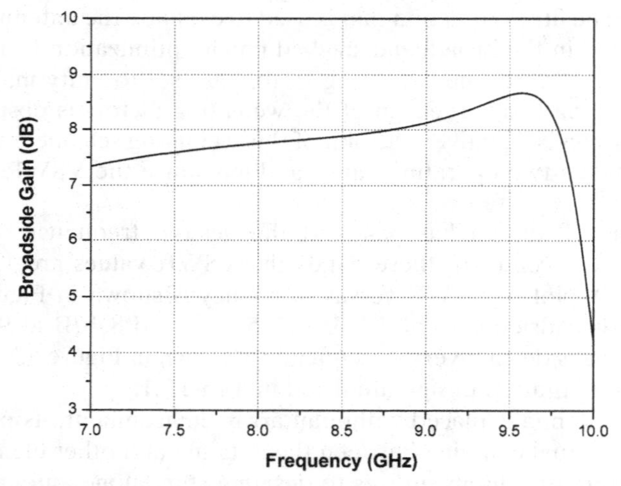

Figure 7.10. *Broadside gain versus frequency for the optimized dual-band stacked patch antenna.*

A binary-coded micro–genetic algorithm (MGA) was utilized to increase the convergence rate of the optimization. The MGA uses tournament selection and single-point crossovers with a crossover rate of 50%. Elitism is used to retain the best population members. The MGA uses a population size of 50 and it is set to optimize through 100 generations.

The MGA was parallelized to reduce the overall simulation time of the stacked patch optimization. The parallelization of the MGA is based on a "master–slave" model. The model uses a single population with selection and mating controlled globally by the "master" processor. Evaluations of the fitness function of the population members are performed in parallel by the "slave" processors. Communication between the "master" and "slaves" occurs only when the "slaves" receive the population members to evaluate and when the fitness values are returned to the "master" processor. The communication between the processors is controlled using the message-passing interface (MPI) library.

The goal of the optimization is to create an array with a broadband response that is centered at 8.5 GHz. For this design, the element spacing in both directions is fixed at 1.3 cm. The performance of the antenna elements is sampled at three frequency points ($N = 3$): 7.5, 8.5, and 9.5 GHz. The relative merit of the antennas is evaluated using the following cost function:

$$cost = \sum_{i=1}^{N} P_i \tag{7.12}$$

where

$$P_i = \begin{cases} S_{11}(\text{dB}) + 10\ \text{dB} & \text{if} \quad S_{11}(\text{dB}) \geq -10 \\ 0 & \text{if} \quad S_{11}(\text{dB}) \leq -10 \end{cases} \tag{7.13}$$

The parameters of the optimized stacked patch antenna are listed in Table 7.2. A plot of S_{11} versus frequency is shown in Figure 7.11. Radiation pattern cuts of a stacked patch element within the array environment are shown in Figure 7.12.

TABLE 7.2. Parameters of Optimized Stacked Patch Array Elements

Parameters	L_1	W_1	H_1	ε_1	L_2	W_2	H_2	ε_2	F
Optimized values (mm)	11.375	4.875	1.5	1.2	9.75	11.375	1.5	2.4	0

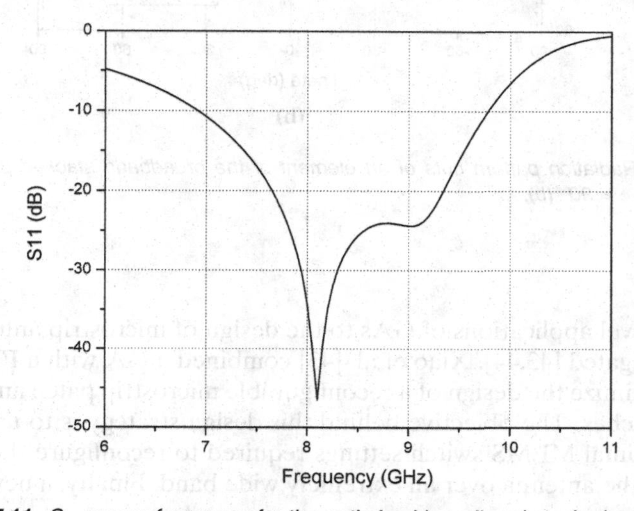

Figure 7.11. *S_{11} versus frequency for the optimized broadband stacked patch array.*

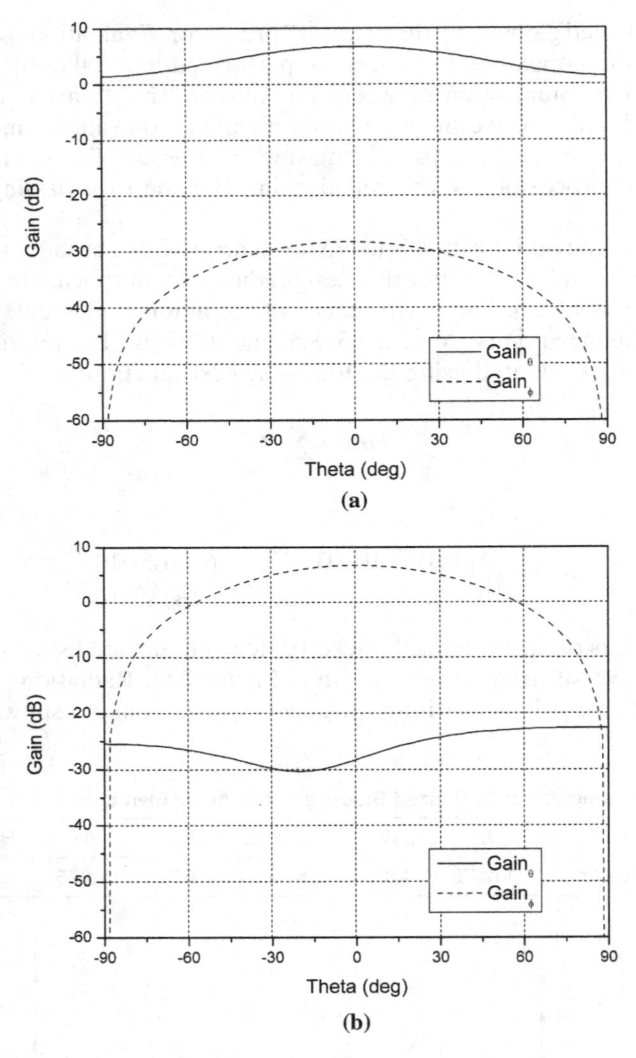

Figure 7.12. *Radiation pattern cuts of an element of the broadband stacked patch array at $\phi = 0°$ (a) and $\phi = 90°$ (b).*

Other novel applications of GAs to the design of microstrip antennas have been investigated [43,44]. Xiao et al. [43] combined a GA with a FDTD technique to optimize the design of a reconfigurable microstrip patch antenna with MEMS switches. The objective behind this design strategy is to use a GA to find the optimal MEMS switch settings required to reconfigure the radiation patterns of the antenna over an extremely wide band. Finally, a new family of microstrip patch antennas introduced by Werner et al. [44] was based on *fractile* (i.e., fractal tile) geometries. It was shown that by utilizing the unique properties of fractiles, it is possible to create high-gain single-feed modular designs for microstrip patch antennas. In this case, the GA is used as an optimization tool for determining the best place to attach a coaxial probe feed to the fractile patch antenna.

REFERENCES

1. P. Blacksmith, Jr., *A Method for Reducing Far-out Sidelobes*, AFCRC-TR-57-115, July 1957, DDC No. AD 133707.
2. C. O. Yowell and R. D. Etcheverry, Millimeter wave isolation measurements of diagonal horns and tunnel antennas, *Proc. 1974 IEEE AP-S Symp.* June 1974, Vol. 12, pp. 57–59.
3. R. C. Johnson and D. W. Hess, Performance of a compact antenna range, *Proc. IEEE AP-S Symp.* June 1975, Vol. 13, pp. 349–352.
4. W. Burnside, M. Gilreath, B. Kent, and G. Clerici, Curved edge modification of compact range reflectors, *IEEE AP-S Trans.* **35**(2):176–182 (Feb. 1987).
5. R. L. Haupt, *Low Sidelobe Resistive Reflector Antenna*, US Patent 5,134,423 (July 29, 1992).
6. M. Shimizu, Pattern tuning of defocus array-fed reflector antennas, *Proc. IEEE AP-S Symp.* June 1998, Vol. 4, pp. 2070–2073.
7. M. Vall-llossera, J. M. Rius, N. Duffo, and A. Cardama, Design of single-shaped reflector antennas for the synthesis of shaped contour beams using genetic algorithms, *Microw. Opt. Technol. Lett.* **27**(5):358–361 (Dec. 5, 2000).
8. Y. Lu, X. Cai, and Z. Gao, Optimal design of special corner reflector antennas by the real-coded genetic algorithm, *Proc. Asia-Pacific Microwave Conf.* Dec. 2000, pp. 1457–1460.
9. K. Barkeshli, F. Mazlumi, and R. Azadegan, The synthesis of offset dual reflector antennas by genetic algorithms, *Proc. IEEE AP-S Symp.* June 2002, Vol. 1, pp. 670–673.
10. S. L. Avila, W. P. Carpes, and J. A. Vasconcelos, Optimization of an offset reflector antenna using genetic algorithms, *IEEE Trans. Magn.* **40**(2):1256–1259 (March 2004).
11. D. G. Kurup, M. Himdi, and A. Rydberg, Design of an unequally spaced reflectarray, *Anten. Wireless Propag. Lett.* **2**(1):33–35 (2003).
12. M. Mussetta, P. Pirinoli, G. Dassano, R. E. Zich, and M. Orefice, Experimental validation of a genetically optimized microstrip reflectarray, *Proc. IEEE AP-S Symp.* June 2004, Vol. 1, pp. 9–12.
13. R. E. Zicht, M. Mussetta, Tovaglieri, P. Pirinoli, and M. Orefice, Frequency response of a new genetically optimized microstrip reflectarray, *Proc. IEEE AP-S Symp.* July 2003, Vol. 1, pp. 173–176.
14. L. Chung, T. Chang, and W. D. Burnside, An ultrawide-bandwidth tapered resistive TEM horn antenna, *IEEE AP-S Trans.* **52**(2):1848–1857 (Feb. 2004).

15. P. L. Garcia-Muller, Optimisation of compact horn with broad sectoral radiation pattern, *Electron. Lett.* **37**(6):337–338 (March 15, 2001).
16. D. Yang, Y. C. Chung, and R. L. Haupt, Genetic algorithm optimization of a corrugated conical horn antenna, *Proc. IEEE AP-S Symp.* June 2002, Vol. 1, pp. 342–345.
17. L. Lucci, R. Nesti, Pelosi, and G. Selleri, Optimization of profiled corrugated circular horns with parallel genetic algorithms, *Proc. IEEE AP-S Symp.* June 2002, Vol. 2, pp. 338–341.
18. L. Lucci, R. Nesti, G. Pelosi, and S. Selleri, NURBS profiled corrugated circular horns, *Proc. IEEE AP-S Symp.* June 2003, Vol. 1, pp. 161–164.
19. S. Sinton, J. Robinson, and Y. Rahmat-Samii, Standard and micro genetic algorithm optimization of profiled corrugated horn antennas, *Micro. Opt. Technol. Lett.* **35**(6):449–453 (Dec. 20, 2002).
20. E. I. Muehldorf, The phase center of horn antennas, *IEEE AP-S Trans.* **18**(6):753–760 (Nov. 1970).
21. C. A. Balanis, *Antenna Theory Analysis and Design*, 2nd ed., Wiley, New York, 1997.
22. M. Hakkak, Design of pyramidal horns for fixed phase center as well as optimum gain, *IEEE AP-S Mag.* **33**(3):53–55 (June 1991).
23. D. Lee and S. Lee, Design of a coaxially fed circularly polarized rectangular microstrip antenna using a genetic algorithm, *Microw. Opt. Technol. Lett.* **26**(5):288–291 (Sept. 2000).
24. O. Ozgun, S. Mutlu, M. I. Aksun, and L. Alatan, Design of dual-frequency probe-fed microstrip antennas with genetic optimization algorithm, *IEEE Trans. Anten. Propag.* **51**(8):1947–1954 (Aug. 2003).
25. B. Aljiouri, E. G. Lim, H. Evans, and A. Sambell, Multi-objective genetic algorithm approach for a dual-feed circular polarized patch antenna design, *Electron. Lett.* **36**(12):1005–1006 (June 2000).
26. A. Raychowdhury, B. Gupta, and R. Bhattacharjee, Bandwidth improvement of microstrip antennas through a genetic-algorithm-based design of a feed network, *Microw. Opt. Technol. Lett.* **27**(4):273–275 (Nov. 2000).
27. C. Delabie, M. Villegas, and O. Picon, Creation of new shapes for resonant microstrip structures by means of genetic algorithms, *Electron. Lett.* **33**(18):1509–1510 (Aug. 1997).
28. J. M. Johnson and Y. Rahmat-Samii, Genetic algorithms and method of moments (GA/MOM) for the design of integrated antennas, *IEEE Trans. Anten. Propag.* **47**(10):1606–1614 (Oct. 1999).
29. F. J. Villegas, T. Cwik, Y. Rahmat-Samii, and M. Manteghi, Parallel genetic-algorithm optimization of a dual-band patch antenna for wireless communications, *Proc. 2002 IEEE Antennas and Propagation Society Int. Symp.* June 2002, Vol. 1, pp. 334–337.
30. F. J. Villegas, T. Cwik, Y. Rahmat-Samii, and M. Manteghi, A parallel electromagnetic genetic-algorithm optimization (EGO) application for patch antenna design, *IEEE Trans. Anten. Propag.* **52**(9):2424–2435 (Sept. 2004).
31. H. Choo and H. Ling, Design of multiband microstrip antennas using a genetic algorithm, *IEEE Microw. Wireless Compon. Lett.* (see also *IEEE Microw. Guided Wave Lett.*) **12**(9):345–347 (Sept. 2002).

32. H. Choo and H. Ling, Design of broadband and dual-band microstrip antennas on a high-dielectric substrate using a genetic algorithm, *IEE Proc. Microw. Anten. Propag.* **150**(3):137–142 (June 2003).
33. H. Choo, A. Hutani, L. C. Trintinalia, and H. Ling, Shape optimisation of broadband microstrip antennas using genetic algorithm, *Electron. Lett.* **36**(25):2057–2058 (Dec. 2000).
34. P. Soontornpipit, C. M. Furse, and Y. C. Chung, Miniaturized biocompatible microstrip antenna using genetic algorithm, *IEEE Trans. Anten. Propag.* **53**(6):1939–1945 (June 2005).
35. P. Soontornpipit, C. M. Furse, Y. C. Chung, and B. M. Lin, Optimization of a buried microstrip antenna for simultaneous communication and sensing of soil moisture, *IEEE Trans. Anten. Propag.* **54**(3):797–800 (March 2006).
36. M. Lech, A. Mitchell, and R. B. Waterhouse, Optimization of broadband microstrip patch antennas, *Proc. 2000 Asia Pacific Microwave Conf.*, Sydney, Astralia, Dec. 2000, pp. 711–714.
37. R. Zentner, Z. Sipus, and J. Bartolic, Optimization synthesis of broadband circularly polarized microstrip antennas by hybrid genetic algorithm, *Microw. Opt. Technol. Lett.* **31**(3):197–201 (Nov. 2001).
38. A. Mitchell, M. Lech, D. M. Kokotoff, and R. B. Waterhouse, Search for high-performance probe-fed stacked patches using optimization, *IEEE Trans. Anten. Propag.* **51**(2):249–255 (Feb. 2003).
39. T. G. Spence, D. H. Werner, and R. D. Groff, Genetic algorithm optimization of some novel broadband and multiband microstrip antennas, *Proc. 2004 IEEE Antennas and Propagation Society Int. Symp.* June 2004, Vol. 4, pp. 4408–4411.
40. R. K. Shaw and D. H. Werner, Design of optimal broadband microstrip antenna elements in the array environment using genetic algorithms, *Proc. 2006 Antennas and Propagation Int. Symp.* July 2006, pp. 3727–3730.
41. J. L. Volakis, A. Chatterjee, and L. C. Kempel, *Finite Element Method for Electromagnetics*, IEEE Press, New York, 1998.
42. L. Li and D. H. Werner, Design of all-dielectric frequency selective surfaces using genetic algorithm combined with finite-element boundary-integral method, *Proc. Antennas and Propagation Society Int. Symp.* July 3–8, 2005, Vol. 4A, pp. 376–379.
43. S. Xiao, B.-Z. Wang, X.-S. Yang, and G. Wang, Reconfigurable microstrip antenna design based on genetic algorithm, *Proc. 2003 IEEE Antennas and Propagation Society Int. Symp.* June 2003, Vol. 1, pp. 407–410.
44. T. G. Spence and D. H. Werner, Genetically optimized fractile microstrip patch antennas, *Proc. 2004 IEEE Antennas and Propagation Society Int. Symp.* June 2004, Vol. IV, pp. 4424–4427.

8

Optimization of Scattering

This chapter takes a break from antenna topics and looks at the closely related topics of scattering and radar cross section (RCS). Scattering occurs when an incident electromagnetic wave induces currents on an object and those currents radiate a scattered field. The RCS of an object is how large it appears to a radar. The RCS in three dimensions is given by

$$\sigma_{3D}(\theta, \phi) = \lim_{r \to \infty} 4\pi r^2 \frac{|\mathbf{E}^{\text{scattered}}|^2}{|\mathbf{E}^{\text{incident}}|^2} \tag{8.1}$$

in units of area. If the objects are limited to the two-dimensional x–y plane, then

$$\sigma_{2D}(\phi) = \lim_{\rho \to \infty} 2\pi\rho \frac{|\mathbf{E}^{\text{scattered}}|^2}{|\mathbf{E}^{\text{incident}}|^2} \tag{8.2}$$

in units of length. Frequently, the RCS is expressed in dBsm (dB relative to a square meter) for 3D or dBm (dB relative to a meter) for 2D formats.

For an infinite, planar structure, the scattering from the device can be represented by reflection and transmission coefficients, where the reflection coefficient is the ratio between the total reflected and the incident electric fields. The transmission coefficient would then be the ratio between the total transmitted and the incident electric fields. For a filtering application, such as

Genetic Algorithms in Electromagnetics, by Randy L. Haupt and Douglas H. Werner

described in Section 8.2.2, a planar device is designed to exhibit stopbands, with low transmission magnitude and high reflection magnitude, and/or passbands, with high transmission magnitude and low reflection magnitude. The phases of the transmission and/or reflection coefficients can also be designed to have specific values, such as discussed in Section 8.2.3 for electromagnetic bandgap devices. Here, the reflection phase is optimized to have a phase of 0°, so that the reflected wave is in phase with the incident wave just above the surface of the planar device. In Section 8.3.4, both the magnitude and phase of the reflection coefficient are optimized to be small for absorber applications.

8.1 SCATTERING FROM AN ARRAY OF STRIPS

This section explains how to use a GA to control the scattering patterns from arrays of strips. A strip is defined as having a finite width in the x direction, is infinitely thin in the y direction, and is infinitely long in the z direction. Placing strips side-by-side forms an array or grid. The size and spacing of the strips are variables that affect the scattering pattern when a plane wave is incident. An optimum spacing between strips can be found that results in the lowest maximum backscatter sidelobe level. The model is an array of perfectly conducting strips as shown in Figure 8.1. Each strip is 0.25λ wide. A plane wave in which the electric field is parallel to the edge of the strips and a magnetic field of unity amplitude is incident at an angle ϕ. The spacing between the strips is variable and limited to 0–0.5λ. The currents flow in the z direction and are found for each relevant angle by solving

$$H_z e^{jkx\cos\phi} = \frac{k}{4}\sum_{n=1}^{N}\int_{a_n}^{b_n} J_z(x')H_0^{(2)}(k|x-x'|)dx' \tag{8.3}$$

for the surface current density using the method of moments (MoM). The left-hand side of this equation is the incident magnetic field. A physical optics (PO) formulation for the current density is

$$J_z(x') = 2H_z \sin\phi e^{jkx'\cos\phi} \tag{8.4}$$

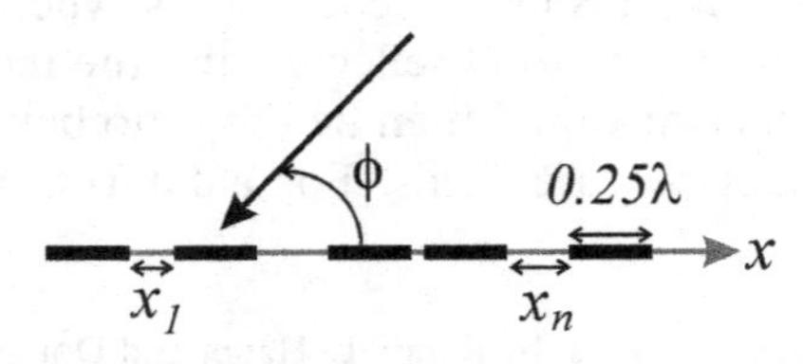

Figure 8.1. *Diagram of an array of unequally spaced strips that are 0.25λ wide.*

The surface current density is substituted into the expression for the backscattering RCS

$$\sigma(\phi) = \frac{k}{4}\left|\sum_{n=1}^{N}\int_{x_n}^{x_n+\Delta} J_z(x')e^{jkx'\cos\phi}dx'\right|^2 \tag{8.5}$$

where N = number of strips
$J_z(x')$ = surface current density in z direction
x_n = beginning of strip n
Δ = strip width = 0.25λ
$H_z = 1$ = magnitude of incident magnetic field
$H_0^{(2)}(\cdot)$ = zero-order Hankel function of the second kind

Backscattering assumes that the incident angle and the observation angle are equal.

The goal here is to minimize the maximum sidelobe level by finding the optimum spacing between strips. Eight minimization techniques

- Nelder Mead downhill simplex
- BFGS (Broyden–Fletcher–Goldfarb–Shannon)
- DFP (Davidon–Fletcher–Powell)
- Steepest-descent method
- Random search
- Binary GA
- Continuous GA
- Hybrid GA

were applied to minimizing the cost function

$$cost = \max\{sidelobe \text{ level}[\sigma(\phi)]\} \tag{8.6}$$

for $N = 10, 40$. The local optimizers started at 20 independent random points. All optimizers were terminated after 1000 cost function evaluations. Better results might have been found with more cost function evaluations. The binary GA has a population size of 16, a 3% mutation rate, and a 50% discard rate. The continuous parameter GA has a population size of 8, an 8% mutation rate, and a 50% discard rate. The hybrid GA combines the strengths of the GA and a local optimizer. It begins the optimization with a continuous parameter GA. The local optimizer kicks in after 700 function evaluations.

Table 8.1 displays the results from optimizing (8.6) using the 8 different approaches and averaging over 20 independent runs (i.e., each run starts with a new random seed). An independent run consists of a new random starting point or population. The hybrid GA produces the best results. Surprisingly,

TABLE 8.1. Backscattering Maximum Sidelobe Level (dB) after Optimizing Spacing between 10 Strips with 8 Different Optimization Methods and Averaging over 20 Independent Runs Bold Font Indicates Best Value in Column. (Lowest Cost in Boldface)

	Method of Moments			Physical Optics		
	Maximum	Minimum	Mean	Maximum	Minimun	Mean
Nelder–Mead	−13.7	−15.8	−15.0	−14.3	−19.9	−16.9
BFGS	−13.3	−16.4	−14.8	−14.4	−19.4	−17.0
DFP	−12.0	−16.2	−14.3	−12.2	−20.4	−16.9
Steepest descent	−12.6	**−17.8**	−15.4	−13.6	−19.9	−16.8
Random search	−13.7	−15.5	−14.4	−12.5	−19.0	−16.3
Binary GA (6-bit)	−14.8	−16.0	−15.5	−17.7	−21.1	−19.2
Continuous GA	−13.3	−16.3	−15.0	−16.6	−20.8	−18.6
Hybrid GA	**−15.3**	−16.7	**−16.2**	**−18.0**	**−21.3**	**−19.5**

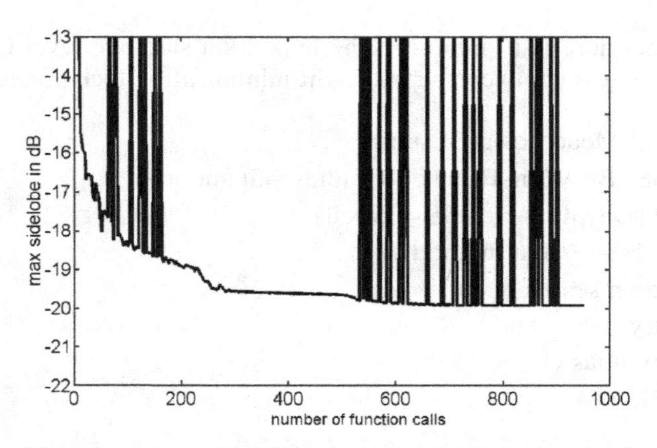

Figure 8.2. *Convergence of the Nelder–Mead downhill simplex algorithm.*

steepest descent found the lowest minimum when the method of moments cost function was used. The binary and continuous GAs performed quite well. These results show that this cost function is multimodal with many local minima. Convergence for the best run by the Nelder–Mead algorithm for the physical optics cost function is shown in Figure 8.2. Spikes in the convergence are due to bad movements by the algorithm. Convergence for the best runs by the three derivative-based local optimizers appears in Figure 8.3. DFP worked the best, while the steepest descent came in third. Figure 8.4 shows the convergence for the best runs by the three GA algorithms. The hybrid GA is equivalent to the continuous GA up to 700 function calls. At that point, a Nelder–Mead algorithm completes the remaining 300 function calls before stopping. This hybrid GA resulted in the best performance. Figure 8.5 displays the best grid spacing found by each of the eight optimization algorithms.

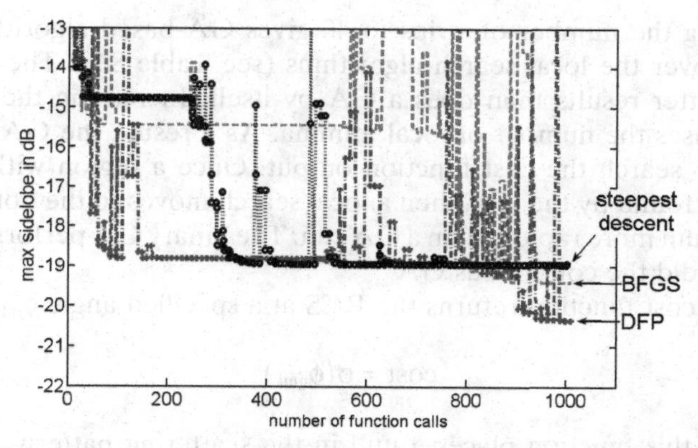

Figure 8.3. *Convergence of the steepest descent, BFGS, and DFP algorithms.*

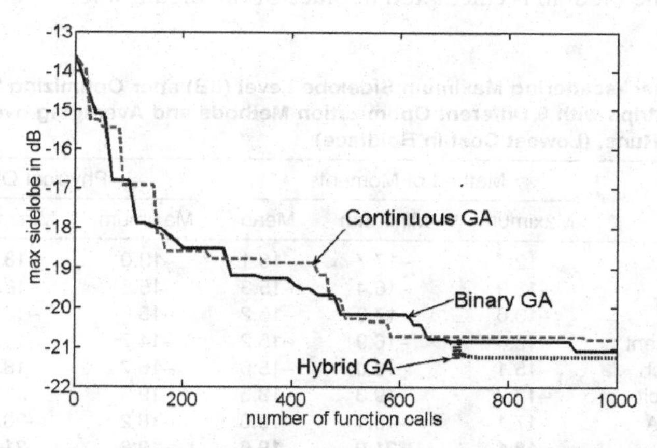

Figure 8.4. *Convergence of the continuous, binary, and hybrid GAs.*

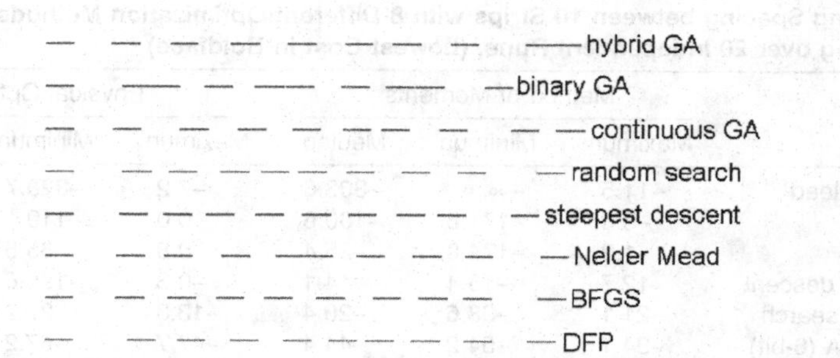

Figure 8.5. *Optimum strip configurations found by the eight different optimization methods for the method of moments cost function.*

Increasing the number of strips to 40 gives GA-based algorithms a clear advantage over the local search algorithms (see Table 8.2). The hybrid GA provides better results than does a GA by itself. Increasing the number of strips increases the number of local minima. As a result, the GAs are better equipped to search the cost function output. Once a region with a suitable minimum is found by the GA, then a local search moves to the bottom of that local minimum more rapidly than a GA can. The binary GA performed slightly better than did the continuous GA.

Another cost function returns the RCS at a specified angle ϕ_{full}:

$$\text{cost} = \sigma(\phi_{null}) \tag{8.7}$$

Minimizing this function places a null in the scattering pattern. Finding the spacing of 10 strips to place a null at $\phi = 73.7°$ is the next example. Table 8.3 summarizes the results for each algorithm. The hybrid GA proved best this time, too. The median is calculated in place of the mean, since an output of $-\infty$

TABLE 8.2. Backscattering Maximum Sidelobe Level (dB) after Optimizing Spacing between 40 Strips with 8 Different Optimization Methods and Averaging over 20 Independent Runs. (Lowest Cost in Boldface)

	Method of Moments			Physical Optics		
	Maximum	Minimum	Mean	Maximum	Minimun	Mean
Nelder–Mead	−12.7	−17.7	−16.1	−13.0	−18.9	−16.2
BFGS	−13.9	−16.4	−15.3	−15.5	−18.7	−16.9
DFP	−13.6	−17.0	−15.2	−15.0	−18.1	−16.6
Steepest descent	−13.4	−16.9	−15.2	−14.7	−18.6	−16.7
Random search	−15.1	−16.0	−15.6	−16.7	−18.0	−17.4
Binary GA (6-bit)	−17.7	−19.3	−18.5	−19.3	−21.4	−20.0
Continuous GA	−17.1	−19.1	−18.0	−18.2	−20.9	−19.3
Hybrid GA	**−18.1**	**−21.0**	**−19.6**	**−19.6**	**−21.6**	**−20.2**

TABLE 8.3. Null Depth (dB) in Backscattering Pattern at $\phi = 73.7°$ Created by Optimizing Spacing between 10 Strips with 8 Different Optimization Methods and Averaging over 20 Independent Runs. (Lowest Cost in Boldface)

	Method of Moments			Physical Optics		
	Maximun	Minimun	Median	Maximun	Minimun	Median
Nelder–Mead	−11.5	$-\infty$	−303.6	−7.2	**−328.7**	−205.3
BFGS	−8.6	−171.8	−100.6	0.0	−119.1	−9.4
DFP	1.2	−174.0	−25.4	0.8	−65.5	−10.7
Steepest descent	−12.7	−15.4	−14.1	−0.3	−195.6	−8.3
Random search	−21.1	−38.6	−26.4	−13.3	−27.2	−20.9
Binary GA (6-bit)	−34.1	−54.2	−44.4	−47.7	−67.2	−55.1
Continuous GA	−32.9	−56.4	−43.5	−41.8	−67.7	−52.7
Hybrid GA	**−305.5**	−329.2	**−314.2**	**−307.9**	−326.0	**−318.0**

TABLE 8.4. Null Depth (dB) in Backscattering Pattern at $\phi = 86°$ Created by Optimizing Spacing between 40 Strips with 8 Different Optimization Methods and Averaging over 20 Independent Runs. (Lowest Cost in Boldface)

	Method of Moments			Physical Optics		
	Maximun	Minimun	Median	Maximun	Minimun	Median
Nelder–Mead	14.5	−85.3	−49.4	7.1	−92.6	−64.9
BFGS	9.1	−23.3	−4.5	2.4	−34.0	−12.0
DFP	5.9	−26.8	−2.5	4.7	−29.3	−9.5
Steepest descent	7.9	**−193.4**	−89.9	2.6	**−181.3**	−11.9
Random search	−10.0	−27.9	−18.9	−18.8	−51.8	−25.4
Binary GA (6-bit)	−24.8	−51.0	−31.8	−38.2	−60.7	−46.3
Continuous GA	−22.9	−41.9	−31.3	−32.5	−55.0	−41.2
Hybrid GA	**−140.6**	−151.1	**−144.3**	**−144.0**	−141.7	**−142.1**

is possible. The hybrid GA had the best maximum and median values, but the Nelder–Mead algorithm had the lowest median null depth for MoM and PO.

Table 8.4 shows the results for 40 strips. The median results for the binary and continuous GAs were consistently good, but one of the local optimizers did better. The hybrid GA gave the lowest maximum and median values by far. Some of the local optimizers were able to find very low minima in one or more of the runs.

Additional results and information on optimizing (8.6) and (8.7) can be found in Ref. 1. Other researchers have optimized scattering from strips. In Ref. 2, the reflection and transmission properties of multilayered dielectric structures periodically loaded with metal strips are optimized. The chromosomes consisted of binary and continuous values. The choice of dielectric constant and the presence or absence of a strip were represented with binary numbers. On the other hand, the thicknesses of the dielectric layers were represented by continuous values. The GA was able to find configurations that allowed reflection and transmission for several different frequency bands. A GA has also been used to solve inverse scattering problems. A GA is used to find the widths and locations of strips given the scattered field at various angles [3]. Other objects have also been imaged from scattered field data using a GA. The shape and location of a cylinder are found in Ref. 4. A GA using a method of moments cost function was used to predict the shape of objects from measured data in Ref. 5.

8.2 SCATTERING FROM FREQUENCY-SELECTIVE SURFACES

Frequency-selective surfaces (FSSs) are planar structures consisting of a doubly periodic metallic screen usually backed by one or more dielectric layers [6]. A schematic diagram of a FSS screen and dielectric substrate is shown in Figure 8.6. Additional substrates or superstrates can be incorporated into

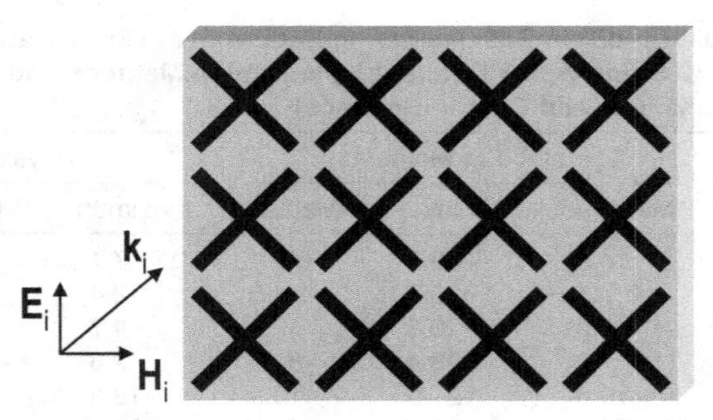

Figure 8.6. Diagram of a frequency-selective surface (FSS) composed of a periodic array of metallic crossed dipoles printed on a dielectric substrate.

the structure to enhance the desired scattering performance of the device or to protect the metallic screen elements from damage. A ground plane may be added to the back of the device for certain applications such as constructing electromagnetic bandgap (EBG) surfaces as discussed in Section 8.2.3.

An FSS can be designed to behave as a lowpass, highpass, bandpass, stop-band, or multiband filter for electromagnetic waves. When illuminated by electromagnetic radiation, wavelengths at which the metallic screen patches resonate are reflected back from the FSS. The FSS screen is transparent to other wavelengths of electromagnetic radiation. For example, the crossed dipole metallic patches depicted in Figure 8.6 are resonant at a wavelength approximately twice the crossed dipole armlength in the dielectric [6]. Thus, this FSS filter exhibits a single stopband at the resonant frequency of the crossed dipole screen elements.

In practice, there are many variables, such as metallic element shape, spacing, orientation, and dielectric layer properties, which must be adjusted simultaneously to achieve a desired high-performance filter response. Hence, GAs are a natural choice for synthesizing FSS designs with optimal performance. Representation of the design variables within binary chromosomes is straightforward, as will be shown in the next section. Calculating the gradient of the scattering coefficients in terms of the design variables is quite difficult, and not worthwhile, as there are many local minima in which a local optimization could potentially become stuck. A robust global optimization scheme is needed for FSS synthesis. In addition, the desired scattering properties can be cast in terms of a cost function, as will be shown in the examples to follow.

GA techniques have successfully optimized the designs for several types of FSS devices. The first device type, which will be discussed in more detail in Section 8.2.1, is an FSS filter. For this application, the FSS scattering variables are optimized for minimum transmission at desired stopband frequencies and maximum transmission at desired passband frequencies.

Another interesting twist is to add reconfigurability into the FSS structure. Reconfigurable frequency-selective surfaces (RFSS) have been studied more recently, where a static metallic FSS screen geometry, such as an array of linear or crossed dipoles, is interconnected with switches that can be independently turned on or off [7]. The switch settings are then optimized via a GA to achieve a desired scattering response. This topic is discussed further in Section 8.2.2.

Metamaterials that possess an effective negative refractive index (Re $[n] < 0$) are currently receiving much attention due mainly to the promise they hold for realizing a "perfect lens," which can resolve features smaller than the wavelength of light [8]. One of the most critical issues with realizing a flat negative-index metamaterial (NIM) lens capable of imaging beyond the diffraction limit has been overcoming the relatively high losses inherent in these materials. To this end, a planar NIM synthesis technique has been developed by using a GA to optimize the scattering from planar FSS screens to exhibit negative-index behavior as well as extremely low losses (i.e., low reflection mismatch together with low effective material losses) [9]. A similar technique has also been used to synthesize zero-index metamaterials (ZIM) [10,11].

EBG devices represent another important class of FSS-based metamaterials that have been given a lot of attention in recent years. These are made by placing a metallic ground plane on the bottom of the FSS substrate. By doing so, the phase of the reflection coefficient from the FSS passes through 0° at its resonant frequencies. This surface would then act like an artificial magnetic conductor (AMC) under the 0° reflection phase condition, since the reflected wave is in phase with the incident wave. A perfect electric conductor (PEC), on the other hand, has a reflection phase of 180°, such that the reflected wave is exactly out of phase with the incident wave. According to image theory, a horizontal dipole antenna should be placed a quarter-wavelength above a PEC ground plane in order for the reflected image of the antenna to radiate in phase with the physical antenna. Theoretically, an antenna could then be placed extremely close to the AMC surface, as opposed to requiring a quarter-wavelength standoff in the case of a conventional ground plane made from a PEC. EBG FSS metamaterials have been synthesized using GA optimization to possess a variety of advantageous properties, including multiband response, angular stability, and small size. This topic is discussed further in Section 8.2.3. EBG FSS metamaterials have also been used in electromagnetic absorber applications, which are described in Section 8.3.

8.2.1 Optimization of FSS Filters

FSS are used as spatial filters at microwave frequencies for a variety of applications, including radomes, waveguide filters [12], and reflectors for satellite communication [13]. At infrared (IR) and visible wavelengths, FSS filters with micrometer- and nanoscale features are being synthesized for signal processing applications [14–17] and as IR emitters [18]. Since metallic losses become significant at optical wavelengths, all-dielectric FSS (DFSS) are also being studied for use in filtering applications. DFSS are composed of multiple dielectric materials with no metallic screens and are more advantageous at optical wavelengths where a variety of dielectric materials are available with extremely low losses [16,19].

For these applications, the design goal is to achieve a set of desired filter characteristics. This includes specified frequency points for stopbands or passbands and can also include shaping the transitions between stopbands and passbands. In terms of scattering variables, S_{21} should be minimized (S_{11} maximized) for stopbands, and S_{21} should be maximized (S_{11} minimized) for passbands. Typically, the mean-squared error (MSE) is minimized between the calculated S parameters and the desired transmission or reflection value at each specified frequency point. Specific cost functions will be given for two synthesis examples presented later in this section.

Various studies on GA synthesis of FSS and DFSS filters have been published [12–17,20–24]. While most of these studies generalize the metallic screen geometry to a pixelized grid that can be optimized by the GA, two approaches are based on choosing a fixed screen geometry a priori and then optimizing a smaller subset of FSS variables to achieve a desired filter response. For example, it is well known that an FSS with a conventional square array of crossed dipoles has a stopband that changes its center frequency for different angles of incidence. However, it has been shown [21] that a GA could be used to optimize the spacing and orientation of the crossed dipole elements in the FSS to achieve a design with a robust frequency response that is essentially independent of the incidence angle. In a second study [16], a GA is used to optimize the variables of a specific FSS unit cell geometry. In this case a DFSS with dielectric blocks embedded in a slab of a different dielectric material is optimized to have a strong stopband at a desired frequency. This is accomplished by using the GA to evolve the best size and periodicity of the embedded dielectric blocks.

A number of other studies consider pixilated FSS screens, which allow the screen geometry to be included as an additional variable for the GA to optimize. Typically, an 8-fold symmetry condition is imposed on the unit cell geometry as shown in Figure 8.7 to ensure that the FSS filter response is the same for both vertically and horizontally polarized incident waves (i.e., its response is polarization-independent). The unit cell is divided into eight triangular folds, such that a single fold is parameterized in the GA and the remaining unit cell is generated by mirroring the geometry across the diagonal, horizontal, and vertical lines shown in Figure 8.7.

The individual pixels have binary values with 1 representing the presence of metal and 0 representing the absence of metal. For a binary-coded GA, it is straightforward to represent the screen geometry with 8 genes. The gene corresponding to the first row of the pixelized screen geometry would be

$$gene = [b_1^1 b_2^1 b_3^1 b_4^1 b_5^1 b_6^1 b_7^1 b_8^1] \tag{8.8}$$

A chromosome is formed by placing the genes for each row of the geometry together in sequence followed by the cell size, substrate variables, and any

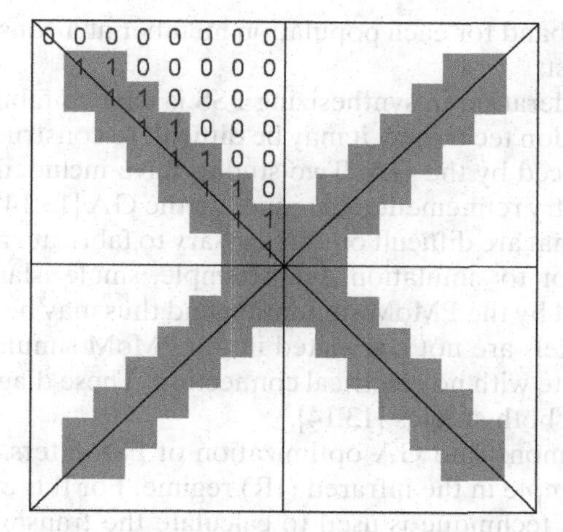

Figure 8.7. *Unit cell of a FSS screen with 8-fold symmetry enforced. Chromosome values are shown over one triangular fold of the screen.*

other variables to be optimized by the GA. If we encode the geometry shown in Figure 8.7, then the chromosome would be

$$\begin{aligned} chromosome &= [b_1^1 b_2^1 b_3^1 b_4^1 b_5^1 b_6^1 b_7^1 b_8^1 b_1^2 b_2^2 b_3^2 b_4^2 b_5^2 b_6^2 b_7^2 b_1^3 b_2^3 b_3^3 b_4^3 b_5^3 b_6^3 b_1^4 \\ &\quad b_2^4 b_3^4 b_4^4 b_5^4 b_1^5 b_2^5 b_3^5 b_4^5 b_1^6 b_2^6 b_3^6 b_1^7 b_2^7 b_1^8] \\ &= [00000000 \mid 1100000 \mid 110000 \mid 11000 \mid 1100 \mid 110 \mid 11 \mid 1] \end{aligned} \tag{8.9}$$

The transmission and reflection from the FSS screen is typically calculated using a periodic method of moments (PMoM) full-wave simulation code [25,26]. Because of the computational complexity of using a full-wave simulation tool to calculate the scattering variables for every population member at multiple frequencies over many generations, several schemes have been investigated in the literature for synthesizing FSS filters that speed up the convergence of the GA. A hybrid GA has been presented [22] that uses a local optimizer between generations to optimize variables other than the screen geometry, upon which the cost function has a smooth dependence. The local optimizer uses an interpolation scheme, reducing the number of times that the PMoM code must be run. A similar scheme called a *hierarchical GA* which also allows the number of screens in the FSS to be chosen by the GA has been reported [23]. A micro–genetic algorithm (MGA) approach that uses a small population size and includes no mutation operator has also been considered [24]. While the MGA sparsely samples the cost function, it has the advantage of converging more rapidly to a local solution, which may be sufficient to meet the intended design requirements. Another method for speeding up the GA optimization interpolates the impedance matrices across a wide frequency band in the PMoM analysis [27]. This provides a fine frequency resolution over a wide frequency band for each population member at a considerably reduced computational cost.

Another consideration in synthesizing FSS is ease of fabrication. Depending on the fabrication technique, it may be difficult to construct the synthesized FSS screen produced by the GA. Two studies have included fabrication constraints, or geometry refinement techniques, in the GA [13,14]. In both of these studies, features that are difficult or unnecessary to fabricate are removed from the geometry prior to simulation. For example, single island pixels are not taken into account by the PMoM simulation and thus may be removed. Diagonally touching pixels are not connected in the PMoM simulation and may be difficult to fabricate with no electrical connection. These diagonal connections are suppressed in both studies [13,14].

In order to demonstrate GA optimization of FSS filters, we will consider here a design example in the infrared (IR) regime. For this example, a PMoM full-wave analysis technique is used to calculate the transmission and reflection from the FSS screen [25,26]. The current basis functions on the screen geometry are represented by "rooftops," which conveniently allow the screen to be generalized to pixels [27]. For IR wavelengths, the loss in the metallic screen begins to impact the performance of the FSS and must be taken into account. Frequency-dependent loss models in the optical and IR regimes have been published for common metals [29]. The loss models presented in Ref. 29 can be incorporated into the PMoM code in the form of a surface impedance given by

$$Z_s = \frac{Z_0}{\sqrt{\varepsilon_1 - j\varepsilon_2}} \tag{8.10}$$

where $Z_0 \approx 377\,\Omega$ is the intrinsic impedance of free space [19]. A 4-μm-thick flexible slab of polyimide is used as the dielectric substrate for this FSS filter. The permittivity of polyimide is estimated to be $\varepsilon_r = 3.0 - 0.3j$.

The unit cell geometry is represented by a 16 × 16 array of binary pixels. As shown in Figure 8.7, 8-fold symmetry is enforced to achieve polarization insensitivity. Because the substrate variables are chosen a priori, the chromosome for this optimization consists of one triangular fold of the geometry as shown in (8.9) and an 8-bit string representing the unit cell dimension. Fabrication constraints as described earlier were enforced during optimization, so that each population member was modified for ease of fabrication prior to PMoM simulation [17].

The cost function used for this optimization is given by

$$\mathrm{FF} = \sum_{\text{pass}} |\Gamma|^2 + 3\sum_{\text{stop}} |T|^2 \tag{8.11}$$

where Γ is the reflection coefficient and T is the transmission coefficient. A weighting coefficient of 3 is used to achieve greater attenuation in the specified stopbands. There were three stopband frequencies specified around 2.7 THz and three stopband frequencies specified around 5.9 THz. Additionally, 18 passband frequencies were specified elsewhere from 1 to 10 THz.

A population of 75 was optimized over 300 generations with a crossover probability of 0.5 and a mutation probability of 0.1%. The synthesized geometry shown in Figure 8.8a has a cell size of 35.9 μm on a side. This design was fabricated using standard optical photolithography. An optical microscope image of the fabricated FSS shown in Figure 8.8b demonstrates that the geometry was able to be easily and accurately fabricated. The simulated and measured transmission spectra in Figure 8.9 show excellent agreement between the design goals and the response of the fabricated structure.

As a second example, consider the design of a DFSS filter. The analysis of DFSS can be carried out by the finite-element boundary–integral (FEBI)

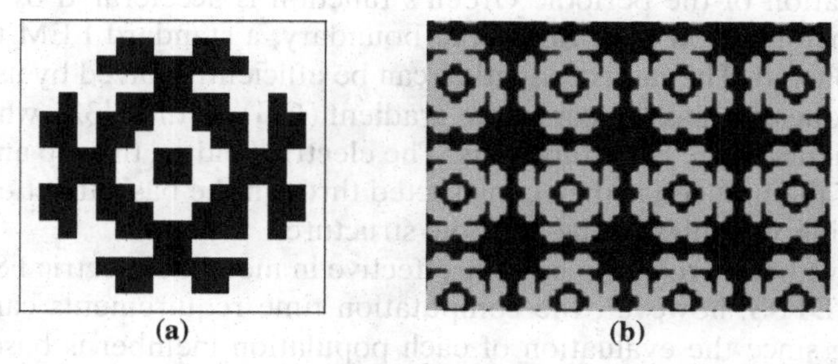

Figure 8.8. IR FSS filter designed to have stopbands at 2.7 and 5.9 THz: (a) unit cell geometry is shown along with (b) optical microscope image of fabricated FSS structure.

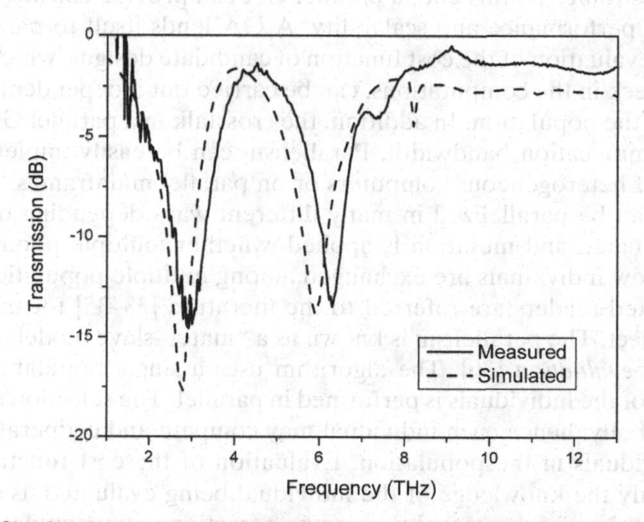

Figure 8.9. Simulated and measured transmission spectra for the dual-band IR FSS filter shown in Figure 8.8.

method, which can efficiently model the material inhomogeneities and can simulate periodic boundaries without any difficulty. Moreover, it can be used in conjunction with a GA to provide a powerful tool for the design optimization of DFSS.

The FEBI method, a hybridization of the traditional MoM with the finite-element method (FEM), is one of the most powerful and efficient full-wave numerical methods for handling the analysis of inhomogeneous problems [30,31]. For periodic structures, the method begins with a functional description of the field problem where only a single unit cell of the array is considered. This unit cell is meshed with three-dimensional cubic elements and the electric field intensity is discretized with edge-based basis functions. Periodic boundary conditions are enforced on the opposite sidewalls of the unit cell. On the top and bottom planar surfaces of the structure, the FEM computation domain is terminated by applying the mixed potential integral equation (MPIE) [31]. The calculation of the periodic Green's function is accelerated by using the Ewald transformation [31]. Inside the boundary, a standard FEM technique [30] is employed. The matrix equation can be efficiently solved by using iterative techniques such as the conjugate gradient (CG) method [32], which yields the coefficients of the basis functions. The electric field on the top and bottom surfaces of the unit cell is then constructed through the basis functions to find the scattering variables of the periodic structure.

GAs have been proven to be very effective in metallodielectric FSS design. As to the DFSS, however, the computation time requirements can be very demanding since the evaluation of each population member is based on the volumetric discretization of the entire unit cell structure. Even with a numerical technique as efficient as FEBI, finding a way to shorten the design cycle is highly desirable. To this end, a parallel GA can provide considerable gains in terms of performance and scalability. A GA lends itself to parallelization in that the evaluation of the cost function of candidate designs, which is usually the bottleneck in the computations, can be carried out independently for each member of the population. In addition, the crosstalk in a parallel GA requires a low communication bandwidth. Parallelism can be easily implemented on networks of heterogeneous computers or on parallel mainframes.

A GA can be parallelized in many different ways depending on how the cost is evaluated and mutation is applied, whether multiple populations are used and how individuals are exchanged among multiple populations, and so on. Interested readers are referred to the literature [33–35] for more details on this subject. The parallelism is known as a "mater–slave model", or *distributed fitness evaluation* [36]. The algorithm uses a single population and the evaluation of the individuals is performed in parallel. The selection and mating is done globally; hence each individual may compete and cooperate with any other individuals in the population. Evaluation of the cost functions, which requires only the knowledge of the individual being evaluated, is distributed to multiple "slave" processors by assigning a fraction of the population to each processor available. Communication occurs only when each "slave" processor receives the individual (or subset of individuals) to evaluate and when the "slaves" return the cost values to the "master" processor, which performs mutation (sometimes with crossover), selection, and recombination. The evolution and the convergence of the parallel GA implemented in this way are exactly the same as those in a traditional sequential GA. However, the time savings brought about by the parallelization is roughly proportional to the number of processors involved in the calculation.

With this parallel GA scheme, DFSS optimization can be done much more efficiently. At each generation, FEBI is used to analyze each candidate design. The distribution of the materials, the number and the thickness of the layers, and the electromagnetic properties of the materials can be optimized independently or simultaneously to achieve the desired frequency response.

The FSS considered in this section consists of an inhomogeneous dielectric slab made up of two different materials. The unit cell of the DFSS is volumetrically pixelized into 16×16 small cubes, as shown in Figure 8.10, and a GA is used to optimize the distribution of the two materials in these small cubes so that the desired frequency response can be achieved. The dielectric constant of the two materials, the periodicity of the DFSS and the thickness of the dielectric layer are chosen in advance. In order to get a polarization insensitive frequency response from the DFSS, a rotationally symmetric scheme is enforced on the material distribution, which means that only a quarter of the unit cell needs to be optimized. To improve the speed of convergence, a MGA is used in this example with a population size of 50. A crossover probability of 0.5 is used with no mutations. The iteration is terminated after completion of 100 generations. The goal of the GA in this example is to determine the optimal material distribution in the dielectric slab to provide a single stopband at 100 THz. The cost function of the GA, given by

$$\mathrm{FF} = 2 \sum_{\substack{\text{stopband(s)} \\ T>0.1}} (|T| - 0.1)^{3/2} + \sum_{\substack{\text{passband(s)} \\ \Gamma>0.2}} (|\Gamma| - 0.2)^{3/2} \tag{8.12}$$

is designed to simultaneously minimize the transmission coefficient T in the stopband and reflection coefficient Γ in the passband. The geometry of the DFSS as well as the corresponding scattering response of the periodic structure are shown in Figures 8.11 and 8.12, respectively. Note that the scattering

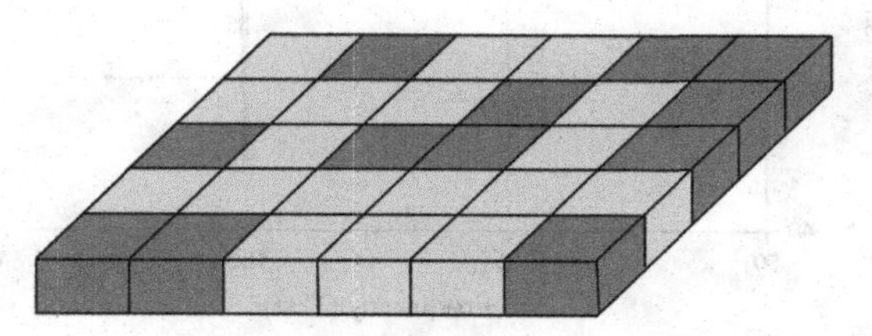

Figure 8.10. Pixelization of the DFSS unit cell.

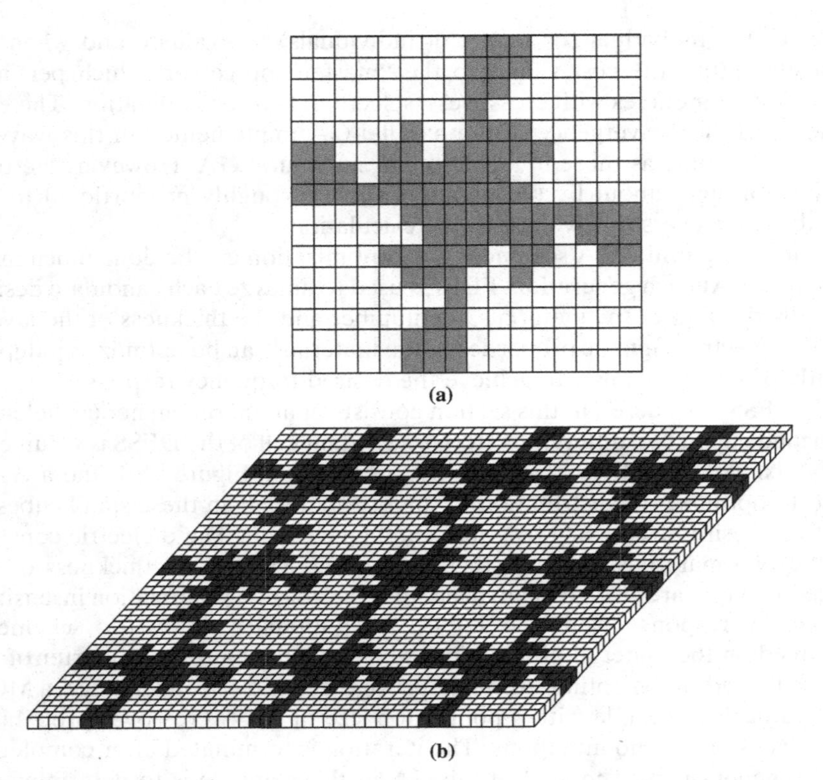

Figure 8.11. Material distribution inside the DFSS unit cell with scattering response shown in Figure 8.12. The dark pixels represent material with relative dielectric constant 1.82, and the light pixels represent material with relative dielectric constant 2.82: (a) unit cell geometry; (b) DFSS screen geometry.

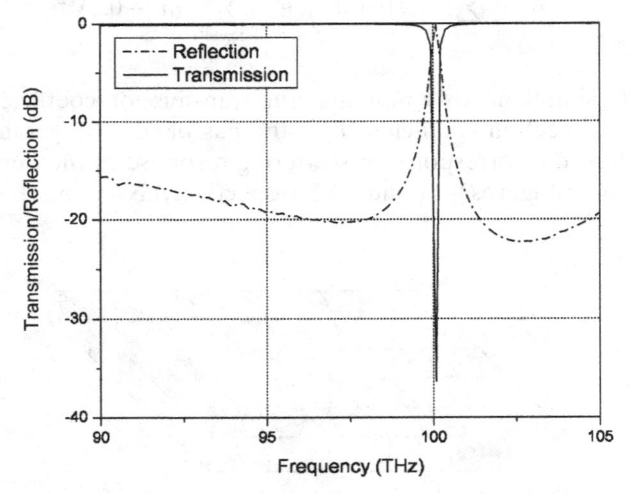

Figure 8.12. Frequency response for both TE and TM polarizations of the DFSS with the unit cell structure shown in Figure 8.11.

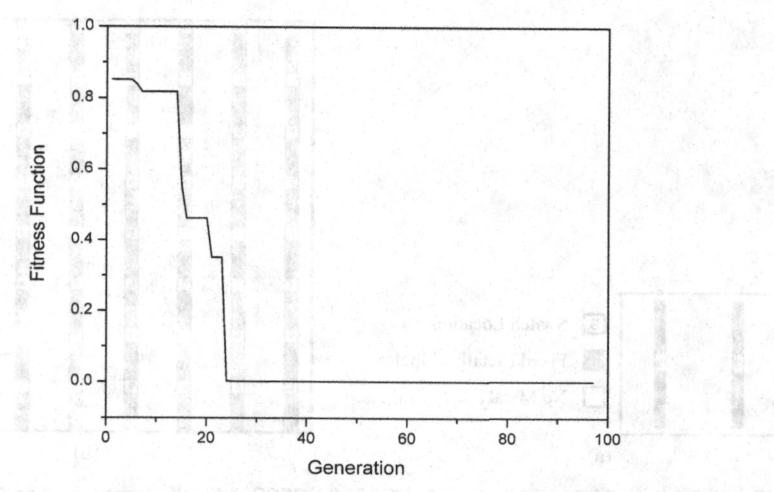

Figure 8.13. *Convergence plot for the MGA optimization of the single stopband filter shown in Figure 8.11.*

responses of the DFSS are the same for both TE and TM (transverse electric and magnetic) polarizations. The periodicity of the DFSS is 2.29 μm, and the thickness of the DFSS is 0.97 μm. The relative dielectric constants of the two materials are 1.82 and 2.82, respectively. Finally, Figure 8.13 shows a convergence plot for the MGA optimization used in this case.

8.2.2 Optimization of Reconfigurable FSSs

For some applications, it is desirable to construct a FSS filter that is capable of changing its filtering characteristics in real time. A methodology for accomplishing this using a FSS screen with fixed metallic features interconnected by switches has been presented [7]. Consider the cell geometry shown in Figure 8.14a. If the short dipoles in the unit cell are connected by ideal switches, then the stopband of the FSS screen will change in frequency when the switches are toggled off and on. When the switches are off, or open, the unit cell has four shorter dipoles, producing a stopband for linearly (e.g., vertically) polarized incident waves at a wavelength approximately twice the length of the dipoles. When the switches are on, or closed, the unit cell consists of two longer dipoles. In this case, the screen will be resonant at a lower frequency when the wavelength is approximately twice the length of the longer dipoles. Thus, the filter response of this FSS is reconfigurable. If we generalize this simple unit cell geometry to a larger array of 8×8 dipoles, connected by a grid of 8×8 independent ideal switches as shown in Figure 8.14b, we can potentially reconfigure the switches to produce a wide variety of desired filter responses for linearly polarized incident waves.

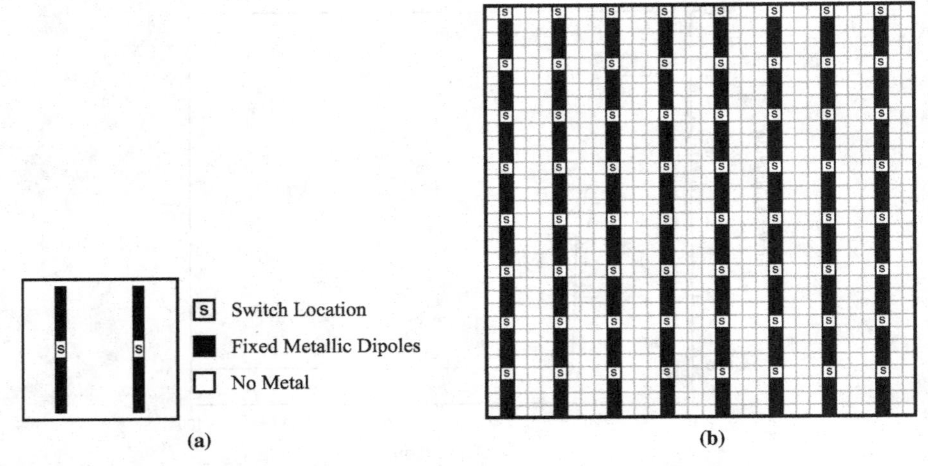

Figure 8.14. *(a) Unit cell for a simple RFSS; (b) RFSS unit cell composed of an 8 × 8 array of linear dipoles (J. A. Bossard et al., © IEEE, 2005 [7]).*

The RFSS illustrated in Figure 8.14b lends itself to binary GA optimization. The only variables in the RFSS that are not fixed are the switch settings. These switch settings are encoded into a chromosome such that a 1 corresponds to the switch being on, or closed, and a 0 corresponds to the switch being off, or open. Thus, the chromosome for the RFSS in Figure 8.14b has 64 bits corresponding to the 64 switches in the unit cell. The dimensions of the RFSS unit cell are chosen to be 32 mm on a side, while the dipoles are 3 mm long and 1 mm wide with a 1 mm gap between each dipole where a switch is placed. The substrate thickness and relative permittivity are assumed to be 0.2 mm and 2, respectively. The scattering from the RFSS is calculated at specified desired stopband and passband frequencies by the PMoM code described in Section 8.2.1. The scattering variables are then used to calculate the cost of each design according to the cost function given in (8.11).

In order to demonstrate the flexibility of this RFSS concept, let us consider the design of a triband filter with the optimal switch settings determined using a MGA. Three desired stopband frequencies were specified at 4, 7, and 9 GHz. In addition, nine passband frequencies were specified elsewhere throughout the band from 1 to 10 GHz. For the MGA, a population of 50 was optimized over 100 generations with a crossover probability of 0.5 and no mutation. The RFSS geometry with optimized switch settings is shown in Figure 8.15. The simulated transmission and reflection spectra shown in Figure 8.16 demonstrate that the MGA was able to evolve a set of switch settings that produce the desired triband response.

This RFSS concept can also be used with alternate geometries such as metallic crossed dipole elements that can produce either the same or different

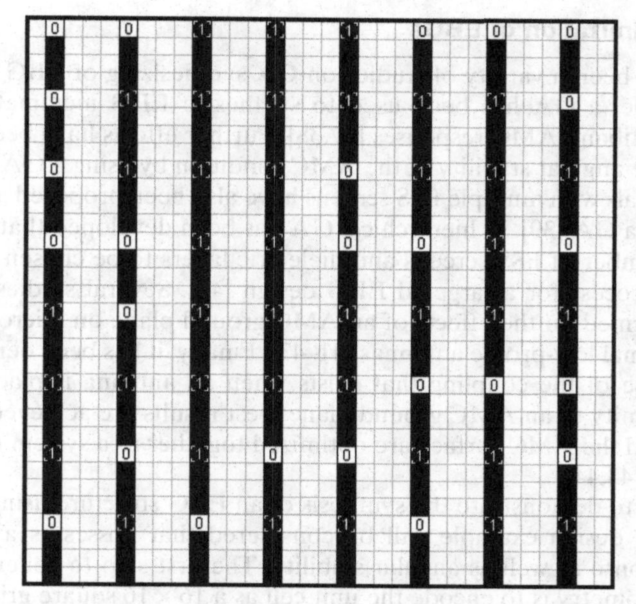

Figure 8.15. *RFSS optimized for triband response (J. A. Bossard et al., © IEEE, 2005 [7]).*

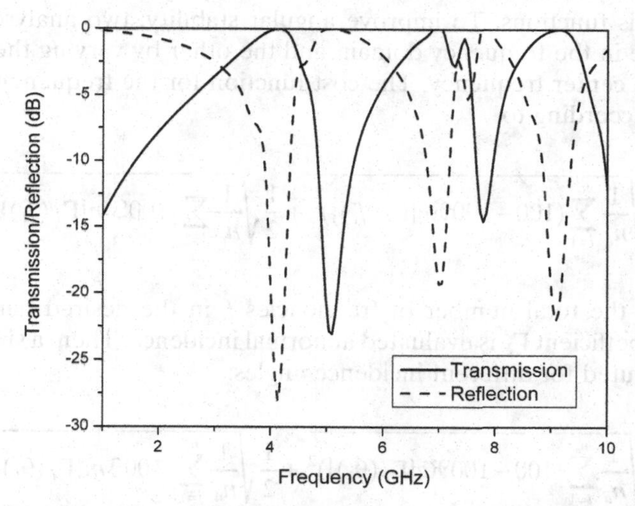

Figure 8.16. *Transmission and reflection spectra for the RFSS unit cell geometry shown in Figure 8.15 (J. A. Bossard et al., © IEEE, 2005 [7]).*

filter responses depending on the polarization of the incident wave (i.e., vertical or horizontal polarizations) as discussed in Ref. 7. Several other examples have also been presented [7] that demonstrate the considerable flexibility of this concept.

8.2.3 Optimization of EBGs

There have been a variety of studies on GA synthesizing of EBG structures. For example, a GA has been used to synthesize EBG metamaterials that exhibit multiband AMC responses [37,38]. Further efforts have been made to improve the angular stability of the AMC condition by using a GA [38]. EBG metamaterials with multiple FSS screens have also been proposed and synthesized using a GA [39]. A hierarchical GA has been developed that allows the optimal number of FSS screens and dielectric layers to be chosen during the synthesis process for a targeted EBG design [40]. Several studies have also been performed on the effects of an AMC ground plane on microstrip patch and conformal low-profile antennas [41–43]. Finally, it has been demonstrated that because of the coupling that exists when an antenna is placed in very close proximity to an AMC ground plane, best results are achieved when the antenna and the AMC surface are optimized together as a system rather than separately [43,44].

In order to demonstrate the synthesis of an EBG structure using GA optimization, a design example will be considered that possesses a multiband AMC response as well as angular stability. The first step to generalizing the unit cell geometry is to encode the unit cell as a 16×16 square grid of pixels, each representing either the presence or absence of metal. The unit cell is analyzed using a full-wave PMoM code that employs Floquet's theorem and rooftop basis functions. To improve angular stability, two analyses are performed: one in the frequency domain and the other by varying the incidence angle at the center frequency. The cost function for the frequency domain is calculated according to

$$F_{\mathrm{TE,TM}}^{\mathrm{freq}} = \frac{1}{2}\sqrt{\frac{1}{n_c}\sum_{i=1}^{n_c}(100-100\Re e\{\Gamma_E(f_i)\})^2} + \frac{1}{2}\sqrt{\frac{1}{n_c}\sum_{i=1}^{n_c}(100\Im m\{\Gamma_E(f_i)\})^2} \tag{8.13}$$

where n_c is the total number of frequencies f_i in the desired band and the reflection coefficient Γ_E is evaluated at normal incidence. Then, a similar quantity is computed for different incidence angles:

$$F_{\mathrm{TE,TM}}^{\mathrm{inc}} = \frac{1}{2}\sqrt{\frac{1}{n_a}\sum_{i=1}^{n_a}(100-100\Re e\{\Gamma_E(\theta_i)\})^2} + \frac{1}{2}\sqrt{\frac{1}{n_a}\sum_{i=1}^{n_a}(100\Im m\{\Gamma_E(\theta_i)\})^2} \tag{8.14}$$

where n_a is the total number of incidence angles θ_i at the central frequency of the desired band. It is useful to note that the separation of the reflection coefficient into real and imaginary parts allows for the maximization of the amplitude of Γ_E in the presence of losses in the dielectric materials or the metallic conducting FSS screen. For both of these cases, the desired value for the reflection coefficient Γ_E is $1 + 0j$.

To obtain the cost for each frequency band, the TE and TM responses are averaged, while a weighted mean is used for the frequency and incidence angle cost data. The cost is calculated as

$$\mathrm{FF} = \sum_{j=1}^{N_{\mathrm{bands}}} \alpha F_j^{\mathrm{freq}} + (1-\alpha) F_j^{\mathrm{inc}} \tag{8.15}$$

where N_{bands} is the number of frequency bands of the EBG that exhibit AMC resonances and $0 \leq \alpha \leq 1$ is the weight assigned to the frequency cost function. The average of the TE and TM responses for the different frequencies of interest and angles of incidence are represented by F_j^{freq} and F_j^{inc}, respectively. The frequency bands optimized for this design include the GPS (Global Positioning System) L_1 frequency of 1.575 GHz and the cellular band covering 1.93–1.99 GHz. In the actual design, the cell phone band is approximated by its center frequency of 1.96 GHz. Five frequency points ($n_c = 5$) were chosen around each band, and three incidence angles ($n_a = 3$) were chosen between normal incidence ($\theta = 0°$) and grazing ($\theta = 90°$). The weighting coefficient for the frequency function was chosen to be $\alpha = 0.65$. A dielectric material with permittivity $\varepsilon_r = 13$ and a thickness value $t = 0.508$ cm has also been imposed, corresponding to a commercially available substrate. Eight-fold symmetry is enforced on the FSS screen as shown in Figure 8.7 in order to ensure the same response for vertical and horizontal polarizations.

A GA was used to optimize a population of 200 members over 200 generations, successfully synthesizing the desired dual-band AMC structure. A unit cell of the FSS screen is shown along with a 5 × 5 tiling of the periodic structure in Figure 8.17. The dimensions of the unit cell were found to be $T_x = T_y = 2.6254$ cm. The

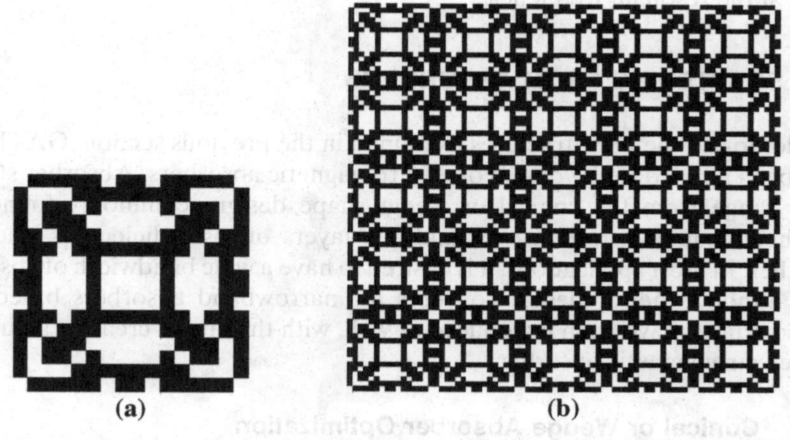

Figure 8.17. EBG AMC surface optimized for triband response; (a) unit cell geometry with black as "metal" and white as "no metal"; (b) 5 × 5 tiling of screen geometry (D. J. Kern et al., © IEEE, 2005 [38]).

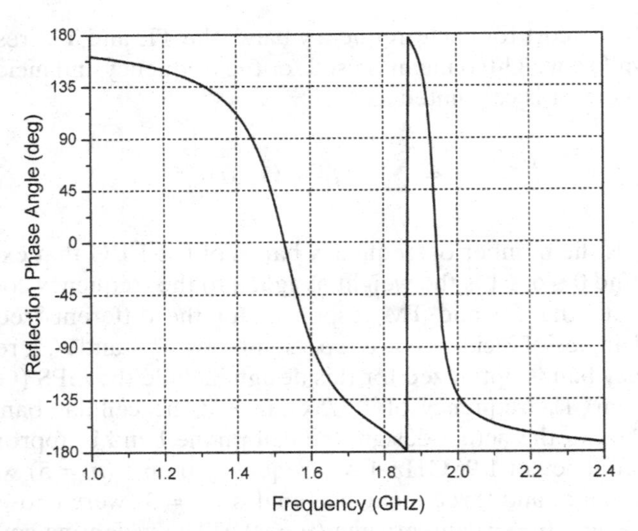

Figure 8.18. *Frequency response of EBG AMC surface shown in Figure 8.17 for normal incidence (D. J. Kern et al., © IEEE, 2005 [38]).*

frequency response shown in Figure 8.18 indicates an AMC condition at the targeted bands of 1.575 and 1.96 GHz. Since the system contains no loss, only the phase of the reflection coefficient is shown. The relative bandwidths, defined by the ±90° phase points, are 8.5% in the first band and 2.14% in the second. The angular performance of the dual-band AMC structure is shown in Figure 8.19. A properly tuned GA optimization procedure can provide very robust solutions with respect to the variation of the incidence angle; in this case, phase values below ±45° were obtained up to an incidence angle of $\theta = 60°$ at the resonant frequencies.

8.3 SCATTERING FROM ABSORBERS

In addition to the FSS structures examined in the previous section, GAs have also been applied to the design of electromagnetic absorbers. Absorber structures range from the conical or wedge-shape designs commonly found in anechoic chambers to single or multiple layers of lossy dielectrics stacked together. In most applications, it is desired to have a wide bandwidth of absorption, although newer design concepts for narrowband absorbers based on EBG surfaces have been introduced as well, with the aim of creating an ultrathin electromagnetic absorber.

8.3.1 Conical or Wedge Absorber Optimization

The effect of a wedge or conical absorber on electromagnetic absorption is dependent on the shape and dielectric properties of the material in the

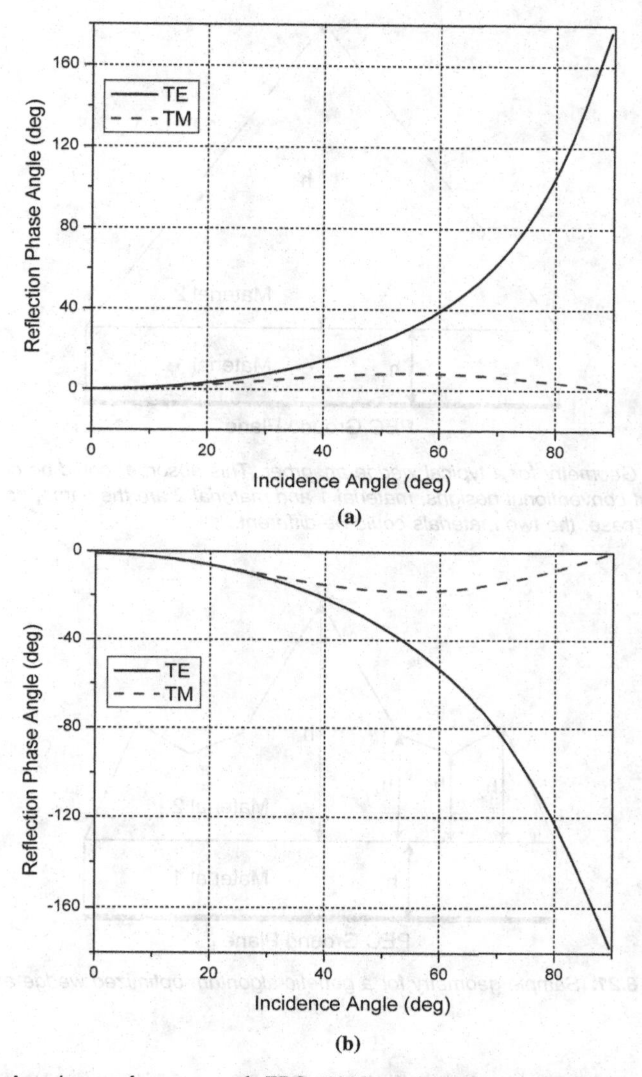

Figure 8.19. *Angular performance of EBG AMC surface shown in Figure 8.17 at (a) f = 1.575 GHz and (b) f = 1.96 GHz (D. J. Kern et al., © IEEE, 2005 [38]).*

frequency band of interest. By altering the shape and dielectric material properties, one can significantly modify the frequency response. A typical wedge absorber is shown in Figure 8.20. In general, the wedge absorber may consist of two dielectric layers with different properties. However, in most conventional designs, the two materials are considered to be the same. A GA is useful in this design because it allows for the optimization of the dielectric properties for maximum absorption, as well as optimizing the total height and width of the wedge shape itself.

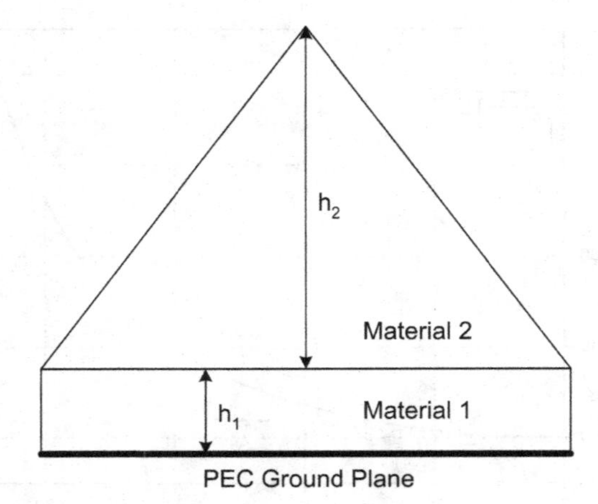

Figure 8.20. *Geometry for a typical wedge absorber. This absorber could be conical or pyramidal. In most conventional designs, material 1 and material 2 are the same; however, for the more general case, the two materials could be different.*

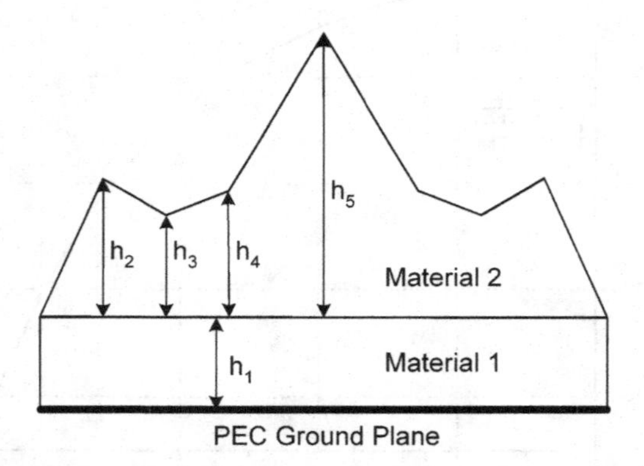

Figure 8.21. *Sample geometry for a genetic-algorithm-optimized wedge absorber.*

An alternative method of utilizing a GA for optimizing wedge absorbers is to allow the wedge shape itself to be modified [45,46]. Rather than using a standard wedge or conical shape, the angle of the wedge can be changed at many points along the absorber. An example of this type of design is shown in Figure 8.21. As can be seen, varying the height of the wedge at various locations allows for a completely different shape of absorber to be designed. This structure can be optimized for performance similar to that of the conventional wedge absorber, but with a smaller maximum height, a narrower width, or both. It is useful to note that the design presented in Figure 8.21 is far from intuitive, which highlights another advantage of the GA when designing these structures. A conical absorber shape that is seemingly random in structure in fact allows for improved absorption, and it would be nearly impossible to arrive at this design shape by means of trial and error. It has been shown that such an absorber designed with the help of a GA can achieve comparable performance to the conventional conical absorber [46].

8.3.2 Multilayer Dielectric Broadband Absorber Optimization

In addition to the conventional wedge absorber designs, it is also possible to create a planar electromagnetic absorbing structure capable of obtaining a reasonably wide bandwidth [47]. Two methods have been used to obtain this type of design: a multilayer dielectric substrate solution consisting of *N* layers of varying dielectrics backed by a PEC ground plane and a multilayer dielectric substrate absorber with multiple FSS screens placed between the dielectric layers.

For the designs utilizing only dielectric substrate layers backed with a ground plane, the optimization procedure is fairly straightforward. The number of layers can be fixed or included as a variable within the GA, and the dielectric variables can be stored in a lookup table for future optimization. The thickness of each layer can be fixed or variable as well. Therefore, by encoding these variables into the GA, a full description of the absorber can be obtained. This type of solution is ideally suited for a GA because the various dielectric layers can be chosen at random for each member of the population. The evolutionary procedure of the GA results in a drastic reduction of time required as compared to testing every possible configuration of dielectric layers. In fact, since there are no gradients or integrals to compute for the choice of the next members of the population, the GA will outperform most other optimization methods. The result is a robust solution to the problem of a broadband, conformal electromagnetic absorber [48].

Additionally, it has been shown that a GA is successful in optimizing a multilayer dielectric absorber with multiple FSS screens embedded throughout the structure. The GA can be used to determine the FSS unit cell size, FSS cell symmetry if desired, the number of dielectric substrates, dielectric properties of each layer such as permittivity and thickness, and the total thickness of the absorber [49]. By incorporating FSS screens within the multiple dielectric layers, an overall reduced thickness of the absorber structure is possible. Furthermore, the ability to optimize multiple FSS screens embedded in multiple dielectrics provides a solid foundation for analyzing absorbers with only one or two layers and FSS screens for narrowband operation.

8.3.3 Ultrathin Narrowband Absorber Optimization

In many practical design situations the physical thickness of the absorber itself is of critical importance. For instance, in many military or radar applications it is desirable to have a radar-absorbing material (RAM) coating for an aircraft to avoid detection [50]. However, when coating any object with an additional material, the thickness and weight of the coating, as well as the durability, are principal concerns. Conical absorbers would be obviously impractical for such applications. While the multilayer dielectric coatings would not be as susceptible to aerodynamic and weather effects, these designs still have considerable weight issues to be dealt with for any application such as RAM coatings.

Therefore, it becomes necessary to investigate the potential solution to such a problem, as well as the significant tradeoffs in the design procedure to allow for a thin, lightweight, conformal RAM coating to be developed. A typical method of reducing the RCS of a structure is to coat the surface with some type of lossy material that can increase the absorption of electromagnetic waves at the operating frequency of interest. A classical absorber design of this type is the Salisbury screen [51], which consists of a resistive metallic screen placed a quarter-wavelength above a ground plane, separated by a dielectric spacer. While such a structure is very simple to build, it suffers from the quarter wavelength restriction on the minimum thickness. A typical Salisbury screen and its corresponding frequency response are shown in Figure 8.22.

It has been shown [52] that the absorber thickness can be considerably reduced by utilizing a superdense dipole surface (also called a "gangbuster FSS" [53]) and a resistive sheet placed above the ground plane. When the FSS is placed correctly above the PEC ground plane the surface acts as an artificial magnetic conductor (AMC). The resistive sheet is then placed directly above the gangbuster FSS to provide loss. However, the complexity of such a design can be reduced by incorporating the resistive sheet directly into the FSS itself [54,55]. By allowing the metallic FSS screen to have large enough loss, it can act in the same manner as the gangbuster FSS with resistive sheet.

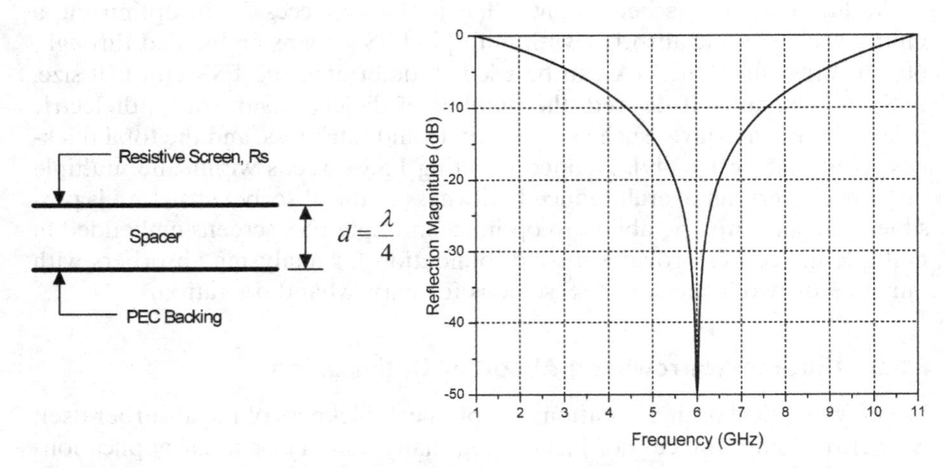

Figure 8.22. *Conventional Salisbury screen with quarter wavelength thickness and the corresponding frequency response when $R_s = 377\,\Omega$.*

The EBG designs from the previous section are of interest in these types of narrowband absorber structures because the optimization procedure is quite similar. Instead of optimizing lossless structures for an AMC surface, the GA must now optimize for minimum reflection phase at the desired resonant frequency to ensure the AMC condition, as well as a minimum in the reflection magnitude to act as an effective absorber. By allowing the FSS screen to have significant loss, the reflection phase curve is quite similar to that of a typical EBG surface described in Section 8.2, yet with a minimum reflection magnitude at the AMC resonance. Hence, effective designs for ultrathin electromagnetic absorbers can be realized through this GA optimization approach. An example cost function for such a design can be expressed as

$$FF = \frac{1}{0.2|\varphi_{max}/180| + 0.8|\Gamma_{max}|} \tag{8.16}$$

where φ_{max} (in degrees) and Γ_{max} are the maximum reflection coefficient phase and magnitude, respectively. The weighting of 0.2 and 0.8 was found to be the most effective in generating the desired results, but other values could also be used.

The question remains as to how much of an improvement such a design actually provides over the conventional Salisbury screen. In many applications, a quarter-wavelength is not a desirable thickness, yet a slightly thinner design will result in significantly degraded performance. For this type of structure to be useful and worth the additional complexity, a dramatic reduction of thickness is required. The exact thickness necessary depends on the specific engineering solution; however, a design that is at least 4 times thinner would represent a reasonable improvement.

The design shown in Figure 8.23 has a total thickness of $\lambda/50$ at the resonant frequency, where λ is the wavelength in the dielectric. This is a reduction of thickness by more than 12 compared to that of the conventional Salisbury screen. The dielectric used in this design has a permittivity of $\varepsilon_r = 1.044$, with a total thickness for the design of $h = 0.952$ mm. Thus, by using a substrate with dielectric properties close to that of air, a dramatic reduction in total thickness is possible. The unit cell geometry and a portion of the FSS screen evolved by the GA are shown in Figure 8.24, where the black and white portions of the screen correspond to the presence and absence of metal, respectively.

Increasing the dielectric permittivity could potentially reduce the overall thickness even further, with the unwanted effect of reducing the already narrow bandwidth. However, an alternative solution for reduced thickness exists. It has been shown [55] that by including a relatively small amount of magnetic loading within the dielectric substrate, it is possible to further reduce the overall thickness without causing a loss of bandwidth or reduced absorption. It is important to note here that the magnetic permeability must not be too large compared to the permittivity of the substrate, as that would result in an unrealistic substrate material. Rather, the dielectric permittivity is

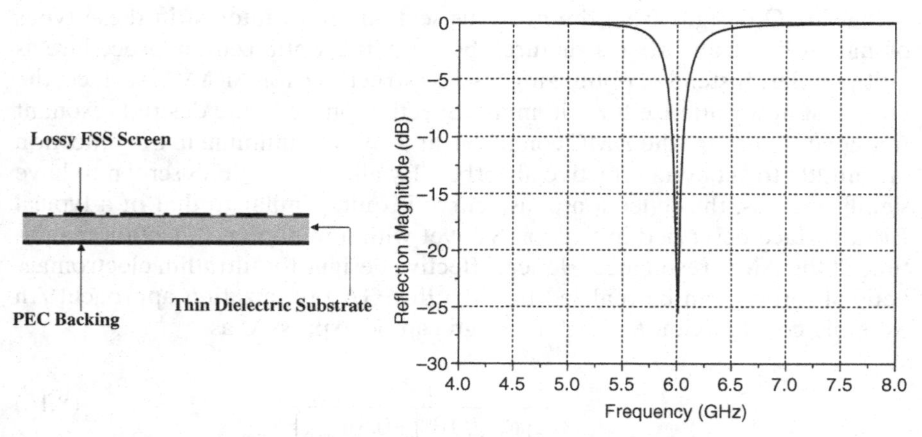

Figure 8.23. Example of a GA-designed EBG absorber and corresponding reflection magnitude versus frequency.

Figure 8.24. Unit cell geometry and a portion of the FSS screen for the sample EBG absorber.

assumed to be considerably greater than the permeability, which is quite common in many high-frequency substrates.

The next design demonstrates this ability by including magnetic permeability as well as dielectric permittivity within the substrate. This design was optimized using a GA with the cost function of (8.16), for a resonant frequency of 2 GHz. The permittivity, permeability, thickness, unit cell size, and resistance of the FSS screen were all included in the optimization variables for the GA. The optimal material properties were found to be unit cell size = 1.45 cm

$\varepsilon_r = 15.12$

$\mu_r = 2.13$

$h = 1.98\,\text{mm}\ (\lambda/76)$

$R = 1.866\,\Omega$

where R is the resistance of the FSS screen. The magnitude and phase of the frequency response are shown in Figure 8.25. It is important to note here that the resonant frequency of 2 GHz is also the frequency of AMC operation, as can be seen by the reflection phase of zero degrees. The unit cell and sample screen geometries are shown in Figure 8.26.

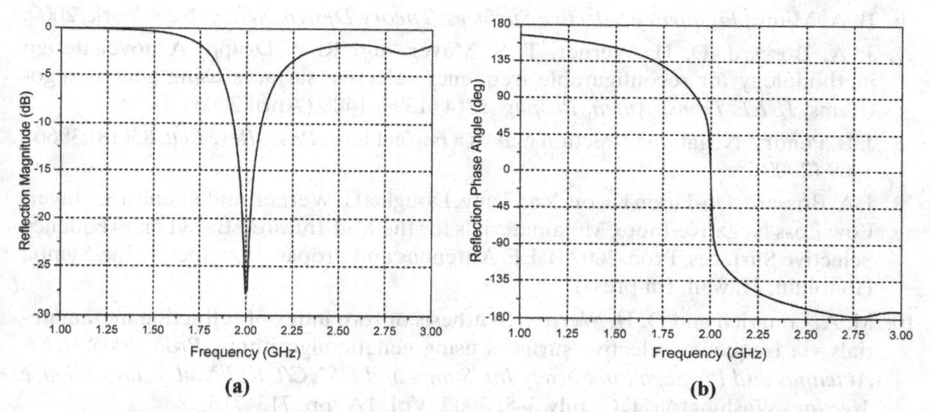

Figure 8.25. Reflection coefficient magnitude and phase response versus frequency.

Figure 8.26. Unit cell geometry and a portion of the FSS screen.

REFERENCES

1. R. Haupt and Y. C. Chung, Optimizing backscattering from arrays of perfectly conducting strips, *IEEE AP-S Maga.* **45**(5):26–33 (Oct. 2003).
2. K. Aygun, D. S. Weile, and E. Michielssen, Design of multilayered periodic strip gratings by GAs, *Microw. Opt. Technol. Lett.* **14**(2):81–85 (Feb. 5, 1997).
3. R. Takenaka, Z. Q. Meng, T. Tanaka, and W. C. Chew, Local shape function combined with genetic algorithm applied to inverse scattering for strips, *Microw. Opt. Technol. Lett.* **16**(6):337–341 (Dec. 20, 1997).
4. A. Qing, An experimental study on electromagnetic inverse scattering of a perfectly conducting cylinder by using the real-coded genetic algorithm, *Microw. Opt. Technol. Lett.* **30**(5):315–320 (Sept. 5, 2001).
5. Y. Zhou and H. Ling, Electromagnetic inversion of Ipswich objects with the use of the genetic algorithm, *Microw. Opt. Technol. Lett.* **33**(6):457–459 (June 20, 2002).
6. B. A. Munk, *Frequency Selective Surfaces: Theory Design*, Wiley, New York, 2000.
7. J. A. Bossard, D. H. Werner, T. S. Mayer, and R. P. Drupp, A novel design methodology for reconfigurable frequency selective surfaces using genetic algorithms, *IEEE Trans. Anten. Propag.* **53**(4):1390–1400 (April 2005).
8. J. B. Pendry, Negative refraction makes a perfect lens, *Phys. Rev. Lett.* **85**(18):3966–3969 (2000).
9. J. A. Bossard, Do-Hoon Kwon, Yan Tang, Douglas H. Werner, and Theresa S. Mayer, Low Loss Negative Index Metamaterials for the Mid-Infrared Based on Frequency Selective Surfaces, Proc. 2007 IEEE Antennas and Propagation Society Int. Symp., Honolulu, Hawaii, (in press).
10. M. A. Gingrich and D. H. Werner, Synthesis of zero index of refraction metamaterials via frequency-selective surfaces using genetic algorithms, *Proc. 2005 IEEE Antennas and Propagation Society Int. Symp. and USNC/URSI Natl. Radio Science Meeting*, Washington, DC, July 3–8, 2005, Vol. 1A, pp. 713–716.
11. M. A. Gingrich and D. H. Werner, Synthesis of low/zero index of refraction metamaterials from frequency selective surfaces using genetic algorithms, *Electron. Lett.* **41**(23):13–14 (Nov. 2005).
12. A. Monorchio, G. Manara, U. Serra, G. Marola, and E. Pagana, Design of waveguide filters by using genetically optimized frequency selective surfaces, *IEEE Microw. Wireless Compon. Lett.* **15**(6):407–409 (June 2005).
13. M. Ohira, H. Deguchi, M. Tsuji, and H. Shigesawa, Multiband single-layer frequency selective surface designed by combination of genetic algorithm and geometry-refinement technique, *IEEE Trans. Anten. Propag.* **52**(11):2925–2931 (Nov. 2004).
14. J. A. Bossard, J. A. Smith, D. H. Werner, T. S. Mayer, and R. P. Drupp, Multiband planar infrared metallodielectric photonic crystals designed using genetic algorithms with fabrication constraints, *Proc. 2005 IEEE Antennas and Propagation Society Int. Symp. and USNC/URSI Natl. Radio Science Meeting*, Washington, DC, July 3–8, 2005, Vol. 1A, pp. 705–708.
15. R. P. Drupp, J. A. Bossard, D. H. Werner, and T. S. Mayer, Single-layer multi-band infrared metallodielectric photonic crystals designed by genetic algorithm optimization, *Appl. Phys. Lett.* **86**:811021–811023 (Feb. 2005).

16. L. Li and D. H. Werner, Design of all-dielectric frequency selective surfaces using genetic algorithms combined with the finite element-boundary integral method, *Proc. 2005 IEEE Antennas and Propagation Society Int. Symp. and USNC/URSI Natl. Radio Science Meeting*, Washington, DC, July 3–8, 2005, Vol. 4A, pp. 376–379.
17. J. A. Bossard, D. H. Werner, T. S. Mayer, J. A. Smith, Y. U. Tang, R. P. Drupp, and L. Li, The design and fabrication of planar multiband metallodielectric frequency selective surfaces for infrared applications, *IEEE Trans. Anten. Propag.* **54**(4):1265–1276 (April 2006).
18. B. Monacelli, J. B. Pryor, B. A. Munk, D. Kotter, and G. D. Boreman, Infrared frequency selective surface based on circuit-analog square loop design, *IEEE Trans. Anten. Propag.* **53**(2):745–752 (Feb. 2005).
19. J. E. Raynolds, B. A. Munk, J. B. Pryor, and R. J. Marhefka, Ohmic loss in frequency-selective surfaces, *J. Appl. Phys.* **93**(9):5346–5358 (May 2003).
20. G. Manara, A. Monorchio, and R. Mittra, Frequency selective surface design based on genetic algorithm, *Electron. Lett.* **35**(17):1400–1401 (Aug. 1999).
21. E. A. Parker, A. D. Chuprin, J. C. Batchelor, and S. B. Savia, GA optimization of crossed dipole FSS array geometry, *Electron. Lett.* **37**(16):996–997 (Aug. 2001).
22. D. S. Weile and E. Michielssen, The use of domain decomposition genetic algorithms exploiting model reduction for the design of frequency selective surfaces, *Comput. Meth. Appl. Mech. Eng.* **186**(4):439–458 (June 2000).
23. Y. Yuan, C. H. Chan, K. F. Man, and R. Mittra, A genetic algorithm approach to FSS filter design, *Proc. 2001 IEEE Antennas and Propagation Society Int. Symp. and USNC/URSI Natl. Radio Science Meeting*, Boston, MA, July 8–13, 2001, Vol. 4, pp. 688–691.
24. S. Chakravarty and R. Mittra, Application of the micro-genetic algorithm to the design of spatial filters with frequency-selective surfaces embedded in dielectric media, *IEEE Trans. Electromagn. Compat.* **44**(2):338–346 (May 2002).
25. T. K. Wu, ed., *Frequency Selective Surface and Grid Array*, Wiley, New York, 1995.
26. R. Mittra, C. H. Chan, and T. Cwik, Techniques for analyzing frequency selective surfaces—a review, *Proc. IEEE*, **76**(12):1593–1615 (Dec. 1988).
27. L. Li, D. H. Werner, J. A. Bossard, and T. S. Mayer, A model-based parameter estimation technique for wideband interpolation of periodic moment method impedance matrices with application to genetic algorithm optimization of frequency selective surfaces, *IEEE Trans. Anten. Propag.* **54**(3):908–924 (March 2006).
28. C. H. Chan and R. Mittra, On the analysis of frequency selective surfaces using subdomain basis functions, *IEEE Trans. Anten. Propag.* **AP-38**:40–50 (Jan. 1990).
29. A. D. Rakić, A. B. Djurišić, J. M. Elazar, and M. L. Majewski, Optical properties of metallic films for vertical-cavity optoelectronic devices, *Appl. Opt.* **37**(22):5271–5283 (Aug. 1998).
30. J. L. Volakis, A. Chatterjee, and L. C. Kempel, *Finite Element Method for Electromagnetics*, IEEE Press, New York, 1998.

31. T. F. Eibert, J. L. Volakis, D. R. Wilton, and D. R. Jackson, Hybrid FE/BI modeling of 3-D doubly periodic structures utilizing triangular prismatic elements and an MPIE formulation accelerated by the Ewald transformation, *IEEE Trans. Anten. Propag.* **47**(5):843–850 (May 1999).
32. T. K. Sarkar, *Application of Conjugate Gradient Method to Electromagnetics Signal Analysis*, Elsevier, 1991.
33. E. Cantu-Paz, *A Survey of Parallel Genetic Algorithms*, IlliGAL Report 97003, Univ. Illinois, 1997 (available online at fttp://ftp-illigal.ge.uiuc.edu/pub/papers/IlliGALs/97003.ps.Z).
34. D. E. Goldberg, *Genetic Algorithm in Search, Optimization and Machine Learning*, Addison-Wesley, New York, 1989.
35. T. C. Fogarty and R. Huang, Implementing the genetic algorithm on transputer based parallel processing systems, in *Parallel Problem Solving from Nature*, Springer-Verlag, Berlin, 1991.
36. E. Cantu-Paz, *Designing Efficient Master-Slave Parallel Genetic Algorithms*, IlliGAL Report 97004, Univ. Illinois, 1997 (available online at fttp://ftp-illigal.ge.uiuc.edu/pub/papers/IlliGALs/97004.ps.Z).
37. D. J. Kern, D. H. Werner, and M. J. Wilhelm, Genetically engineered multiband high-impedance frequency selective surfaces, *Microw. Opt. Technol. Lett.* **38**(5):400–403 (Sept. 2003).
38. D. J. Kern, D. H. Werner, A. Monorchio, L. Lanuzza, and M. J. Wilhelm, The design synthesis of multiband artificial magnetic conductors using high impedance frequency selective surfaces, *IEEE Trans. Anten. Propag.* **53**(1):8–17 (Jan. 2005).
39. A. Monorchio, G. Manara, and L. Lanuzza, Synthesis of artificial magnetic conductors by using multilayered frequency selective surfaces, *IEEE Anten. Wireless Propag. Lett.* **1**(1):196–199 (2002).
40. Y. Yuan, C. H. Chan, K. F. Man, and K. M. Luk, Meta-material surface design using the hierarchical genetic algorithm, *Microw. Opt. Technol. Lett.* **39**(3):226–230 (Nov. 2003).
41. G. Qiang, D.-B. Yan, Y.-Q. Fu, and N.-C. Yuan, A novel genetic-algorithm-based artificial magnetic conductor structure, *Microw. Opt. Technol. Lett.* **47**(1):20–22 (Oct. 2005).
42. J. Yeo, J.-F. Ma, and R. Mittra, GA-based design of artificial magnetic ground planes (AMGs) utilizing frequency-selective surfaces for bandwidth enhancement of microstrip antennas, *Microw. Opt. Technol. Lett.* **44**(1):6–13 (Jan. 2005).
43. D. J. Kern, T. G. Spence, and D. H. Werner, The design optimization of antennas in the presence of EBG AMC ground planes, *Proc. 2005 IEEE Antennas and Propagation Society Int. Symp. and USNC/URSI Natl. Radio Science Meeting*, Washington, DC, July 3–8, 2005, Vol. 3A, pp. 10–13.
44. D. H. Werner, D. J. Kern, and M. G. Bray, Advances in EBG design concepts based on planar FSS structures, *Proc. 2005 Loughborough Antennas and Propagation Conf.* Loughborough Univ. Loughborough, Leicestershire, UK, April 4–5, 2005, pp. 259–262.
45. S. Cui, A. Mohan, and D. S. Weile, Pareto optimal design of absorbers using a parallel elitist nondominated sorting genetic algorithm and the finite element-boundary integral method, *IEEE Trans. Anten. Propag.* **53**(6):2099–2107 (2005).

46. S. Cui and D. S. Weile, Robust design of absorbers using genetic algorithms and the finite element-boundary integral method, *IEEE Trans. Anten. Propag.* **51**:3249–3258 (2003).
47. B. Chambers and A. Tennant, Optimised design of Jaumann radar absorbing materials using a genetic algorithm, *IEEE Proc. Radar Sonar Nav.* **143**(1):23–30 (1996).
48. E. Michielssen, J. Sajer, S. Ranjithan, and R. Mittra, Design of lightweight, broadband microwave absorbers using genetic algorithms, *IEEE Trans. Microw. Theory Tech.* **41**(6/7):1024–1031 (1993).
49. S. Chakravarty, R. Mittra, and N. R. Williams, Application of a microgenetic algorithm (MGA) to the design of broad-band microwave absorbers using multiple frequency selective surface screens buried in dielectrics, *IEEE Trans. Anten. Propag.* **50**(3):284–296 (2002).
50. K. J. Vinoy and R. M. Jha, *Radar Absorbing Materials: From Theory to Design* and *Characterization*, Kluwer, Boston, 1996.
51. R. L. Fante and M. T. McCormack, Reflection properties of the Salisbury screen, *IEEE Trans. Anten. Propag.* **36**(10):1443–1454 (Oct. 1988).
52. N. Engheta, Thin absorbing screens using metamaterial surfaces, *Proc. IEEE AP-S/URSI Symp.* June 2002, Vol. 2, pp. 392–395.
53. S. W. Schneider and B. A. Munk, The scattering properties of "super dense" arrays of dipoles, *IEEE Trans. Anten. Propag.* **42**(4):463–472 (April 1994).
54. D. J. Kern and D. H. Werner, A genetic algorithm approach to the design of ultra-thin electromagnetic bandgap absorbers, *Microw. Opt. Technol. Lett.* **38**:61–64 (2003).
55. D. J. Kern and D. H. Werner, Ultra-thin electromagnetic bandgap absorbers synthesized via genetic algorithms, *Proc. IEEE AP-S/URSI Symp.* June 2003, Vol. 2, pp. 1119–1122.

9

GA Extensions

This chapter provides additional information relevant to GAs. Selecting GA parameters, such as population size and mutation rate, is controversial. Appropriate GA parameters depend on the particular implementation of the algorithm, the final goal, and the cost function. Thus, recommendations vary widely. We present a brief introduction to the subject. Since the GA is a random search, averaging results to arrive at any conclusions about adjusting parameters is mandatory. A brief introduction to particle swarm optimization (PSO) is also given. PSO offers a possible alternative to the GA. Finally, the very important subject of multiple objective optimization is addressed. Many times, there is more than one cost or objective associated with a cost function.

9.1 SELECTING POPULATION SIZE AND MUTATION RATE

The mutation rate and population size for a GA are the major contributions to the convergence speed of a GA. Other operators and parameters affect GA convergence but to a lesser extent. De Jong did the first study of GA parameters versus performance [1]. He tested his GAs on the five cost functions in Table 9.1. After trying various mutation rates, population sizes, crossover rates, and replacement percentages, he deduced that

- A small population size improved performance in early generations, while a large population size improved performance in later generations.

TABLE 9.1. De Jong Test Functions

Function	Limits
$\sum_{n=1}^{3} x_n^2$	$-5.12 \le x_n \le 5.12$
$100(x_1^2 - x_2)^2 + (1 - x_1)^2$	$-2.048 \le x_n \le 2.048$
$\sum_{n=1}^{5} round(x_n)$	$-5.12 \le x_n \le 5.12$
$\sum_{n=1}^{30} nx_n^4 + randn$	$-1.28 \le x_n \le 1.28$
$0.002 + \sum_{m=1}^{25} \frac{1}{m + (x_1 - a_{1m})^6 + (x_2 - a_{2m})^6}$	$-65.536 \le x_n \le 65.536$

- A high mutation rate improved offline performance, while a low mutation rate improved online performance. Offline performance is the best cost found up to the present generation. Online performance is the average of all costs up to the present generation.
- The best crossover rate is approximately 100%.
- The type of crossover was not a factor.

Grefenstette used a GA to optimize parameters for GAs [2]. The six parameters he varied were mutation rate, population size, crossover rate, number of chromosomes replaced, cost normalization, and whether elitism is used. The cost function was a GA that ran until 5000 cost function evaluations were performed on one of the De Jong test functions. After finding the 20 best parameter settings for GAs, he unleashed those GAs on De Jong's test functions over five independent runs. The best online performance resulted when the population size was 30, the crossover rate 0.95, the mutation rate 1%, the cost function was scaled, and elitism was used. He also found that a wide range of parameter settings gave excellent performance.

Another study added five more cost functions and used 8400 possible combinations of parameter settings [3]. The GAs terminated after 10,000 function evaluations and results were averaged over 10 independent runs. The best parameter values for this study were a population size of 20 or 30, a crossover rate between 0.75 and 0.95, a mutation rate of 0.005 or 0.01, and two-point crossover.

Thomas Back postulated that the optimal mutation rate is $1/\ell$, where ℓ is the length of the chromosome [4]. He also found that convergence improved when starting the GA with large mutation rates of up to 50% and gradually decreasing to $1/\ell$. Gao showed that the larger the probability of mutation and the smaller the population, the faster the GA should converge for short-term performance [5]. Iterative approaches where mutation rate varies over the course of a run such as done by Back [6,7] and Davis [8] are probably best, but there is no acceptable strategy as to how to adaptively change the parameters.

Another GA parameter study was done on antenna array factors [9], and the results are summarized here. The objective function returns the maximum sidelobe level of an 18-element array with $d = 0.5\lambda$ spacing for a given set of weights

$$f = \max \text{ sidelobe of} \left\{ \sum_{n=1}^{N} a_n e^{j\delta_n} e^{j(n-0.5)kdu} \right\} \tag{9.1}$$

where N = number of array elements
$a_n e^{j\delta_n}$ = element weights
d = element spacing
u = angle variable

The amplitude taper is symmetric about the center of the array with the two center elements having an amplitude of 1. The algorithm stops whenever the maximum sidelobe level is 25 dB below the peak of the mainbeam or after 50,000 function calls. The results were averaged for population sizes, 4, 8, 12, . . . , 64, and mutation rates 0.01, 0.02, 0.03, . . . , 0.4 after 20 independent runs. Figure 9.1 is a plot of the number of function calls required versus the population size and mutation rate. The best results occur when the population size is between 4 and 16 and the mutation rate is between 0.1 and 0.2. Table 9.2 presents the minimum, maximum, and mean numbers of function calls when the GA is averaged 200 times for eight different combinations of population size and mutation rate. A population size of 4 with a mutation rate of 15% produced the best average results.

The next example finds a low-sidelobe symmetric phase taper for a linear array. Table 9.3 shows the results after running a GA 100 times to find the best

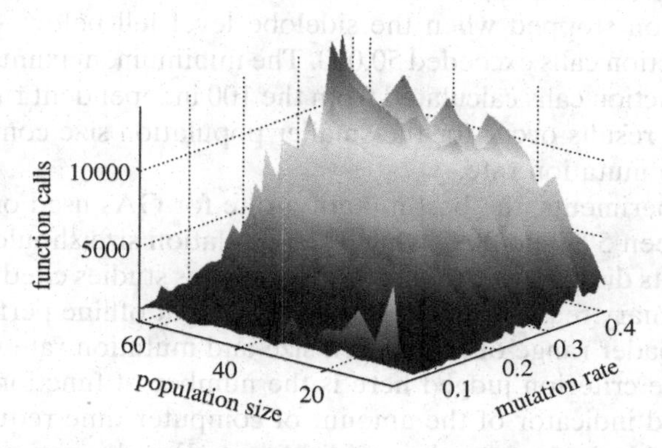

Figure 9.1. Number of function calls needed for the GA to finish for many different combinations of population sizes and mutation rates.

TABLE 9.2. Minimum, Maximum, and Mean Numbers of Function Calls from Running a GA 200 Times to Find Optimum Amplitude Taper for an 18-Element Array that Minimizes the Maximum Sidelobe Level[a]

Mutation Rate	Population Size	Minimum	Maximum	Mean
0.15	4	26	3,114	398
0.20	4	110	50,002	7,479
0.15	8	60	2,457	461
0.20	8	49	2,624	654
0.01	64	300	50,031	1,158
0.02	64	277	11,818	1,028
0.01	128	393	2,535	1,410
0.02	128	1215	50,071	10,208

[a] A single GA run stopped when the sidelobe level went below −25 dB or the number of function calls exceeded 50,000.

TABLE 9.3. Minimum, Maximum, and Mean Numbers of Function Calls from Running a GA 100 Times to Find the Optimum Phase Taper that Minimizes the Maximum Sidelobe Level of a 40-Element Array[a]

Mutation Rate	Population Size	Minimum	Maximum	Mean
0.15	4	134	50,002	2,973
0.20	4	163	50,000	5,232
0.15	8	168	8,223	1,827
0.20	8	124	21,307	3,220
0.01	64	614	50,024	7,914
0.02	64	546	50,036	6,624
0.01	128	955	50,043	4,791
0.02	128	933	50,033	3,942

[a] A single GA run stopped when the sidelobe level went below −14 dB or the number of function calls exceeded 50,000.

phase taper that minimizes the highest sidelobe level of a 40-element array. A GA optimization stopped when the sidelobe level fell below −14 dB or the number of function calls exceeded 50,000. The minimum, maximum, and mean numbers of function calls calculated from the 100 independent runs are shown here. The best results occur for the smaller population size combined with a relatively large mutation rate.

In these experiments, the best mutation rate for GAs used on these problems lies between 5% and 20% while the population size should be less than 16. These results disagree with some of the previous studies cited and common usage. The primary reasons for these results are that offline performance was used and a broader range of population size and mutation rate was included. In addition, the criterion judged here is the number of function evaluations, which is a good indicator of the amount of computer time required to solve the problem. Also, most of the published studies do not perform many independent runs and average the results. Since the GA is a random search, performance measures make sense only after observing many independent runs. An extensive investigation into GA parameters is found in Ref. 10. Another difference here is that the GA stopped when the cost fell below a target value, rather than running the GA for a very large fixed number of function calls. There is no foolproof acceptable way to evaluate GA performance. Counting function evaluations is important in electromagnetics, because calculating the cost tends to take considerable computer time. Be wary of results that prove a certain GA works better than others if the authors do not average results over many independent runs and do not count function evaluations.

9.2 PARTICLE SWARM OPTIMIZATION (PSO)

PSO was inspired by the social behavior of animals, such as bird flocking or fish schooling [11,12]. Like a GA, the PSO begins with a random population matrix. Unlike the GA, PSO does not have operators such as crossover and mutation. The rows in the matrix are called *particles* instead of *chromosomes*. They contain the continuous (not binary) variable values. Each particle moves about the search space with a designated velocity. Updates to the velocities and positions of each particle are based on the local and global best solutions

$$v_{m,n}^{\text{new}} = v_{m,n}^{\text{old}} + \ell_1 \times r_1 \times (p_{m,n}^{\text{local best}} - p_{m,n}^{\text{old}}) + \ell_2 \times r_2 \times (p_{m,n}^{\text{global best}} - p_{m,n}^{\text{old}}) \tag{9.2}$$

$$p_{m,n}^{\text{new}} = p_{m,n}^{\text{old}} + v_{m,n}^{\text{new}} \tag{9.3}$$

where $v_{m,n}$ = particle velocity
$p_{m,n}$ = particle variables
r_1, r_2 = independent uniform random numbers
$\ell_1 = \ell_2$ = learning factors = 2
$p_{m,n}^{\text{local best}}$ = best local solution
$p_{m,n}^{\text{global best}}$ = best global solution

The PSO algorithm updates the velocity vector for each particle and then adds that velocity to the particle position or values. Velocity updates are influenced by both the best global solution associated with the lowest cost ever found by a particle and the best local solution associated with the lowest cost in the present population. If the best local solution has a cost less than the cost of the current global solution, then the best local solution replaces the best global solution. Since velocity is the derivative of position, the particle velocity is similar to the derivative used in local minimizers. The constant ℓ_1 is called the "cognitive parameter." The constant ℓ_2 is called the "social parameter." The advantages of PSO are that it is easy to implement and there are few parameters to adjust.

The PSO is able to tackle tough cost functions with many local minima, just like a GA. Consider the problem of finding the nonuniform spacing of 40 elements in a linear array that yields the lowest maximum sidelobe level. The array lies along the x axis and is symmetric about its center. The objective function is given by

$$af(x) = \text{max sidelobe of}\left\{\sum_{n=1}^{20}\cos(kx_n u)\right\} \tag{9.4}$$

Figure 9.2 shows plots of $p_{m,n}^{\text{local best}}$ and $p_{m,n}^{\text{global best}}$ as well as the population average as a function of generation. $p_{m,n}^{\text{global best}}$ serves the same function as elitism strategy in the GA. The resulting optimum array is illustrated in Figure 9.3. Its corresponding array factor is shown in Figure 9.4. The maximum sidelobe level is −19.95 dB below the peak of the main beam.

Several researchers have used PSO to tackle various electromagnetics problems. Finding optimum sidelobe levels for linear phased arrays [13–15], optimum design of corrugated horns [16], and to search for miniaturized designs for Yagi–Uda arrays [17] are three excellent applications. So far, though, there is no compelling reason to use PSO over a GA other than perhaps for simplicity of the algorithm.

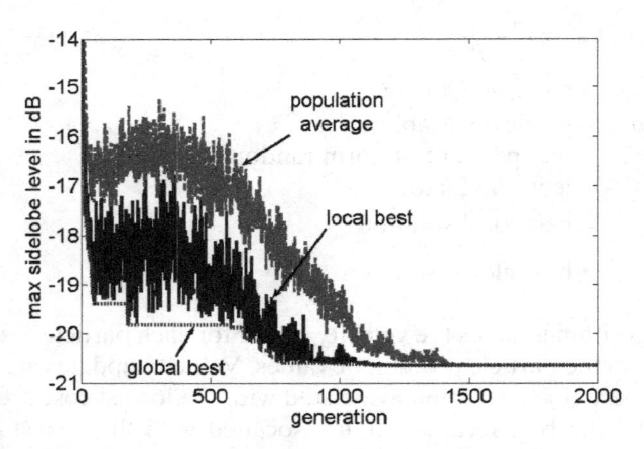

Figure 9.2. Convergence of PSO. The top solid line is the local best, and the bottom solid line is the global best particle after each generation. The bottom dotted line is the best particle found so far by the algorithm.

o o o o oooooooooooooooooooooooooooooooo o o o o ⟶ x in λ

Figure 9.3. The optimum element spacing found by the PSO.

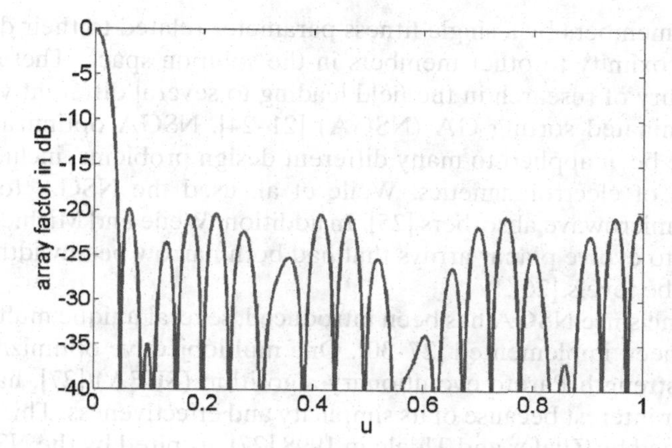

Figure 9.4. The array factor of the optimized array.

9.3 MULTIPLE-OBJECTIVE OPTIMIZATION

9.3.1 Introduction

GAs, simulated annealing, and other global optimization techniques evolve solutions based on a single objective; however, many complex design problems require the optimization of more than one variable. Determining which solutions are best in a multidimensional solution space can be difficult, especially if one does not understand the tradeoffs between the solutions. One of the more common techniques for comparing cases involves specifying a line in the solution space and basing the fitness on an orthographic projection of the solution on that line. This is achieved simply by using a weighted sum of the parameters for the fitness. This method is ideal if a specific ratio or balance between these parameters is desired, such as the methods used to optimize adaptive arrays [18]. In addition, more complicated functions can be used to determine the global fitness, essentially creating a contour mapping of the solution space. These techniques can be very effective if one can fully anticipate the nature of the solution space and understand what tradeoffs are more important.

The main drawback with creating single objective optimization problems from multiobjective problems is that one limits their search to a small sector of the solution space. In an ideal case, engineers would like to know all the solutions that lie on the Pareto front (i.e. the set of all nondominated solutions) so that they can compare the results and choose the case which is best to use. The first attempt at a multiple objective optimization algorithm was the vector-evaluated GA (VEGA), introduced by Schaffer in 1984 [19]. Instead of finding the entire Pareto front, VEGA often finds groups of solutions along the front. The next major advances in this field were the concepts of nondominated sorting and fitness sharing, introduced by Goldberg in 1989 [20]. Nondominated sorting and fitness sharing are used together to ultimately judge population members by a single fitness parameter related to their dominance over and proximity to other members in the solution space. These concepts sparked a fury of research in the field leading to several different versions of the nondominated sorting GA (NSGA) [21–24]. NSGA optimization techniques have been applied to many different design problems, including some in the field of electromagnetics. Weile et al. used the NSGA to optimize broadband microwave absorbers [25]. In addition, Weile and Michielssen used the NSGA to evolve planar arrays that had both narrow beamwidths and low peak sidelobe levels [26].

In the time since NSGA has been introduced, several unique multiobjective GAs have been implemented [27–30]. One multiobjective optimization technique, the strength Pareto evolutionary algorithm (SPEA) [27], has become of particular interest because of its simplicity and effectiveness. This algorithm was developed by Zitzler and Thiele in 1998 [27], inspired by the NSGA work of Horn et al. [23,24] a few years earlier. Like the NSGA, SPEA also ultimately evaluates members by a single fitness value, which they called the "strength"; however, this strength is not based on individual fitness values or proximity to other members but solely on the domination of members of the population over others. In this section, we will introduce SPEA and apply the algorithm to optimize both planar thinned periodic arrays and planar polyfractal arrays.

9.3.2 Strength Pareto Evolutionary Algorithm—Strength Value Calculation

The strength Pareto evolutionary algorithm sorts all solutions into two groups: one group containing the Pareto front (solutions not dominated by any solution) and the other containing all solutions dominated by at least one solution. A member of the population A is considered dominated by another solution B when

$$\text{Fitness}_i(A) < \text{fitness}_i(B) \text{ for every value } i \tag{9.5}$$

After dividing the solutions into groups, the strength of each member of the Pareto front can be found. The strength of a Pareto optimal solution is determined by the number of solutions in the general population that it dominates. The following equation is used to calculate the value of the strength:

$$S_{\text{Pareto}} = \frac{\text{number of population members dominated}}{\text{total number of population members} + 1} \tag{9.6}$$

Finally, the strength of each member of the general population can be found. The strength of a dominated population member can be calculated by knowing the strengths of each Pareto optimal solution that dominates the population member. The calculation of this strength can be represented by

$$S_{\text{population}} = \sum_{\substack{\text{dominating} \\ \text{Pareto solutions}}} S_{\text{Pareto}} + 1 \tag{9.7}$$

Through this process, the solutions that have the smallest strength value are said to be the most optimal. More specifically, a Pareto optimal solution that dominates only a few solutions is ranked better than one that dominates a majority of the solutions. In addition, a solution in the general population that is dominated by a Pareto solution with a small strength is more optimal than one dominated by a Pareto optimal solution with a large strength. The goal of this process is to reduce the number of dominated solutions bunched together by decreasing the fitness of the solutions that are bunched together. The strength values for a sample Pareto front and population are illustrated in Figure 9.5. In this example, adapted from [27], strength values are calculated from two fitness values evaluating each member of the population. The three solutions not dominated by any possible solution make up the Pareto front and the other eight solutions make up the general population.

9.3.3 Strength Pareto Evolutionary Algorithm—Pareto Set Clustering

During an optimization procedure, all members of the Pareto set are carried over to the next generation. This is done because the Pareto set contains the most desirable solutions. In addition, because their fitness values have already been evaluated, it costs nothing computationally to keep these solutions. But because no Pareto optimal solution is thrown away, the Pareto set can become very large and densely populated, especially when the algorithm begins to converge. In such a case, the strength value calculation becomes ineffective in

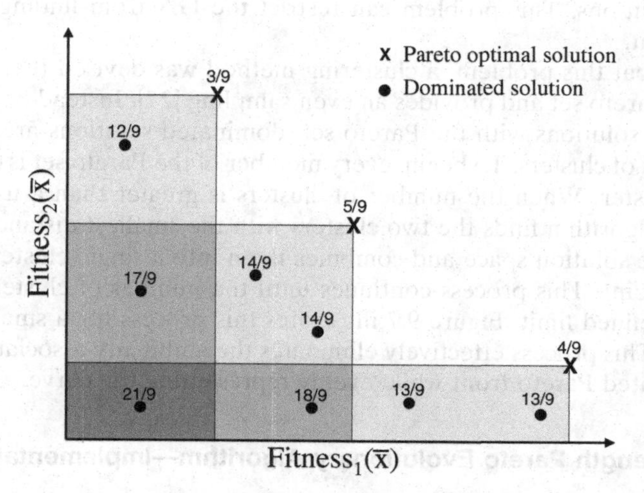

Figure 9.5. Strength values for a sample population and Pareto front evaluated by two fitnesses [27].

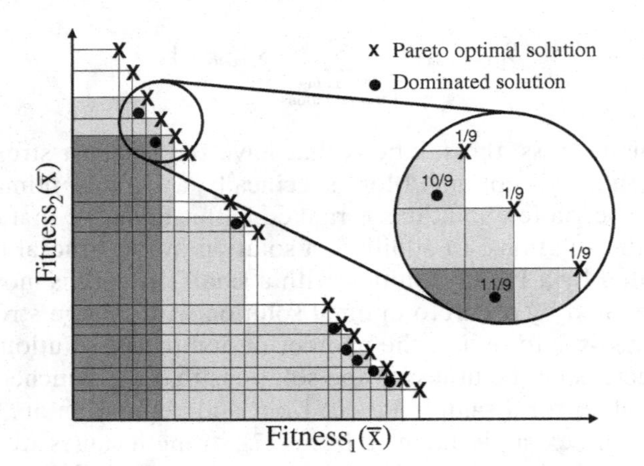

Figure 9.6. *Strength values for a sample population and Pareto front when the Pareto front becomes large in size.*

ranking Pareto optimal and dominated population members. Figure 9.6 illustrates such a case. In this example, there is a large group of dominated solutions near the bottom of the figure, each of which is given the strength of $\frac{10}{9}$. In addition, near the top of the figure are two dominated solutions. Ideally, a few of the bottom solutions would be removed so that the cluster near the top can expand, possibly to fill the gap in the Pareto front directly below it. Instead, it is likely that a second solution near the top with a strength of $\frac{11}{9}$ will be removed because it is the only member of the population dominated by two solutions. This problem can restrict the GA from finding a smooth Pareto front.

To combat this problem, a clustering method was devised that artificially thins the Pareto set and provides an even sampling [27]. Instead of comparing dominated solutions with the Pareto set, dominated solutions are compared with the set of clusters. To begin, every member of the Pareto set is considered to be a cluster. When the number of clusters is greater than a user-defined limit, the algorithm finds the two clusters with the smallest distance between them in the solution space and combines them into a single cluster placed at their midpoint. This process continues until the number of clusters is under the user-defined limit. Figure 9.7 illustrates this process for a small group of solutions. This process effectively eliminates the ambiguity associated with an overpopulated Pareto front while evenly representing the curve.

9.3.4 Strength Pareto Evolutionary Algorithm—Implementation

The description of the strength Pareto evolutionary algorithm's operation begins with the Pareto set and the general population of dominated solutions.

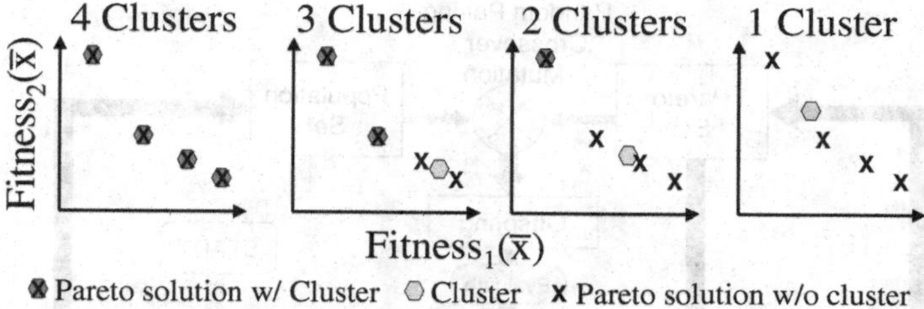

Figure 9.7. *Example of the clustering process introduced in [27].*

These sets of solutions come from the previous generation of the evolutionary algorithm. The first step the algorithm takes is to create a new set of solutions. This is accomplished by first randomly pairing members from the population and Pareto set. After that, the crossover and mutation functions are performed on the chosen members at specified rates. These solutions are then evaluated and placed in an unsorted population of solutions. The Pareto set from the previous generation is also copied to the unsorted population; however, these solutions do not need to be reevaluated because their fitness parameters are known. If this is the initial generation, the unsorted population is filled with random solutions. Next, SPEA determines which members of this unsorted population belong to the Pareto set and which are dominated by Pareto members. These solutions are separated into two different groups, one for the Pareto members and one for the dominated population members. At this point, the Pareto set is copied to create the set of clusters. If there are too many Pareto optimal solutions, the clustering procedure is performed. Finally, the strength of the dominated solutions is evaluated. The size of the population of dominated solutions is reduced to its original size through a natural selection based on the strength parameter. This process, outlined in Figure 9.8, is repeated until the Pareto set no longer improves.

9.3.5 SPEA-Optimized Planar Arrays

The strength Pareto evolutionary algorithm has been used to evolve thinned periodic planar arrays. The optimized arrays are based on a 20×20 square periodic lattice, with 0.5λ spacing between the lattice sites. The strength values are based on three fitness values. The first fitness value is equal to the peak sidelobe level of the array. The second fitness value is based on the conformity of the widest mainbeam cross section to, in this example, a cosine pattern with a 12° half-power beamwidth. The third fitness value is based on the conformity of the narrowest mainbeam cross section to another cosine pattern with an 8.4° half-power beamwidth. This smaller cross section is 70% the width of the maximum cross section. A cosine function is often used to model an element pattern in a linear array

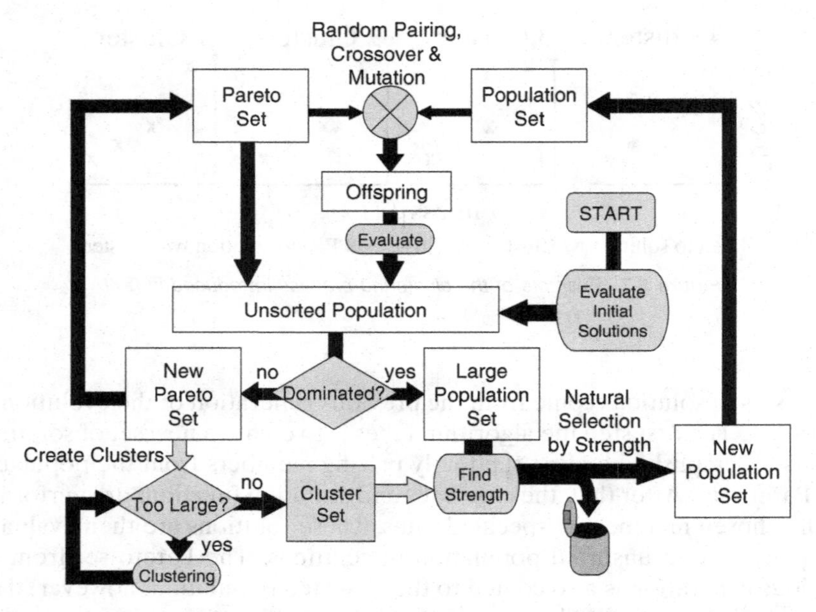

Figure 9.8. *Flowchart of the strength Pareto evolutionary algorithm (SPEA).*

$$f(\theta) = \begin{cases} \cos\dfrac{\pi}{2}\left(\dfrac{\theta}{\delta}\right), & \text{where} \quad \theta \in [-\delta, \delta] \\ 0, & \text{elsewhere} \end{cases} \tag{9.8}$$

where 2δ is the first null beamwidth of the array. SPEA optimizes the population of arrays over 100 generations, adding 300 new arrays to the population on each generation and keeping only 100 dominated solutions for the next generation. The arrays are compared against a maximum of 75 clusters. The final population is illustrated at several angles in Figure 9.9.

From the population of thinned periodic planar arrays shown in Figure 9.9, two Pareto optimal solutions are chosen for discussion. The first example is an optimized 178-element array with a −20.92 dB peak sidelobe level, a maximum half-power beamwidth of 11.5°, and a minimum half-power beamwidth of 7.5°. This solution had the best sidelobe level suppression in the population. The geometry of this array and its respective radiation pattern are shown in Figure 9.10, and the performance parameters of this array are summarized in Table 9.4. The second example is an optimized 164-element array with a peak sidelobe level of −18.97 dB, a maximum half-power beamwidth of 12.0°, and a minimum half-power beamwidth of 7.93°. This array was chosen because the shape of the mainbeam is closest to the desired design goals. In fact, the maximum half-power beamwidth is exactly equal to the design goal of 12°. This array's geometry and its respective radiation pattern are shown in Figure

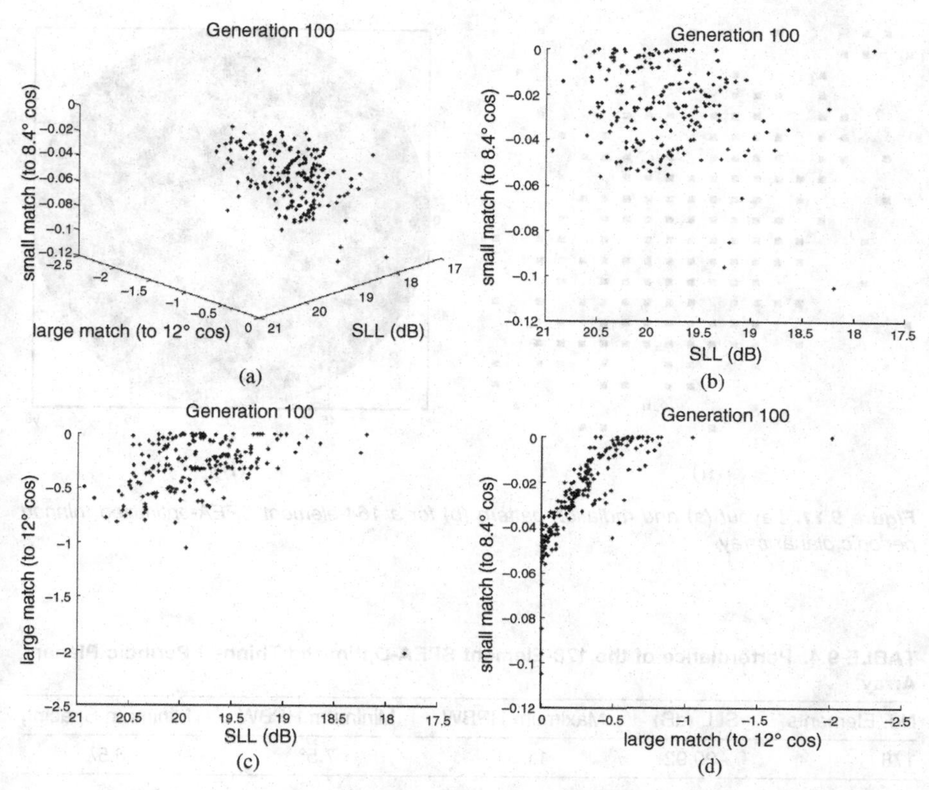

Figure 9.9. *Final three-dimensional solution space of optimized thinned periodic planar arrays found by the strength Pareto evolutionary algorithm. Image (a) shows the solution space in three dimensions, while images (b), (c), and (d) show two-dimensional cross sections of the solution space.*

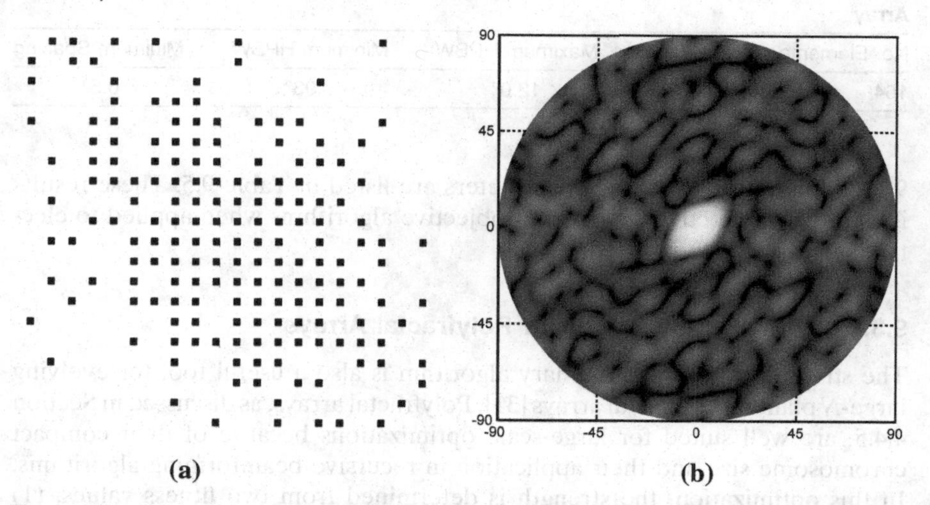

Figure 9.10. *Layout (a) and radiation pattern (b) for a 178-element SPEA-optimized thinned periodic planar array.*

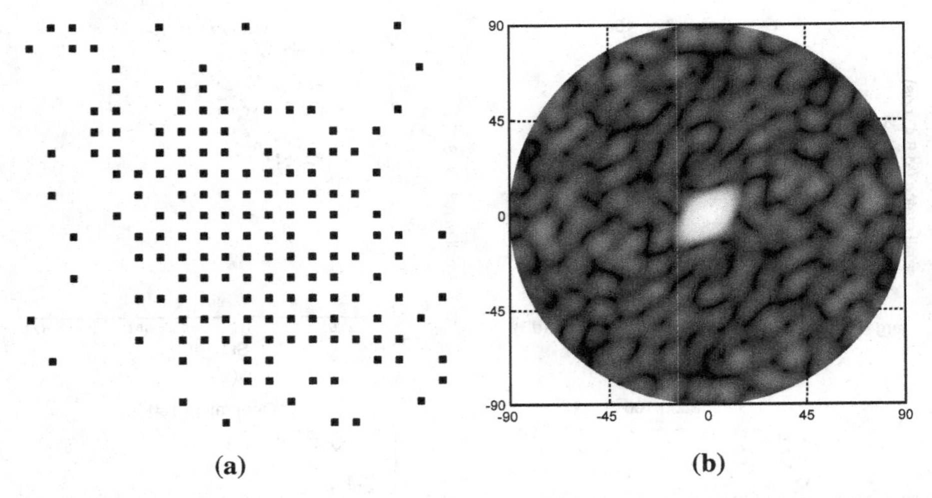

Figure 9.11. Layout (a) and radiation pattern (b) for a 164-element SPEA-optimized thinned periodic planar array.

TABLE 9.4. Performance of the 178-Element SPEA-Optimized Thinned Periodic Planar Array

No. Elements	SLL (dB)	Maximum HPBW	Minimum HPBW	Minimum Spacing
178	–20.92	11.5°	7.5°	0.5λ

TABLE 9.5. Performance of the 164-Element SPEA-Optimized Thinned Periodic Planar Array

No. Elements	SLL (dB)	Maximum HPBW	Minimum HPBW	Minimum Spacing
164	–18.97	12.0°	7.93°	0.5λ

9.11, while the performance parameters are listed in Table 9.5. These results illustrate the effectiveness of multiobjective algorithms when applied to electromagnetic design problems.

9.3.6 SPEA-Optimized Planar Polyfractal Arrays

The strength Pareto evolutionary algorithm is also a useful tool for evolving large-N planar–polyfractal arrays [31]. Polyfractal arrays, as discussed in Section 4.4.5, are well suited for large-scale optimizations because of their compact chromosome size and their application in recursive beamforming algorithms. In this optimization, the strength is determined from two fitness values: (1)

fitness equal to the peak sidelobe level of the array and (2) fitness equal to the ratio of the maximum and minimum half-power beamwidths, with the objective of having this ratio equal to unity. This fitness function keeps the mainbeam highly circular. For the examples considered here, the optimization began with an initial population of two-generator polyfractal arrays. These initial arrays are based on a 6561-element periodic array with 0.5λ interelement spacing. During each generation, 200 new population members are created. A maximum of 30 clusters are used to calculate the strength of each array. Finally, the resulting set of dominated solutions is reduced back to an original value of 100 members. During the optimization process, the number of generators of the arrays is strategically increased several times so that every member of the final Pareto front is constructed from 8 generators. The final Pareto front, illustrated in Figure 9.12, results after 300 generations and consists of arrays between 2115 and 2652 elements.

Two examples of planar polyfractal arrays are chosen from the Pareto front for discussion. The first example is an optimized 2542-element planar polyfractal array with a −18.11 dB sidelobe level and a beam ratio of 93%. This solution was chosen from the part of the Pareto front with the best sidelobe level suppression. This polyfractal array geometry is shown in Figure 9.13 along with its radiation pattern. Performance parameters are listed in Table 9.6. When a more circular beam is desired, a 2192-element array can be chosen from the Pareto front that has a sidelobe level of −16.62 dB and a beam ratio of 96%. The geometry and radiation pattern of this polyfractal array are shown in Figure 9.14, and the performance parameters are listed in Table 9.7. Both arrays have a 0.5λ minimum spacing between elements. Finally, the radiation

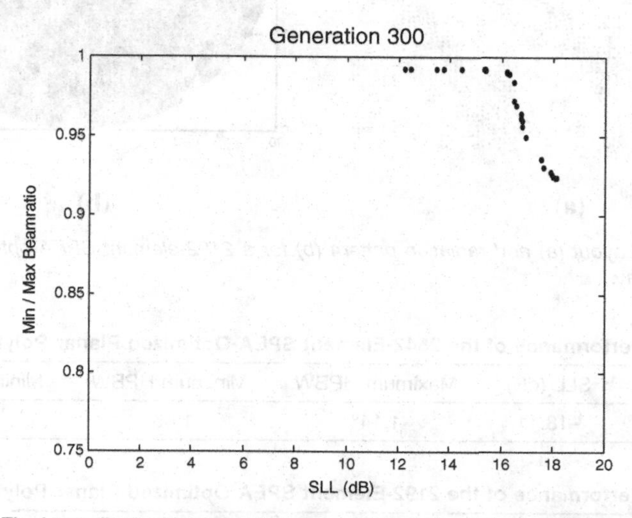

Figure 9.12. *Final two-dimensional solution space of optimized planar polyfractal arrays found by the strength Pareto evolutionary algorithm (SPEA).*

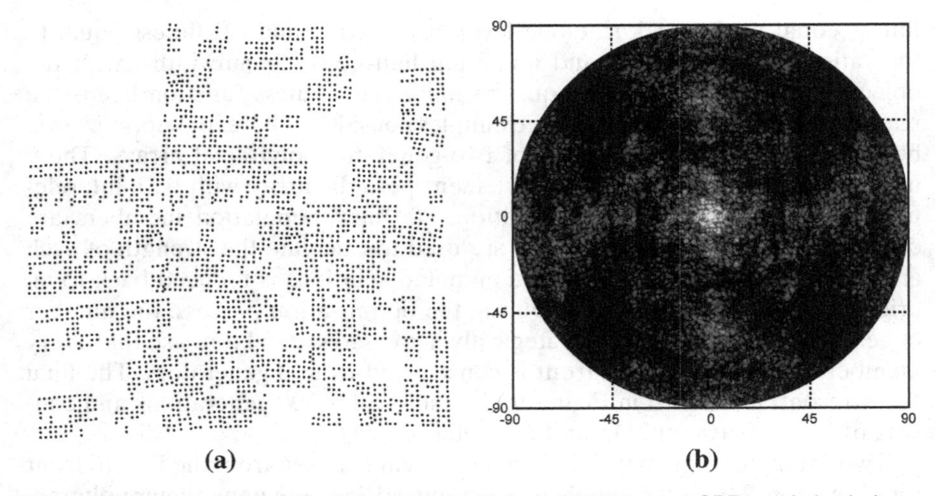

Figure 9.13. Layout (a) and radiation pattern (b) for a 2542-element SPEA-optimized planar polyfractal array.

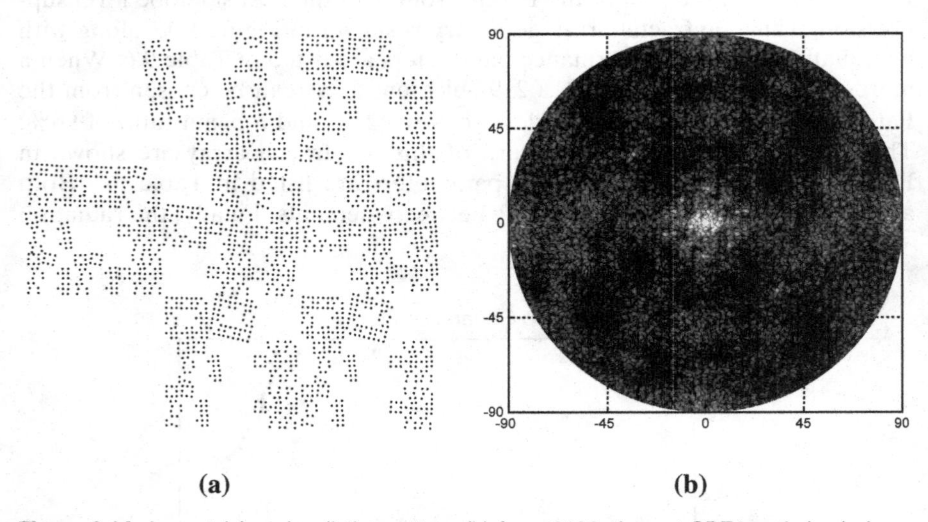

Figure 9.14. Layout (a) and radiation pattern (b) for a 2192-element SPEA-optimized planar polyfractal array.

TABLE 9.6. Performance of the 2542-Element SPEA-Optimized Planar Polyfractal Array

No. Elements	SLL (dB)	Maximum HPBW	Minimum HPBW	Minimum Spacing
2542	–18.11	1.14°	1.06°	0.5λ

TABLE 9.7. Performance of the 2192-Element SPEA Optimized Planar Polyfractal Array

No. Elements	SLL (dB)	Maximun HPBW	Minimum HPBW	Minimum Spacing
2192	–16.62	1.08°	1.03°	0.5λ

patterns of polyfractal arrays can be calculated much more rapidly by taking into account their self-similar structure. For instance, the arrays with two generators of this size can be evaluated with approximately 100 times the speed using recursive beamforming as compared to conventional DFT methods. These results translate into better overall performance and faster convergence of the SPEA algorithm.

REFERENCES

1. K. A. De Jong, *Analysis of the Behavior of a Class of Genetic Adaptive Systems*, PhD dissertation, Univ. Michigan, Ann Arbor, MI, 1975.
2. J. J. Grefenstette, Optimization of control parameters for genetic algorithms, *IEEE Trans. Syst. Man Cybern.* **SMC 16**: 128. (Jan./Feb. 1986)
3. J. D. Schaffer, R. A. Caruana, L. J. Eshelman, and R. Das, A study of control parameters affecting online performance of genetic algorithms for function optimization, *Proc. 3rd Int. Conf. Genetic Algorithms*, Los Altos, CA, June 1989 (D. Schaffer, ed.), Morgan Kaufmann, 1989, pp. 51–60.
4. T. Back, Optimal mutation rates in genetic search, *Proc. 5th Int. Conf. Genetic Algorithms* (S. Forrest, ed.), Morgan Kaufmann, 1993, pp. 2–9.
5. Y. Gao, An upper bound on the convergence rates of canonical genetic algorithms, *Complexity Int.* **5** (1998).
6. T. Back and M. Schutz, Intelligent mutation rate control in canonical genetic algorithms, *Proc. Foundations of Intelligent Systems 9th Int. Symp.* (Z. W. Ras and M. Michalewicz, eds.), Springer-Verlag, Berlin, 1996, pp. 158–167.
7. T. Back, Evolution strategies: An alternative evolutionary algorithm, in *Artificial Evolution*, J. M. Alliot et al., eds., Springer-Verlag, Berlin, 1996, pp. 3–20.
8. L. Davis, Parameterizing a genetic algorithm, in *Handbook of Genetic Algorithms*, L. Davis, ed., Van Nostrand Reinhold, New York, 1991.
9. R. L. Haupt and S. E. Haupt, Optimum population size and mutation rate for a simple real genetic algorithm that optimizes array factors, *Appl. Comput. Electromagn. Soc. J.* **15**(2):94–102, (July 2000).
10. R. L. Haupt and S. E. Haupt, *Practical Genetic Algorithms*, 2nd ed., Wiley, New York, 2004.
11. J. Kennedy and R. C. Eberhart, Particle swarm optimization, *Proc. IEEE Int. Conf. Neural Networks*, IV, IEEE Service Center, Piscataway, NJ, 1995, pp. 1942–1948.
12. J. Kennedy and R. C. Eberhart, *Swarm Intelligence*, Morgan Kaufmann, San Francisco, 2001.
13. D. W. Boeringer and D. H. Werner, Particle swarm optimization versus genetic algorithms for phased array synthesis, *IEEE AP-S Trans.* **52**(3):771–779, (March 2004).
14. D. W. Boeringer and D. H. Werner, Efficiency-constrained particle swarm optimization of a modified Bernstein polynomial for conformal array excitation amplitude synthesis, *IEEE AP-S Trans.* **53**(8):2662–2673, (Aug. 2005).
15. M. M. Khodier and C. G. Christodoulou, Linear array geometry synthesis with minimum sidelobe level and null control using particle swarm optimization, *IEEE AP-S Trans.* **53**(8):2674–2679, (Aug. 2005).

16. J. Robinson and Y. Rahmat-Sami, Particle swarm optimization in electromagnetics, *IEEE AP-S Trans.* **52**(2):397–407 (Feb. 2004).
17. Z. Bayraktar, D. H. Werner, and P. L. Werner, The design of miniature three-element stochastic Yagi-Uda arrays using particle swarm optimization, *IEEE Anten. Wireless Propag. Lett.* **5**(1):22–26 (Dec. 2006).
18. J. M. Johnson, Genetic algorithm design of a switchable shaped beam linear array with phase only control, *Proc. 1999 IEEE Aerospace Conf.* March 1999, Vol. 3, pp. 297–303.
19. J. D. Schaffer, *Multiple Objective Optimization with Vector Evaluated Genetic Algorithms*, PhD thesis, Vanderbilt Univ., 1984.
20. D. E. Goldberg, *Genetic Algorithms in Search, Optimization & Machine Learning*, Addison-Wesley, Reading, MA, 1989.
21. C. M. Fonseca and P. J. Fleming, Genetic algorithms for multiobjective optimization: formulation, discussion and generalization, *Proc. 5th Int. Conf. Genetic Algorithms*, Univ. Illinois at Urbana–Champaign and Morgan Kauffman, San Mateo, CA, 1993, pp. 416–423.
22. N. Srinivas and K. Deb, Multiobjective optimization using nondominated sorting in genetic algorithms, *Evolut. Computa.* **2**:221–248 (1994).
23. J. Horn, *The Nature of Niching, Genetic Algorithms and the Evolution of Optimal Cooperative Populations*, PhD thesis, Dept. Mathematics and Computer Science, Univ. Illinois and Urbana–Champaign, 1997.
24. J. Horn, N. Nafpliotis, and D. E. Goldberg, A niched Pareto genetic algorithm for multiobjective optimization, *Proc. 1st IEEE Conf. Evolutionary Computation*, IEEE World Congress on Computational Intelligence, Piscataway, NJ, June 1994, Vol. 1, pp. 82–87.
25. D. S. Weile, E. Michielssen, and D. E. Goldberg, Genetic algorithm design of Pareto optimal broadband microwave absorbers, *IEEE Trans. Electromagn. Compat.* **38**(3):518–525 (Aug. 1996)
26. D. S. Weile and E. Michielssen, Integer coded Pareto genetic algorithm design of constrained antenna arrays, *IEE Electron. Lett.* **32**(19):1744–1745 (Sept. 1996).
27. E. Zitzler and L. Thiele, *An Evolutionary Algorithm for Multiobjective Optimization: The Strength Pareto Approach*, TIK-Report, Swiss Federal Institute of Technology, Zurich, May 1998.
28. J. Knowles and D. Corne, The Pareto archived evolution strategy: A new baseline algorithm for Pareto multiobjective optimisation, *Proc. 1999 Congress on Evolutionary Computation*, Washington, DC, July 1999, pp. 98–105.
29. D. A. Van Veldhuizen, *Multiobjective Evolutionary Algorithms: Classifications, Analyses, and New Innovations*, PhD dissertation, Air Force Institute of Technology, June 1999.
30. M. Laumanns, G. Rudolph, and H. Schwefel, A spatial predator-prey approach to multi-objective optimization: A preliminary study, in *Parallel Problem Solving from Nature—PPSN V*, A. E. Eiben, M. Schoenauer, and H.-P. Schwefel, eds., Springer-Verlag, Amsterdam, 1998, pp. 241–249.
31. J. S. Petko and D. H. Werner. Pareto optimization of planar fractal-random arrays using the strength pareto evolutionary algorithm, Proc. 2005 IEEE Antennas and Propagation Society International Symposium, Washington, DC, vol. 2A, pp. 57–60, July 3–8, 2005.

Appendix

MATLAB Code

There are two GAs in this appendix. The first is a binary GA using uniform crossover and tournament selection. The second is a continuous GA using single-point crossover and roulette wheel selection. You are encouraged to try other crossover and mutation operators. The population size and mutation rates are easily changed. Stopping criteria include the maximum number of iterations, maximum number of function calls, and a minimum cost.

The cost function can be changed by placing the name of the MATLAB® function between the single quotes in the line `ff=''`. It returns a column cost vector of costs. Three cost functions are provided here. The first two are the mathematical functions

$$f_1(\mathbf{x}) = \sum_{n=1}^{nvar} x_n \tag{A.1}$$

$$f_2(\mathbf{x}) = 60 + \sum_{n=1}^{nvar} \left[x_n^2 - 10\cos(2\pi x_n) \right] \tag{A.2}$$

Both have a minimum of zero when all $x_n = 0$. The third cost function returns the maximum sidelobe level for an amplitude weighted array factor. Its equation is written as

$$af(u) = \sum_{n=1}^{N} a_n e^{jkx_n u} \tag{A.3}$$

Genetic Algorithms in Electromagnetics, by Randy L. Haupt and Douglas H. Werner

For this function the amplitude weights are assumed symmetric, so *nvar* = *N*/2. This cost function may be tested using thinning and the binary GA or amplitude tapering and the continuous GA.

The MATLAB code is as follows:

```
% bga
%
% This is a typical GA that works with binary
variables
% Uses - uniform crossover
%      - tournament selection
%
% Randy Haupt
% 11/21/05

clear
global funcount
funcount=0;

% Defining cost function
nvar=6;
nbits=3;
Nt=nvar*nbits;
ff='testfun3';

% GA parameters
npop=8; % population size
mutrate=0.2; % mutation rate
el=1; % number of chromosomes not mutated
Nmut=ceil(mutrate*((npop-el)*Nt)); %# mutations
% stopping criteria
maxgen=400; % max # generations
maxfun=2000; % mas # function calls
mincost=-50; % acceptable cost

% initial population
P=round(rand(npop,Nt));

% cost function
cost=feval(ff,P);
[c,in]=min(cost);
tp=P(1,:); tc=cost(1);
P(1,:)=P(in,:); cost(1)=cost(in);
P(in,:)=tp; cost(in)=tc;
minc(1)=min(cost); % best cost in each generation
```

```
for gen=1:maxgen

  % Natural selection
  indx=find(cost<=mean(cost));
  keep=length(indx);
  cost=cost(indx); P=P(indx,:);
  M=npop-keep;

  % Create mating pool using tournament selection
  Ntourn=2;
  for ic=1:M
    rc=ceil(keep*rand(1,Ntourn));
    [c,ci]=min(cost(rc)); % indicies of mother
    ma=rc(ci);
    rc=ceil(keep*rand(1,Ntourn));
    [c,ci]=min(cost(rc)); % indicies of father
    pa=rc(ci);
    % generate mask
    mask=round(rand(1,Nt));
    % crossover
    P(keep+ic,:)=mask.*P(ma,:)+not(mask).*P(pa,:);
  end

  % Mutation
  elP=P(el+1:npop,:);
  elP(ceil((npop-el)*Nt*rand(1,Nmut)))=round(rand(1,
   Nmut));
  P(el+1:npop,:)=elP;

  % cost function
  cost=feval(ff,P);
  [c,in]=min(cost);
  tp=P(1,:); tc=cost(1);
  P(1,:)=P(in,:); cost(1)=cost(in);
  P(in,:)=tp; cost(in)=tc;

  minc(gen+1)=cost(1);
  [gen cost(1)]
  % Convergence check
  if funcount>maxfun | gen>maxgen | minc(gen+1)
   <mincost
    break
  end

end
```

```
% Present results
day=clock;
disp(datestr(datenum(day(1),day(2),day(3),day(4),day(5),day(6)),0))
disp(['optimized function is ' ff])
format short g
disp(['# variables = ' num2str(nvar) ' # bits = '
  num2str(nbits)])
disp(['min cost = ' num2str(mincost)])
disp(['best chromosome = ' num2str(P(1,:))])

figure(1)
plot([0:gen],minc)
xlabel('generation');ylabel('cost')
```

```
% cga
%
% This is a typical GA that works with continuous
variables
% Uses - single point crossover
%      - roulette wheel selection
%
% Randy Haupt
% 11/21/05

clear
global funcount
funcount=0;

% Defining cost function
nvar=10;
ff='testfun3';

% GA parameters
npop=8; % population size
mutrate=0.15; % mutation rate
natsel=npop/2; % #chromosomes kept
M=npop-natsel; % #chromosomes discarded
el=1; % number of chromosomes not mutated
Nmut=ceil(mutrate*((npop-el)*nvar)); %# mutations
parents=1:natsel; % indicies of parents
prob=parents/sum(parents); % prob assigned to parents
odds=[0 cumsum(prob)]; Nodds=length(odds); % cum prob
% stopping criteria
```

```
maxgen=500; % max # generations
maxfun=2000; % mas # function calls
mincost=-50; % acceptable cost

% initial population
P=rand(npop,nvar);

% cost function
cost=feval(ff,P);
% Natural selection
[cost ind]=sort(cost);
P=P(ind(1:natsel),:);
cost=cost(1:natsel);

minc(1)=min(cost); % best cost in each generation

for gen=1:maxgen

  % Create mating pool
  for ic=1:2:M
      r=rand;ma=max(find(odds<r)); % indicies of mother
      r=rand;pa=max(find(odds<r)); % indicies of father
      xp=ceil(rand*nvar);          % crossover point
      r=rand;                      % mixing parameter
      xy=P(ma,xp)-P(pa,xp);        % mix from ma and pa
      % generate masks
      mask1=[ones(1,xp) zeros(1,nvar-xp)];
      mask2=not(mask1);
      % crossover
      P(natsel+ic,:)=mask1.*P(ma,:)+mask2.*P(pa,:);
      P(natsel+ic+1,:)=mask2.*P(ma,:)+mask1.*P(pa,:);
      % create single point crossover variable
      P(natsel+ic,xp)=P(natsel+ic,xp)-r*xy;
      P(natsel+ic+1,xp)=P(natsel+ic+1,xp)+r*xy;
  end

  % Mutation
  elP=P(el+1:npop,:);
  elP(ceil((npop-el)*nvar*rand(1,Nmut)))=rand(1,Nmut);
  P(el+1:npop,:)=elP;

  % cost function
  cost=feval(ff,P);
  % Natural selection
  [cost ind]=sort(cost);
```

```
   P=P(ind(1:natsel),:);
   cost=cost(1:natsel);

   minc(gen+1)=cost(1);
   [gen cost(1)]
   % Convergence check
   if funcount>maxfun | gen>maxgen | minc(gen+1)<
     mincost
      break
   end

end

% Present results
day=clock;
disp(datestr(datenum(day(1),day(2),day(3),day(4),day(5),day(6)),0))
disp(['optimized function is ' ff])
format short g
disp([' # par = ' num2str(nvar)])
disp(['min cost = ' num2str(mincost)])
disp(['best chromosome = ' num2str(P(1,:))])

figure(1)
plot([0:gen],minc)
xlabel('generation');ylabel('cost')

% a test function with one local minima at xn=0
%
% Randy Haupt
% 11/21/05

function sll=testfun1(chrom)
global funcount

[nr,nc]=size(chrom);
funcount=funcount+nr; % keeps counting number of
  function calls
bb=10*(chrom-.5); % transforms chromosome variables

sll(:,1)= bb.^2*ones(nc,1);

% a test function with many local minima at xn=0
%
% Randy Haupt
% 11/21/05
```

```
function sll=testfun2(chrom)
global funcount

[nr,nc]=size(chrom);
funcount=funcount+nr; % keeps counting number of
  function calls
bb=10*(chrom-.5); % transforms chromosome variables

sll(:,1)=10*nc+[bb.^2-10*cos(2*pi*bb)]*ones(nc,1);


% a test function for cga - place a null with phase
only weighting
%
% Randy Haupt
% 11/21/05

function sll=testfun3(chrom)
global funcount

[nr,nc]=size(chrom);
funcount=funcount+nr; % keeps counting number of
  function calls

k=2*pi; % wavenumber
d=0.5; % element spacing
N=2*nc; % number of elements in array
x=(0:(N-1))*d; % element spacing
u=0:2/10/N:1; % u=cos(phi)
Q=exp(j*k*x'*u); % phase

for ic=1:nr
    w=[fliplr(chrom(ic,:)) chrom(ic,:)]; % amplitude
      weights
    af=20*log10(abs(w*Q)).'; % array factor in dB
    af=af-max(af); % normalize array factor
    saf=flipud(sort(af));
    ind=min(find(saf>af));
    sll(ic,1)=saf(ind(1)); % max sidelobe level
end
```

Bibliography

Bibliography is arranged chronologically by topic.

1. Antenna Arrays

R. L. Haupt, J. J. Menozzi, and C. J. McCormack, Thinned arrays using genetic algorithms, *Proc. IEEE AP-S Int. Symp.*, June 1993, pp. 712–715.

R. L. Haupt, Thinned arrays using genetic algorithms, *IEEE AP-S Trans.* **42**:993–999 (July 1994).

A. Tennant, M. M. Dawoud, and A. P. Anderson, Array pattern nulling by element position perturbations using a genetic algorithm, *Electron. Lett.* **30**:174–176 (Feb. 1994).

M. Shimizu, Determining the excitation coefficients of an array using genetic algorithms, *Proc. IEEE AP-S Int. Symp.*, June 1994, pp. 530–533.

D. J. O'Neill, Element placement in thinned arrays using genetic algorithms, *Proc. OCEANS '94*, Sept. 1994, pp. 301–306.

R. L. Haupt, An introduction to genetic algorithms for electromagnetics, *IEEE AP-S Mag.* **37**:7–15 (April 1995).

R. Haupt, Optimization of subarray amplitude tapers, *Proc. IEEE AP-S Int. Symp.*, June 1995, pp. 1830–1833.

R. L. Haupt, Optimum quantised low sidelobe phase tapers for arrays, *Electron. Lett.* **31**:1117–1118 (July 1995).

D. Marcano, F. Duran, and O. Chang, Synthesis of multiple beam linear antenna arrays using genetic algorithms, *Proc. IEEE AP-S Int. Symp.*, July 1995, pp. 328–332.

B. Chambers, A. P. Anderson, and R. J. Mitchell, Application of genetic algorithms to the optimization of adaptive antenna arrays and radar absorbers, *Proc. 1st Int.*

Genetic Algorithms in Engineering Systems: Innovations and Applications, Sept. 1995, pp. 94–99.

D. Marcano, M. Jimenez, F. Duran, and O. Chang, Synthesis of antenna arrays using genetic algorithms, *Proc. IEEE Int. Caracus Conf. Devices, Circuits, and Systems*, Dec. 1995, pp. 328–332.

F. Ares, S. R. Rengarajan, E. Villaneuva, E. Skochinski, and E. Moreno, Application of genetic algorithms and simulated annealing technique in optimising the aperture distributions of antenna array patterns, *Electron. Lett.* **32**:148–149 (Feb. 1, 1996).

R. L. Haupt, Genetic algorithm design of antenna arrays, *Proc. IEEE Aerospace Applications Conf.*, Feb. 1996, pp. 103–109.

D. Marcano, M. Jiminez, and O. Chang, Synthesis of linear array using Schelkunoff's method and genetic algorithms, *Proc. IEEE AP-S Int. Symp.*, July 1996, pp. 814–817.

F. Ares, S. R. Rengarajan, E. Villaneuva, E. Skochinski, and E. Moreno, Application of genetic algorithms and simulated annealing technique in optimizing the aperture distributions of antenna arrays, *Proc. IEEE AP-S Int. Symp.*, July 1996, pp. 806–809.

M. J. Buckley, Linear array synthesis using a hybrid genetic algorithm, *Proc. IEEE AP-S Int. Symp.*, July 1996, pp. 584–587.

D. S. Weile and E. Michielssen, Integer coded Pareto genetic algorithm design of constrained antenna arrays, *Electron. Lett.* **32**:1744–1745 (Sept. 12, 1996).

R. J. Mitchell, B. Chambers, and A. P. Anderson, Array pattern synthesis in the complex plane optimised by a genetic algorithm, *Electron. Lett.* **32**:1843–1845 (Sept. 26, 1996).

A. Alphones and V. Passoupathi, Null steering in phased arrays by positional perturbations: A genetic algorithm approach, *Proc. 3rd Int. Conf. High Performance Computing*, Dec. 1996, pp. 4–9.

R. L. Haupt and S. E. Haupt, Phase-only adaptive nulling with a genetic algorithm, *Proc. IEEE Aerospace Applications Conf.*, Feb. 1997, pp. 151–160.

R. J. Mitchell, B. Chambers, and A. P. Anderson, Array pattern control in the complex plane optimised by a genetic algorithm, *Proc. 10th Int. Conf. Antennas and Propagation*, April 1997, pp. 330–333.

R. L. Haupt, Phase-only adaptive nulling with a genetic algorithm, *IEEE AP-S Trans.* **45**:1009–1015 (June 1997).

W. P. Liao and F. L. Chu, Array pattern nulling by phase and position perturbations with the use of the genetic algorithm, *Microw. Opti. Technol. Lett.* **15**:251–256 (July 1997).

Y. Keen-Keong and L. Yilong, Sidelobe reduction in array-pattern synthesis using genetic algorithm, *IEEE AP-S Trans.* **45**:1117–1122 (July 1997).

D. Marcano, Synthesis of linear and planar antenna arrays using genetic algorithms, *Proc. IEEE AP-S Int. Symp.*, July 1997, pp. 1688–1691.

F. Ares, E. Villanueva, J. A. Rodriguez, and S. R. Rengarajan, Application of genetic algorithms in the design and optimization of array patterns, *Proc. IEEE AP-S Int. Symp.*, July 1997, pp. 1684–1687.

D. Marcano, L. Gomez, and O. Sosa, Planar array antenna synthesis using genetic algorithms with a penalty function, *Proc. IEEE Int. Microwave and Optoelectronics Conf.*, Aug. 1997, pp. 285–290.

J. M. Johnson and Y. Rahmat-Samii, Genetic algorithms in engineering electromagnetics, *IEEE AP-S Mag.* **39**:7–21 (Aug. 1997).

N. V. S. N. Sarma and R. Chandrasekharam, Shaped beam radiation pattern synthesis using genetic algorithm, *Proc. Int. Conf. Electromagnetic Interference and Compatibility*, Dec. 1997, pp. 171–174.

K. Markus and L. Vaskelainen, Optimisation of synthesised array excitations using array polynome complex root swapping and genetic algorithms, *IEE Proc. Microw. Anten. Propag.* **145**:460–464 (Dec. 1998).

P. Kozakowski, M. Mrozowski, and W. Zieniutycz, Synthesis of nonuniformly spaced arrays using genetic algorithm, *Proc. 12th Int. Conf. Microwaves and Radar*, May 1998, pp. 340–344.

G. P. Junker, S. S. Kuo, and C. H. Chen, Genetic algorithm optimization of antenna arrays with variable interelement spacings, *Proc. IEEE AP-S Int. Symp.*, June 1998, pp. 50–53.

C. W. Brann and K. L. Virga, Generation of optimal distribution sets for single-ring cylindrical arc arrays, *Proc. IEEE AP-S Int. Symp.*, June 1998, pp. 732–735.

R. L. Haupt and H. L. Southall, Experimental adaptive nulling with a genetic algorithm, *Microw. J.* **42**(1):78–89 (Jan. 99).

R. L. Haupt and H. Southall, Experimental adaptive cylindrical array, *Proc. IEEE Aerospace Applications Conf.*, March 1999, pp. 291–296.

J. M. Johnson, Genetic algorithm design of a switchable shaped beam linear array with phase-only control, *Proc. IEEE Aerospace Applications Conf.*, March 1999, pp. 297–303.

F. J. Ares-Pena, J. A. Rodriguez-Gonzalez, E. Villanueva-Lopez, and S. R. Rengarajan, Genetic algorithms in the design and optimization of antenna array patterns, *IEEE AP-S Trans.* **47**:506–510 (March 1999).

Y. Beng-Kiong and L. Yilong, Array failure correction with a genetic algorithm, *IEEE AP-S Trans.* **47**:823–828 (May 1999).

R. L. Haupt and J. M. Johnson, Dynamic phase-only array beam control using a genetic algorithm, *Proc. 1st NASA/DoD Workshop on Evolvable Hardware*, July 1999, pp. 217–224.

Y. C. Chung and R. L. Haupt, Optimum amplitude and phase control for an adaptive linear array using a genetic algorithm, *Proc. IEEE AP-S Int. Symp.*, July 1999, pp. 1424–1427.

K. F. Sabet, D. P. Jones, Cheng Jui-Ching, L. P. B. Katehi, K. Sarabaudi, and J. F. Harvey, Efficient printed antenna array synthesis including coupling effects using evolutionary genetic algorithms, *Proc. IEEE AP-S Int. Symp.*, July 1999, pp. 2084–2087.

C. Y. Chung and R. L. Haupt, Adaptive nulling with spherical arrays using a genetic algorithm, *Proc. IEEE AP-S Int. Symp.*, July 1999, pp. 2000–2003.

B. P. Kumar and G. R. Branner, Design of low sidelobe circular ring arrays by element radius optimization, *Proc. IEEE AP-S Int. Symp.*, July 1999, pp. 2032–2035.

A. Udina, N. M. Martin, and L. C. Jain, Linear antenna array optimisation by genetic means, *Proc. 3rd Int. Conf. Knowledge-Based Intelligent Information Engineering Systems*, Aug. 1999, pp. 505–508.

Y. H. Lee, A. C. Marvin, and S. J. Porter, Genetic algorithm using real parameters for array antenna design optimisation, *Proc. High Frequency Postgraduate Student Colloquium*, Sept. 1999, pp. 8–13.

T. Fukusako, H. Shiraishi, S. Itakura, Y. Kasano, and N. Mita, Microstrip adaptive array antenna using semiconductor plasma and genetic algorithm, *Proc. Asia Pacific Microwave Conf.*, Dec. 1999, pp. 76–79.

L. Landesa, F. Obelleiro, and J. L. Rodríguez, Practical improvement of array antennas in the presence of environmental objects using genetic algorithms, *Microw. Opti. Technol. Lett.* **23**:324–326 (Dec. 5, 1999).

J. A. Rodriguez, R. Ares, E. Moreno, and G. Franceschetti, Genetic algorithm procedure for linear array failure correction, *Electron. Lett.* **36**:196–198 (Feb. 3, 2000).

K. N. Sherman, Phased array shaped multi-beam optimization for LEO satellite communications using a genetic algorithm, *Proc. IEEE Int. Conf. Phased Array Systems and Technology*, May 2000, pp. 501–504.

Y. Lu and B. K. Yeo, Adaptive wide null steering for digital beamforming array with the complex coded genetic algorithm, *Proc. IEEE Int. Conf. Phased Array Systems and Technology*, May 2000, pp. 557–560.

H. Cheng-Nan, C. Ching-Song, C. Der-Chorng, L. Koong-Jeng, and H. Chia-I, Design of the cross-dipole antenna with near-hemispherical coverage in finite-element phased array by using genetic algorithms, *Proc. IEEE Int. Conf. Phased Array Systems and Technology*, May 2000, pp. 303–306.

E. A. Jones and W. T. Joines, Genetic design of linear antenna arrays, *IEEE AP-S Mag.* **42**:92–100 (June 2000).

D. Marcano and F. Duran, Synthesis of antenna arrays using genetic algorithms, *IEEE AP-S Mag.* **42**:12–20 (June 2000).

A. Armogida, G. Manara, A. Monorchio, P. Nepa, and E. Pagana, Synthesis of point-to-multipoint patch antenna arrays by using genetic algorithms, *Proc. IEEE AP-S Int. Symp.*, July 2000, pp. 1038–1041.

C. You Chung and R. L. Haupt, GAs using varied and fixed binary chromosome lengths and real chromosomes for low sidelobe spherical-circular array pattern synthesis, *Proc. IEEE AP-S Int. Symp.*, July 2000, pp. 1030–1033.

R. Shavit and S. Levy, Improved Orchard-Elliott pattern synthesis algorithm by pseudo-inverse technique and genetic algorithm, *Proc. IEEE AP-S Int. Symp.*, July 2000, pp. 1042–1045.

Y. Kimura and K. Hirasawa, A CMA adaptive array with digital phase shifters by a genetic algorithm and a steepest descent method, *Proc. IEEE AP-S Int. Symp.*, July 2000, pp. 914–917.

R. L. Haupt, Optimum population size and mutation rate for a simple real genetic algorithm that optimizes array factors, *Proc. IEEE AP-S Int. Symp.*, July 2000, pp. 1034–1037.

B. J. Barbisch, D. H. Werner, and P. L. Werner, A genetic algorithm optimization procedure for the design of uniformly excited and nonuniformly spaced broadband low sidelobe arrays, *Appl. Comput. Electromagn. Soc. J.* **15**(2):34–42 (July 2000).

N. N. Jackson and P. S. Excell, Genetic-algorithm optimization of an array for near-field plane wave generation, *Appl. Comput. Electromagn. Soc. J.* **15**(2):61–74 (July 2000).

R. L. Haupt and S. E. Haupt, Optimum population size and mutation rate for a simple real genetic algorithm that optimizes array factors, *Appl. Comput. Electromagn. Soc. J.* **15**(2):94–102 (July 2000).

D. F. Li and Z. L. Gong, Design of hexagonal planar arrays using genetic algorithms for performance improvement, *Proc. 2nd Int. Conf. Microwave and Millimeter Wave Technology*, Sept. 2000, pp. 455–460.

K. C. Lee, Optimization of a finite dipole array with genetic algorithm including mutual coupling effects, *Int. J. RF Microw. Comput. Aid. Eng.* **10**(6):379–382 (Nov. 2000).

A. Petosa and S. Thirakoune, Linear array of dielectric resonator antennas optimized using a genetic algorithm for low-sidelobe applications, *Proc. Asia-Pacific Microwave Conf.*, Dec. 2000, pp. 21–24.

A. Miura and M. Tanaks, A study of array pattern tuning method using hybrid genetic algorithms for figure-8 satellites's earth station antenna, *Proc. Asia-Pacific Microwave Conf.*, Dec. 2000, pp. 325–329.

Y. Yashchyshyn and M. Piasecki, Improved model of smart antenna controlled by genetic algorithm, *Proc. 6th Int. Conf. CAD Systems in Microelectronics*, Feb. 2001, pp. 147–150.

S. E. El-Khamy, M. A. Lotfy, M. H. Ramadan, and A. A. El-Tayeb, Thinned multi-ring arrays using genetic algorithms, *Proc. 18th Natl. Radio Science Conf.*, March 2001, pp. 113–121.

H. M. Elkamchouchi and M. M. Wagib, Failure restoration and array synthesis using genetic algorithms, *Proc. 18th Natl. Radio Science Conf.*, March 2001, pp. 123–130.

P. Karamalis, A. Marousis, A. Kanatas, and P. Constantinou, Direction of arrival estimation using genetic algorithms, *Proc. Vehicular Technology Conf.*, May 2001, pp. 162–166.

C. Chien-Hung and C. Chien-Ching, Novel radiation pattern by genetic algorithms, *Proc. Vehicular Technology Conf.*, May 2000, pp. 8–12.

D. W. Boeringer, D. W. Machuga, and D. H. Werner, Synthesis of phased array amplitude weights for stationary sidelobe envelopes using genetic algorithms, *Proc. IEEE AP-S Int. Symp.*, July 2001, pp. 684–687.

M. G. Bray, D. H. Werner, D. W. Boeringer, and D. W. Machuga, Thinned aperiodic linear phased array optimization for reduced grating lobes during scanning with input impedance bounds, *Proc. IEEE AP-S Int. Symp.*, July 2001, pp. 688–691.

A. Miura and M. Tanaka, An apply of hybrid GA for array pattern control of quasi-zenithal satellite's Earth station antenna, *Proc. IEEE AP-S Int. Symp.*, July 2001, pp. 230–233.

V. R. Mognon, W. A. Artuzi, Jr., and J. R. Descardeci, Tilt angle and sidelobe level control of array antennas by using genetic algorithm, *Proc. SBMO/IEEE MTT-S Int. Microwave and Optoelectronics Conf.*, Aug. 2001, pp. 299–301.

D. S. Weile and E. Michielssen, The control of adaptive antenna arrays with genetic algorithms using dominance and diploidy, *IEEE Trans. Anten. Propag.* **49**:1424–1433 (Oct. 2001).

P. Lopez, J. A. Rodriguez, F. Ares, and E. Moreno, Subarray weighting for the difference patterns of monopulse antennas: joint optimization of subarray configurations and weights, *IEEE AP-S Trans.* **49**:1606–1608 (Nov. 2001).

D. Ansell and E. J. Hughes, Use of multi-objective genetic algorithms to optimise the excitation and subarray division of multifunction radar antennas, in *IEE Multifunction Radar and Sonar Sensor Management Techniques* (Ref. 2001/173), Nov. 2001, pp. 8/1–8/4.

P. Lopez, J. A. Rodriguez, F. Ares, and E. Moreno, Low-sidelobe patterns from linear and planar arrays with uniform excitations except for phases of a small number of elements, *Electron. Lett.* **37**:1495–1497 (Dec. 6, 2001).

A. Lommi, A. Massa, E. Storti, and A. Trucco, Sidelobe reduction in sparse linear arrays by genetic algorithms, *Microw. Opti. Technol. Lett.* **32**(3):194–196 (Feb. 5, 2002).

I. S. Misra, A. Raychowdhury, K. K. Mallik, and M. N. Roy, Design and optimization of a nonplanar multidipole array using genetic algorithms for mobile communications, *Microw. Opti. Technol. Lett.* **32**(4):301–304 (Feb. 20, 2002).

Y. C. Chung and R. L. Haupt, Low-sidelobe pattern synthesis of spherical arrays using a genetic algorithm, *Microw. Opti. Technol. Lett.* **32**:412–414 (March 2002).

M. G. Bray, D. H. Werner, D. W. Boeringer, and D. W. Machuga, Matching network design using genetic algorithms for impedance constrained thinned arrays, *Proc. IEEE AP-S Int. Symp.*, June 2002, pp. 528–531.

S. Mummareddy, D. H. Werner, and P. L. Werner, Genetic optimization of fractal dipole antenna arrays for compact size and improved impedance performance over scan angle, *Proc. IEEE AP-S Int. Symp.*, June 2002, pp. 98–101.

A. A. Varahram and J. Rashed-Mohassel, Sidelobe level optimization using modified genetic algorithm, *Proc. IEEE AP-S Int. Symp.*, June 2002, pp. 742–745.

T. Dong, Y.-y. Li, and X.-w. Xu, Genetic algorithm in the synthesis of low sidelobe antenna array, *Proc. 3rd Int. Conf. Microwave and Millimeter Wave Technology*, Aug. 2002, pp. 751–754.

M. Vitale, G. Vesentini, N. N. Ahmad, and L. Hanzo, Genetic algorithm assisted adaptive beamforming, *Proc. 56th IEEE Vehicular Technology Conf.*, Sept. 2002, pp. 601–605.

T. Dong, Y.-Y. Li, and X.-W. Xu, Genetic algorithm in the synthesis of low sidelobe antenna array, *Proc. 5th Int. Symp. Wireless Personal Multimedia Communications*, Oct. 2002, pp. 757–761.

W. Yan and L. Yilong, The combination of neural networks and genetic algorithm for fast and flexible wide nulling in digital beamforming, *Proc. 9th Int. Conf. Neural Information Processing*, Nov. 2002, pp. 782–786.

M. G. Bray, D. H. Werner, D. W. Boeringer, and D. W. Machuga, Optimization of thinned aperiodic linear phased arrays using genetic algorithms to reduce grating lobes during scanning, *IEEE AP-S Trans.* **50**:1732–1742 (Dec. 2002).

D. W. Boeringer and D. H. Werner, Adaptive mutation parameter toggling genetic algorithm for phase-only array synthesis, *Electron. Lett.* **38**:1618–1619 (Dec. 5, 2002).

S. Caorsi, A. Lommi, A. Massa, S. Piffer, and A. Trucco, Planar antenna array design with a multi-purpose GA-based procedure, *Microw. Opti. Technol. Lett.* **35**(6):428–430 (Dec. 20, 2002).

D. G. Kurup, M. Himdi, and A. Rydberg, Design of an unequally spaced reflectarray, *Anten. Wireless Propag. Lett.* **2**:33–35 (2003).

R. Haupt, Generating a plane wave with a linear array of line sources, *IEEE AP-S Trans.* **51**:273–278 (Feb. 2003).

K. C. Lee, Genetic algorithms based analyses of nonlinearly loaded antenna arrays including mutual coupling effects, *IEEE AP-S Trans.* **51**:776–781 (April 2003).

R. J. Allard, D. H. Werner, and P. L. Werner, Radiation pattern synthesis for arrays of conformal antennas mounted on arbitrarily-shaped three-dimensional platforms using genetic algorithms, *IEEE AP-S Trans.* **51**:1054–1062 (May 2003).

C. Salvatore, M. Donelli, A. Lommi, and A. Massa, A real-time approach to array control based on a learned genetic algorithm, *Microw. Opti. Technol. Lett.* **36**:235–238 (Feb. 20, 2003).

M. A. Mangoud, M. Aboul-Dahab, and M. Sabry, Optimum steering techniques for linear and planar antenna arrays using genetic algorithm, *Proc. 20th Natl. Radio Science Conf.*, March 2003, Vol. B7, pp. 1–8.

D. W. Ansell and E. J. Hughes, Using multi-objective genetic algorithms to optimise the subarray partitions of conformal array antennas, *Proc. 20th Int. Conf. Antennas and Propagation*, March 2003, pp. 151–155.

H. M. Elkamchouchi and M. M. Wagih, Genetic algorithm operators effect in optimizing the antenna array pattern synthesis, *Proc. 20th Natl. Radio Science Conf.*, March 2003, Vol. B12, pp. 1–7.

L. L. Wang, D. G. Fang, and W. X. Sheng, Combination of genetic algorithm (GA) and fast fourier transform (FFT) for synthesis of arrays, *Microw. Opti. Technol. Lett.* **37**:56–59 (April 5, 2003).

F. H. Kashni, F. Arazm, and M. Asgari, The synthesis of super-resolution array through genetic algorithm using CRB, *Proc. 5th European Personal Mobile Communications Conf.*, April 2003, pp. 60–64.

T. Huang and A. S. Mohan, Effects of array mutual coupling on near-field DOA estimation, *Proc. IEEE Canadian Conf. Electrical and Computer Engineering*, May 2003, pp. 1881–1884.

R. Haupt, Synthesis of a plane wave in the near field with a planar phased array, *Proc. IEEE AP-S Int. Symp.*, June 2003, pp. 792–795.

T. Koleck, Active antenna coverage synthesis for GEO satellite using genetic algorithm, *Proc. IEEE AP-S Int. Symp.*, June 2003, pp. 142–144.

C. H. Hsu, R. Jeen-Sheen, K. Kun-Huang, and Y. Kang, Optimizing broadside array antenna with adaptive interference cancellation using amplitude-position perturbations in a linear array, *Proc. IEEE AP-S Int. Symp.*, June 2003, pp. 69–72.

A. T. Bu, S. Xiao-Wei, L. Ying, and X. Liang-Yong, Design of the sector array antenna based on genetic algorithm for smart antenna system front end, *Proc. IEEE AP-S Int. Symp.*, June 2003, pp. 686–689.

D. W. Boeringer and D. H. Werner, A comparison of particle swarm optimization and genetic algorithms for a phased array synthesis problem, *Proc. IEEE AP-S Int. Symp.*, June 2003, pp. 181–184.

D. W. Boeringer and D. H. Werner, Genetic algorithms with adaptive parameters for phased array synthesis, *Proc. IEEE AP-S Int. Symp.*, June 2003, pp. 169–172.

Y. H. Liu, L. An-Shyi, P. Yi-Hsin, and W. Ruey-Beei, Modeling antenna array elements and bandwidth enhanced by genetic algorithm, *Proc. IEEE AP-S Int. Symp.*, June 2003, pp. 884–887.

M. Wang, Z. Qi, Z. Liangjiang, and H. Mengna, The synthesis and optimization of arbitrarily distributed array with circular sparse array, *Proc. IEEE AP-S Int. Symp.*, June 2003, pp. 812–815.

S. Xiao, W. Bing-Zhong, Y. Xue-Song, and W. Gaofeng, Reconfigurable microstrip antenna design based on genetic algorithm, *Proc. IEEE AP-S Int. Symp.*, June 2003, pp. 407–410.

C. H. Hsu, J. S. Row, and K. H. Kuo, Downlink optimal radiation pattern design of smart antennas by phase-amplitude perturbations in a linear array, *Proc. IEEE AP-S Int. Symp.*, June 2003, pp. 80–83.

R. Haupt, Generating a plane wave in the near field with a planar array antenna, *Microw. J.* **46**(9):152–166 (Sept. 2003).

F. H. Wen-Chia Lue, Use of B-spline curves and genetic algorithms to reduce the sidelobe level in array-patterns, *Microw. Opti. Technol. Lett.* **38**:308–311 (Aug. 20, 2003).

W. Ling-Ling and F. Da-Gang, Combination of genetic algorithm and fast Fourier transform for array failure correction, *Proc. 6th Int. Symp. Antennas, Propagation and EM Theory*, Oct. 2003, pp. 234–237.

A. Taskin and C. S. Gurel, Antenna array pattern optimisation in the case of array element failure, *Proc. 33rd European Microwave Conf.*, Oct. 2003, pp. 1083–1085.

W. Qi and G. Zhong Lin, On the performance of genetic algorithm based adaptive beamforming, *Proc. 6th Int. Symp. Antennas, Propagation and EM Theory*, Oct. 2003, pp. 339–343.

G. Golino, A genetic algorithm for optimizing the segmentation in subarrays of planar array antenna radars with adaptive digital beamforming, *Proc. IEEE Int. Symp. Phased Array Systems and Technology*, IEEE, Oct. 2003, pp. 211–216.

R. G. Hohlfeld and N. Cohen, Genetic optimization of sparse, frequency invariant arrays using the HCR principle, *Proc. IEEE Int. Symp. Phased Array Systems and Technology*, Oct. 2003, pp. 588–593.

W. Ling-Ling and F. Da-Gang, Synthesis of nonuniformly spaced arrays using genetic algorithm, *Proc. Asia-Pacific Conf. Environmental Electromagnetics*, Nov. 2003, pp. 302–305.

D. W. Boeringer and D. H. Werner, Particle swarm optimization versus genetic algorithms for phased array synthesis, *IEEE AP-S Trans.* **52**:771–779 (March 2004).

S. Caorsi, A. Lommi, A. Massa, and M. Pastorino, Peak sidelobe level reduction with a hybrid approach based on GAs and difference sets, *IEEE AP-S Trans.* **52**:1116–1121 (April 2004).

S. H. Son, U. H. Park, K. H. Lee, and S. I. Jeon, Mobile phased array antenna design with low sidelobe pattern by genetic algorithm, *Proc. IEEE AP-S Int. Symp.*, June 2004, pp. 4112–4115.

S. E. El-Khamy, Fractal multiband antennas using GA/MOM optimized log periodic dipole arrays, *Proc. IEEE AP-S Int. Symp.*, June 2004, pp. 3433–3436.

A. Erentok and K. L. Melde, Comparison of MATLAB and GA optimization for three-dimensional pattern synthesis of circular arc arrays, *Proc. IEEE AP-S Int. Symp.*, June 2004, pp. 2683–2686.

D. A. Tonn and R. Bansal, Sidelobe minimization in interrrupted phased arrays by means of a genetic algorithm, *Proc. IEEE AP-S Int. Symp.*, June 2004, pp. 531–534.

C. Sacchi, F. De Natale, M. Donelli, A. Lommi, and A. Massa, Adaptive antenna array control in the presence of interfering signals with stochastic arrivals: assessment of a GA-based procedure, *IEEE Trans. Wireless Communi.* 1031–1036 (July 2004).

F. Soltankarimi, J. Nourinia, and C. Ghobadi, Side lobe level optimization in phased array antennas using genetic algorithm, *Proc. 8th IEEE Int. Symp. Spread Spectrum Techniques and Applications*, Aug. 2004, pp. 389–394.

F. Yu, J. Ronghong, W. Zhengyi, L. Bo, and G. Junping, Pattern synthesis of linear arrays using a hybrid optimization algorithm, *Proc. 7th Int. Conf. Signal Processing*, Aug. 2004, pp. 428–430.

A. Massa, M. Donelli, F. G. B. De Natale, S. Caorsi, and A. Lommi, Planar antenna array control with genetic algorithms and adaptive array theory, *IEEE AP-S Trans.* **52**:2919–2924 (Nov. 2004).

M. Donelli, S. Caorsi, F. De Natale, D. Franceschini, and A. Massa, A versatile enhanced genetic algorithm for planar array design, *J. Electromagn. Waves Appl.* **18**(11):1533–1548 (2004).

D. W. Boeringer, D. H. Werner, and D. W. Machuga, A simultaneous parameter adaptation scheme for genetic algorithms with application to phased array synthesis, *IEEE AP-S Trans.* **53**:356–371 (Jan. 2005).

S. H. Zainud-Deen, M. S. Ibrahem, S. M. M. Ibrahem, and H. A. Sharshar, Synthesis of linear arrays with shaped pattern using genetic algorithm and an orthogonal method, *Proc. 22nd Natl. Radio Science Conf.*, March 2005, pp. 89–96.

S. H. Zainud-Deen, E. S. Mady, K. H. Awadalla, and H. A. Sharshar, Adaptive arrays of smart antennas using genetic algorithm, *Proc. 22nd Natl. Radio Science Conf.*, March 2005, pp. 145–154.

J. N. Bogard and D. H. Werner, Optimization of Peano-Gosper fractile arrays using genetic algorithms to reduce grating lobes during scanning, *Proc. IEEE Int. Radar Conf.*, May 2005, pp. 905–909.

S. Tao and H. Ling, Array beamforming in the presence of a mounting tower using genetic algorithms, *IEEE AP-S Trans.* **53**:2011–2019 (June 2005).

Y. B. Tian and J. Qian, Improve the performance of a linear array by changing the spaces among array elements in terms of genetic algorithm, *IEEE AP-S Trans.* **53**:2226–2230 (July 2005).

S. Yang, Y. B. Gan, A. Qing, and P. K. Tan, Design of a uniform amplitude time modulated linear array with optimized time sequences, *IEEE AP-S Trans.* **53**:2337–2339 (July 2005).

R. L. Haupt, Interleaved thinned linear arrays, *IEEE AP-S Trans.* **53**:2858–2864 (Sept. 2005).

J. S. Petko and D. H. Werner, The evolution of optimal linear polyfractal arrays using genetic algorithms, *IEEE Trans. Anten. Propag.* **53**(11):3604–3615 (Nov. 2005).

2. Wire Antennas

D. S. Linden and E. E. Altshuler, Automating wire antenna design using genetic algorithms, *Proc. 1995 Antenna Applications Symp.*, Allerton Park, Monticello, IL, Sept. 1995.

D. S. Linden and E. E. Altshuler, Automating wire antenna design using genetic algorithms, *Microw. J.* **39**:74–86 (March 1996).

A. Boag, E. Michielssen, and R. Mittra, Design of electrically loaded wire antennas using genetic algorithms, *IEEE Trans. Anten. Propag.* **44**(5):687–695 (May 1996).

Z. Altman, R. Mittra, J. Philo, and S. Dey, New designs of ultra-broadband antennas using a genetic algorithm, *Proc. IEEE Antennas and Propagation Society Int. Symp.*, July 1996, Vol. 3, pp. 2054–2057.

E. E. Altshuler and D. S. Linden, Design of a loaded monopole having hemispherical coverage using a genetic algorithm, *IEEE Trans. Anten. Propag.* **AP-45**(1):1–4 (Jan. 1997).

P. L. Werner, Z. Altman, R. Mittra, D. H. Werner, and A. J. Ferraro, Genetic algorithm optimization of stacked vertical dipoles above a ground plane, *Proc. IEEE Antennas and Propagation Society Int. Symp.*, July 1997, Vol. 3, pp. 1976–1979.

D. Linden, Using a real chromosome in a genetic algorithm for wire antenna optimization, *Proc. IEEE Antennas and Propagation Society Int. Symp.*, July 1997, Vol. 3, pp. 1704–1707.

E. Altshuler and D. Linden, Design of a vehicular antenna for GPS/Iridium using a genetic algorithm, *Proc. IEEE Antennas and Propagation Society Int. Symp.*, July 1997, Vol. 3, pp. 1680–1683.

E. E. Altshuler and D. S. Linden, Design of wire antennas using a genetic algorithm, *J. Electron. Defense.* **20**(7):50–52 (July 1997).

E. Altshuler and D. Linden, Wire-antenna designs using genetic algorithms, *IEEE Anten. Propag. Mag.* **39**(2):33–43 (April 1997).

E. Jones and W. Joines, Design of Yagi-Uda antennas using genetic algorithms, *IEEE Trans. Anten. Propag.* **45**(9):1386–1392 (Sept. 1997).

D. Van Veldhuizen, B. Sandlin, E. Marmelstein, G. Lamont, and A. Terzuoli, Finding improved wire-antenna geometries with genetic algorithms, *Proc. IEEE Int. Conf. Computational Intelligence*, May 1998, pp. 102–107.

B. Sandlin and A. Terzuoli, Sensitivity of a genetic algorithm solution for a wire antenna geometry, *Proc. IEEE Antennas and Propagation Society Int. Symp.*, June 1998, Vol. 1, pp. 54–57.

P. L. Werner, Z. Altman, R. Mittra, D. H. Werner, and A. J. Ferraro, Optimization of stacked vertical dipoles above a ground plane using the genetic algorithm, *J. Electromagn. Waves Appl.* **13**:51–66 (1999).

D. Linden and E. Altshuler, Wiring like mother nature [antenna design], *IEEE Potentials* **18**(2):9–12 (April-May 1999).

J. Hagen, P. Werner, R. Mittra, and D. Werner, Toward the synthesis of an artificial magnetic medium, *Proc. IEEE Antennas and Propagation Society Int. Symp.*, July 11-16, 1999, Vol. 1, pp. 430–433.

D. Linden and E. Altshuler, Evolving wire antennas using genetic algorithms: A review, *Proc. 1st NASA/DoD Workshop on Evolvable Hardware*, July 1999, pp. 225–232.

B. Austin and W. Liu, An optimized shaped Yagi-Uda array using the genetic algorithm, *Proc. IEE Nat. Conf. Antennas and Propagation*, Aug. 1999, pp. 245–248.

D. Correia, A. Soares, and M. Terada, Optimization of gain, impedance and bandwidth in Yagi-Uda antennas using genetic algorithm, *Proc. SBMO/IEEE MTT-S, APS and LEOS—IMOC '99, Int. Microwave and Optoelectronics Conf.*, Aug. 1999, Vol. 1, pp. 41–44.

E. E. Altshuler, Design of a vehicular antenna for GPS/Iridium using genetic algorithms, *IEEE Trans. Anten. Propag.* **48**(6):968–972 (June 2000).

E. A. Jones and W. T. Joines, Genetic design of linear antenna arrays, *IEEE Anten. Propag. Mag.* **42**(3):92–100 (June 2000).

B. Long, P. Werner, and D. Werner, Genetic-algorithm optimization of dipole equivalent-circuit models, *Microw. Opti. Technol. Lett.* **27**(4):259–261 (Nov. 2000).

M. Fernandez-Pantoja, A. Monorchio, A. Rubio-Bretones, and R. Gomez-Martin, Direct GA-based optimisation of resistively loaded wire antennas in the time domain, *Electron. Lett.* **36**(24):1988–1990 (Nov. 2000).

D. Linden, Wire antennas optimized in the presence of satellite structures using genetic algorithms, *IEEE Aerospace Conf. Proc.*, March 2000, Vol. 5, pp. 91–99.

D. S. Linden and R. T. MacMillan, Increasing genetic algorithm efficiency for wire antenna design using clustering, *ACES Special J. Genet. Algorithms* **15**(2):75–86 (July 2000).

D. H. Werner and P. L. Werner, Genetically engineered dual-band fractal antennas, *Proc. IEEE Antennas and Propagation Society Int. Symp.*, July 2001, Vol. 3, pp. 628–631.

D. H. Werner, P. L. Werner, and K. Church, Genetically engineered multiband fractal antennas, *Electron. Lett.* **37**(19):1150–1151 (Sept. 2001).

A. Chuprin, J. Batchelor, and E. Parker, Design of convoluted wire antennas using a genetic algorithm, *IEE Proc. Microw. Anten. Propag.* **148**(5):323–326 (Oct. 2001).

E. Altshuler, Electrically small self-resonant wire antennas optimized using a genetic algorithm, *IEEE Trans. Anten. Propag.* **50**(3):297–300 (March 2002).

J. D. Lohn, W. F. Kraus, and D. S. Linden, Evolutionary optimization of a quadrifiler helical antenna, *Proc. IEEE Antennas and Propagation Society Int. Symp.*, June 2002, Vol. 3, pp. 814–817.

P. L. Werner and D. H. Werner, A design optimization methodology for multiband stochastic antennas, *Proc. IEEE Antennas and Propagation Society Int. Symp.*, June 2002, Vol. 2, pp. 354–357.

D. H. Werner, P. L. Werner, J. Culver, S. Eason, and R. Libonati, Load sensitivity analysis for genetically engineered miniature multiband fractal dipole antennas, *Proc. IEEE Antennas and Propagation Society Int. Symp.*, June 2002, Vol. 4, pp. 86–89.

S. Rengarajan and Y. Rahmat-Samii, On the cross-polarization characteristics of crooked wire antennas designed by genetic algorithms, *Proc. IEEE Antennas and Propagation Society Int. Symp.*, June 2002, Vol. 1, pp. 706–709.

H. Choo, R. Rogers, and H. Ling, Design of electrically small wire antennas using genetic algorithm taking into consideration of bandwidth and efficiency, *Proc. IEEE Antennas and Propagation Society Int. Symp.*, June 2002, Vol. 1, pp. 330–333.

G. Marrocco, A. Fonte, and F. Bardati, Evolutionary design of miniaturized meander-line antennas for RFID applications, *Proc. IEEE Antennas and Propagation Society Int. Symp.*, June 2002, Vol. 2, pp. 362–365.

A. Kerkhoff and H. Ling, The design and analysis of miniaturized planar monopoles, *Proc. IEEE Antennas and Propagation Society Int. Symp.*, June 2002, Vol. 4, pp. 30–33.

M. Fernandez Pantoja, F. Garcia Ruiz, A. Rubio Bretones, R. Gomez Martin, J. Gonzalez-Arbesu, J. Romeu, and J. Rius, GA design of wire pre-fractal antennas and comparison with other Euclidean geometries, *Anten. Wireless Propag. Lett.* **2**(1):238–241 (2003).

S. Rogers, C. Butler, and A. Martin, Design and realization of GA-optimized wire monopole and matching network with 20:1 bandwidth, *IEEE Trans. Anten. Propag.* **51**(3):493–502 (March 2003).

D. H. Werner and P. L. Werner, The design optimization of miniature low profile antennas placed in close proximity to high-impedance surfaces, *Proc. IEEE Antennas and Propagation Society Int. Symp.*, June 2003, Vol. 1, pp. 157–160.

P. L. Werner, M. Wilhelm, R. Salisbury, L. Swann, and D. H. Werner, Novel design techniques for miniature circularly-polarized antennas using genetic algorithms, *Proc. IEEE Antennas and Propagation Society Int. Symp.*, June 2003, Vol. 1, pp. 145–148.

K. O'Connor, R. Libonati, J. Culver, D. H. Werner, and P. L. Werner, A planar spiral balun applied to a miniature stochastic dipole antenna, *Proc. 2003 IEEE Int. Symp. Antennas and Propagation and URSI North American Radio Science Meeting*, Columbus, OH, June 2003, Vol. 3, pp. 938–941.

H. Choo and H. Ling, Design of planar, electrically small antennas with inductively coupled feed using a genetic algorithm, *Proc. IEEE Antennas and Propagation Society Int. Symp.*, June 2003, Vol. 1, pp. 300–303.

N. Venkatarayalu, and T. Ray, Single and multi-objective design of Yagi-Uda antennas using computational intelligence, *The 2003 Congress on Evolutionary Computation*, Dec. 2003, Vol. 2, pp. 1237–1242.

S. Zainud-Deen, D. Mohrram, and H. Sharshar, Optimum design of Yagi fractal arrays using genetic algorithms, *Proc. 21st Natl. Radio Science Conf.*, March 2004, Vol. B19, pp. 1–10.

T. F. Kennedy, S. A. Long, and J. T. Williams, Dielectric bead loading for control of currents on electrically long dipole antennas, *Proc. IEEE Antennas and Propagation Society Int. Symp.*, June 2004, Vol. 4, pp. 4420–4423.

E. E. Altshuler, Electrically small genetic antennas immersed in a dielectric, *Proc. IEEE Antennas and Propagation Society Int. Symp.*, June 2004, Vol. 3, pp. 2317–2320.

M. Pantoja, F. Ruiz, A. Bretones, R. Martin, S. Garcia, J. Gonzalez-Arbesu, J. Romeu, J. Rius, D. Werner, and P. Werner, GA design of small wire antennas, *Proc. IEEE Antennas and Propagation Society Int. Symp.*, June 2004, Vol. 4, pp. 4412–4415.

R. L. Haupt, Adaptive crossed dipole antennas using a genetic algorithm, *IEEE Trans. Anten. Propag.* **52**(8):1976–1982 (Aug. 2004).

E. E. Altshuler and D. S. Linden, An ultrawide-band impedance-loaded genetic antenna, *IEEE Trans. Anten. Propag.* **52**(11):3147–3151 (Nov. 2004).

P. L. Werner and D. H. Werner, Design synthesis of miniature multiband monopole antennas with application to ground-based and vehicular communication systems, *Anten. Wireless Propag. Lett.* **4**:104–106 (2005).

H. Choo, R. Rogers, and H. Ling, Design of electrically small wire antennas using a Pareto genetic algorithm, *IEEE Trans. Anten. Propag.* **53**(3):1038–1046 (March 2005).

A. Bretones, C. Moreno de Jong van Coevorden, M. Pantoja, F. Garcia Ruiz, S. Garcia, and R. Gomez Martin, GA design of broadband thin-wire antennas for GPR applications, *Proc. 3rd Int. Workshop on Advanced Ground Penetrating Radar*, May 2005, pp. 143–146.

Y. Kuwahara, Multiobjective optimization design of Yagi-Uda antenna, *IEEE Trans. Anten. Propag.* **53**(6):1984–1992 (June 2005).

T. Maruyama and K. Cho, Novel design method for small multi-band monopole antenna employing maze-generating algorithm for GA, *Proc. 2005 IEEE Antennas and Propagation Society Int. Symp.*, Washington, DC, July 3–8, 2005, Vol. 2A, pp. 49–52.

Z. Bayraktar, P. L. Werner, and D. H. Werner, Miniature three-element stochastic Yagi-Uda array optimization via particle swarm intelligence, *Proc. 2005 IEEE Antennas and Propagation Society Int. Symp.*, Washington, DC, July 3–8, 2005, Vol. 2B, pp. 263–266.

E. Altshuler, A method for matching an antenna having a small radiation resistance to a 50-ohm coaxial line, *IEEE Trans. Anten. Propag.* **53**(9):3086–3089 (Sept. 2005).

X. Chen, K. Huang, and X. Xu, Automated design of a three-dimensional fishbone antenna using parallel genetic algorithm and NEC, *Anten. Wireless Propag. Lett.* **4**:425–428 (2005).

P. K. Varlamos, P. J. Papakanellos, S. C. Panagiotou, and C. N. Capsalis, Multi-objective genetic optimization of Yagi-Uda arrays with additional parasitic elements, *IEEE Anten. Propag. Mag.* **47**(4):92–97 (Aug. 2005).

Z. Bayraktar, P. L. Werner, and D. H. Werner, The design of miniature three-element stochastic Yagi-Uda arrays using particle swarm optimization, *IEEE Anten. Wireless Propag. Lett.* **5**(1):22–26 (Dec. 2006).

3. Reflector Antennas

M. Shimizu, Pattern tuning of defocus array-fed reflector antennas, *Proc. 1998 IEEE Antennas and Propagation Society Int. Symp.*, June 21–26, 1998, Vol. 4, pp. 2070–2073.

W. H. Theunissen, H.-S. Yoon, G. N. Washington, and W. D. Burnside, Reconfigurable contour beam reflector antennas using adjustable subreflector and adjustable single feed, *Microw. Opt. Technol. Lett.* **21**(6):436–446 (June 20, 1999).

B. Lindmark, P. Slattman, and A. Ahlfeldt, Genetic algorithm optimization of cylindrical reflectors for aperture-coupled patch elements, *Proc. 1999 IEEE Antennas and Propagation Society Int. Symp.*, Aug. 1999, Vol. 1, pp. 442–445.

M. Vall-Ilossera, J. M. Rius, N. Duffo, and J. J. Mallorqui, Single reflector synthesis for producing contour radiation pattern and signal null region using genetic algorithms, *Proc. 1999 IEEE Antennas and Propagation Society Int. Symp.*, Aug. 1999, Vol. 4, pp. 2340–2343.

B. V. Sestroretsky, S. A. Ivanov, M. A. Drize, and K. N. Klimov, The genetic concept of topological synthesis of waveguide polarizator with elliptical factor 0.95 for antennas of satellite communication of a C-band 3.7/6.5 GHz, *Proc. 10th Int. Crimean Microwave Conf.*, Sept. 2000, pp. 388–390.

Y. Lu, X. Cai, and Z. Gao, Optimal design of special corner reflector antennas by the real-coded genetic algorithm, *Proc. 2000 Asia-Pacific Microwave Conf.*, Dec. 2000, pp. 1457–1460.

M. Vall-llossera, J. M. Rius, N. Duffo, and A. Cardama, Design of single-shaped reflector antennas for the synthesis of shaped contour beams using genetic algorithms, *Microw. Opt. Technol. Lett.* **27**(5):358–365 (Dec. 2000).

T. Maruyama and T. Hori, Novel dual frequency antenna with same beam width generated by GA-ICT using improved objective function, *Proc. 2001 IEEE Antennas and Propagation Society Int. Symp.*, July 2001, Vol. 4, pp. 668–671.

K. Barkeshli, F. Mazlumi, and R. Azadegan, The synthesis of offset dual reflector antennas by genetic algorithms, *Proc. 2002 IEEE Antennas and Propagation Society Int. Symp.*, June 2002, Vol. 1, pp. 670–673.

S. Sinton, J. Robinson, and Y. Rahmat-Samii, Standard and micro genetic algorithm optimization of profiled corrugated horn antennas, *Microw. Opt. Technol. Lett.* **35**(6):449–453 (Dec. 20, 2002).

S. Chakravarty and R. Mittra, Design of a frequency selective surface (FSS) with very low cross-polarization discrimination via the parallel micro-genetic algorithm (PMGA), *IEEE Trans. Anten. Propag.* **51**(7):1664–1668 (July 2003).

D. G. Kurup, M. Himdi, and A. Rydberg, Design of an unequally spaced reflectarray, *IEEE Anten. Wireless Propag. Lett.* **2**:33–35 (2003).

R. E. Zicht, M. Mussetta, M. Tovaglieri, P. Pirinoli, and M. Orefice, Frequency response of a new genetically optimized microstrip reflectarray, *Proc. 2003 IEEE Antennas and Propagation Society Int. Symp.*, June 22–27, 2003, Vol. 1, pp. 173–176.

S. L. Avila, W. P. Carpes Jr., and J. A. Vasconcelos, Optimization of an offset reflector antenna using genetic algorithms, *IEEE Trans. Magn.* **40**(2):1256–1259 (March 2004).

P. Slattman and J. R. Sanford, Moment method analysis and genetic algorithm design of shaped beam pillbox antennas, *Proc. 2004 IEEE Antennas and Propagation Society Int. Symp.*, June 20–25, 2004, Vol. 4, pp. 4392–4395.

M. Mussetta, P. Pirinoli, G. Dassano, R. E. Zich, and M. Orefice, Experimental validation of a genetically optimized microstrip reflectarray, *Proc. 2004 IEEE Antennas and Propagation Society Int. Symp.*, June 20–25, 2004, Vol. 1, pp. 9–12.

R. L. Haupt, Adaptive nulling with a reflector antenna using movable scattering elements, *IEEE Trans. Anten. Propag.* **53**(2):887–890 (Feb. 2005).

R. L. Haupt, A horn-fed reflector optimized with a genetic algorithm, *Proc. 2005 IEEE/ACES Int. Conf. Wireless Communications and Applied Computational Electromagnetics*, April 3–7, 2005, pp. 515–518.

M. Mussetta, P. Pirinoli, N. Bliznyuk, N. Engheta, and R. E. Zich, Optimization of a gangbuster reflectarray antenna, *Proc. 2005 IEEE Antennas and Propagation Society Int. Symp.*, July 3–8, 2005, Vol. 3A, pp. 626–629.

4. Horn Antennas

Li-Chung, T. Chang, and W. D. Burnside, An ultrawide-bandwidth tapered resistive TEM horn antenna, *IEEE Trans. Anten. Propag.* **48**(12):1848–1857 (Dec. 2000).

P. L. Garcia-Muller, Optimisation of compact horn with broad sectoral radiation pattern, *Electron. Lett.* **37**(6):337–338 (March 15, 2001).

J. Robinson, S. Sinton, and Y. Rahmat-Samii, Particle swarm, genetic algorithm, and their hybrids: Optimization of a profiled corrugated horn antenna, *Proc.*

2002 IEEE Antennas and Propagation Society Int. Symp., June 2002, Vol. 1, pp. 314–317.

L. Lucci, R. Nesti, G. Pelosi, and S. Selleri, Optimization of profiled corrugated circular horns with parallel genetic algorithms, *Proc. 2002 IEEE Antennas and Propagation Society Int. Symp.*, June 2002, Vol. 2, pp. 338–341.

D. Yang, Y. C. Chung, and R. Haupt, Genetic algorithm optimization of a corrugated conical horn antenna, *Proc. 2002 IEEE Antennas and Propagation Society Int. Symp.*, June 2002, Vol. 1, pp. 342–345.

A. Hoorfar, Evolutionary computational techniques in electromagnetics, *Proc. MMET 2002 Int. Conf. Mathematical Methods in Electromagnetic Theory*, Sept. 10–13, 2002, Vol. 1, pp. 54–60.

S. Sinton, J. Robinson, and Y. Rahmat-Samii, Standard and micro genetic algorithm optimization of profiled corrugated horn antennas, *Microw. Opt. Technol. Lett.* **35**(6):449–453 (Dec. 20, 2002).

S. Selleri, P. Bolli, and G. Pelosi, Automatic evaluation of the non-linear model coefficients in passive intermodulation scattering via genetic algorithms, *Proc. 2003 IEEE Antennas and Propagation Society Int. Symp.*, June 22–27, 2003, Vol. 4, pp. 390–393.

L. Lucci, R. Nesti, G. Pelosi, and S. Selleri, NURBS profiled corrugated circular horns, *Proc. 2003 IEEE Antennas and Propagation Society Int. Symp.*, June 22–27, 2003, Vol. 1, pp. 161–164.

D. Yang, Y. C. Chung, and R. Haupt, Genetic algorithm optimization of a multisectional corrugated conical horn antenna, *Microw. Opt. Technol. Lett.*, **38**(5):352–356 (Sept. 5, 2003).

J. Robinson and Y. Rahmat-Samii, Particle swarm optimization in electromagnetics, *IEEE Trans. Anten. Propag.* **52**(2):397–407 (Feb. 2004).

P. Slattman and J. R. Sanford, Moment method analysis and genetic algorithm design of shaped beam pillbox antennas, *Proc. 2004 IEEE Antennas and Propagation Society Int. Symp.*, June 20–25, 2004, Vol. 4, pp. 4392–4395.

C. A. Grosvenor, R. Johnk, D. Novotny, S. Canales, J. Baker-Jarvis, M. Janezic, J. Drewniak, M. Koledintseva, J. Zhang, and P. Rawa, Electrical material property measurements using a free-field, ultra-wideband system [dielectric measurements], *Proc. CEIDP '04. 2004 Annual Report Conf. Electrical Insulation and Dielectric Phenomena*, Oct. 17–20, 2004, pp. 174–177.

R. L. Haupt, A horn-fed reflector optimized with a genetic algorithm, *Proc. 2005 IEEE/ACES Int. Conf. Wireless Communications and Applied Computational Electromagnetics*, April 3–7, 2005, pp. 515–518.

5. Microstrip Antennas

C. Delabie, M. Villegas, and O. Picon, Creation of new shapes for resonant microstrip structures by means of genetic algorithms, *Electron. Lett* **33**(18):1509–1510 (Aug. 1997).

M. Himdi and J. P. Daniel, Synthesis of slot-coupled loaded patch antennas using a genetic algorithm through various examples, *Proc. 1997 IEEE Antennas and Propagation Society Int. Symp.*, July 1997, Vol. 3, pp. 1700–1703.

J. M. Johnson and Y. Rahmat-Samii, Genetic algorithms and method of moments (GA/MOM) for the design of integrated antennas, *IEEE Trans. Anten. Propag.* **47**(10):1606–1614 (Oct. 1999).

B. Aljiouri, E. G. Lim, H. Evans, and A. Sambell, Multi-objective genetic algorithm approach for a dual-feed circular polarized patch antenna design, *Electron. Lett.* **36**(12):1005–1006 (June 2000).

A. Akdagli and K. Güney, Effective patch radius expression obtained using a genetic algorithm for the resonant frequency of electrically thin and thick circular microstrip antennas, *IEE Proc. Microw. Anten. Propag.* **147**(2):156–159 (April 2000).

A. Armogida, G. Manara, A. Monorchio, P. Nepa, and E. Pagana, Synthesis of point-to-multipoint patch antenna arrays by using genetic algorithms, *Proc. 2000 IEEE Antennas and Propagation Society Int. Symp.*, July 2000, Vol. 2, pp. 1038–1041.

D. Lee and S. Lee, Design of a coaxially fed circularly polarized rectangular microstrip antenna using a genetic algorithm, *Microw. Opt. Technol. Lett.* **26**(5):288–291 (Sept. 2000).

A. Raychowdhury, B. Gupta, and R. Bhattacharjee, Bandwidth improvement of microstrip antennas through a genetic-algorithm-based design of a feed network, *Microw. Opt. Technol. Lett.* **27**(4):273–275 (Nov. 2000).

H. Choo, A. Hutani, L. C. Trintinalia, and H. Ling, Shape optimisation of broadband microstrip antennas using genetic algorithm, *Electron. Lett.* **36**(25):2057–2058 (Dec. 2000).

M. Lech, A. Mitchell, and R. B. Waterhouse, Optimization of broadband microstrip patch antennas, *Proc. 2000 Asia Pacific Microw. Conf.*, Sydney, Australia, Dec. 2000, pp. 711–714.

H. Choo and H. Ling, Design of dual-band microstrip antennas using the genetic algorithm, *Proc. 17th Annual Review of Progress in Applied Computational Electromagnetics Society*, Monterey, CA, March 2001, pp. 600–605.

R. Zentner, Z. Sipus, and J. Bartolic, Optimum synthesis of broadband circularly polarized microstrip antennas by hybrid genetic algorithm, *Microw. Opt. Technol. Lett.* **31**(3):197–201 (Nov. 2001).

F. J. Villegas, T. Cwik, Y. Rahmat-Samii, and M. Manteghi, Parallel genetic-algorithm optimization of a dual-band patch antenna for wireless communications, *Proc. 2002 IEEE Antennas and Propagation Society Int. Symp.*, June 2002, Vol. 1, pp. 334–337.

H. Choo and H. Ling, Design of multiband microstrip antennas using a genetic algorithm, *IEEE Microw. Wireless Compon. Lett.* (see also *IEEE Microw. Guided Wave Lett.*) **12**(9):345–347 (Sept. 2002).

A. Mitchell, M. Lech, D. M. Kokotoff, and R. B. Waterhouse, Search for high-performance probe-fed stacked patches using optimization, *IEEE Trans. Anten. Propag.* **51**(2):249–255 (Feb. 2003).

H. Choo and H. Ling, Design of broadband and dual-band microstrip antennas on a high-dielectric substrate using a genetic algorithm, *IEE Proc. Microw. Anten. Propag.* **150**(3):137–142 (June 2003).

S. Xiao, B.-Z. Wang, X.-S. Yang, and G. Wang, Reconfigurable microstrip antenna design based on genetic algorithm, *Proc. 2003 IEEE Antennas and Propagation Society Int. Symp.*, June 2003, Vol. 1, pp. 407–410.

O. Ozgun, S. Mutlu, M. I. Aksun, and L. Alatan, Design of dual-frequency probe-fed microstrip antennas with genetic optimization algorithm, *IEEE Trans. Anten. Propag.* **51**(8):1947–1954 (Aug. 2003).

T. G. Spence and D. H. Werner, Genetically optimized fractile microstrip patch antennas, *Proc. 2004 IEEE Antennas and Propagation Society Int. Symp.*, June 2004, Vol. IV, pp. 4424–4427.

A. A. Lotfi and F. H. Kashani, Bandwidth optimization of the E-shaped microstrip antenna using the genetic algorithm based on fuzzy decision making, *Proc. 2004 IEEE Antennas and Propagation Society Int. Symp.*, June 2004, Vol. 3, pp. 2333–2336.

T. G. Spence, D. H. Werner, and R. D. Groff, Genetic algorithm optimization of some novel broadband and multiband microstrip antennas, *Proc. 2004 IEEE Antennas and Propagation Society Int. Symp.*, June 2004, Vol. 4, pp. 4408–4411.

F. J. Villegas, T. Cwik, Y. Rahmat-Samii, and M. Manteghi, A parallel electromagnetic genetic-algorithm optimization (EGO) application for patch antenna design, *IEEE Trans. Anten. Propag.* **52**(9):2424–2435 (Sept. 2004).

S. J. Najafabadi and K. Forooraghi, Genetic optimization of an aperture-coupled microstrip patch antenna, *Proc. 3rd Int. Conf. Computational Electromagnetics and Its Applications* (ICCEA '04), Nov. 1–4, 2004, pp. 141–144.

6. Frequency-Selective Surfaces

E. Michielssen, J. M. Sajer, and R. Mittra, Design of multilayered FSS and waveguide filters using genetic algorithms, *Proc. IEEE Antennas and Propagation Society Int. Symp.*, June 28–July 2, 1993, Vol. 3, pp. 1936–1939.

K. Aygun, D. S. Weile, and E. Michielssen, Design of multilayered periodic strip gratings by genetic algorithms, *Microw. Opt. Technol. Lett.* **14**(2):81–85 (Feb. 1997).

G. Manara, A. Monorchio, and R. Mittra, A new genetic algorithm-based frequency selective surface design for dual frequency applications, *Proc. IEEE Antennas and Propagation Society Int. Symp.*, July 1999, Vol. 3, pp. 1722–1725.

G. Manara, A. Monorchio, and R. Mittra, Frequency selective surface design based on genetic algorithm, *Electron. Lett.* **35**(17):1400–1401 (Aug. 1999).

D. S. Weile and E. Michielssen, The use of domain decomposition genetic algorithms exploiting model reduction for the design of frequency selective surfaces, *Comput. Meth. Appl. Mech. Eng.* **186**(4):439–458 (June 2000).

Y. Yuan, C. H. Chan, K. F. Man, and R. Mittra, A genetic algorithm approach to FSS filter design, *Proc. 2001 IEEE Antennas and Propagation Society Int. Symp. and USNC/URSI Nat. Radio Science Meeting*, Boston, July 8–13, 2001, Vol. 4, pp. 688–691.

E. A. Parker, A. D. Chuprin, J. C. Batchelor, and S. B. Savia, GA optimization of crossed dipole FSS array geometry, *Electron. Lett.* **37**(16):996–997 (Aug. 2001).

S. Chakravarty and R. Mittra, Application of the micro-genetic algorithm to the design of spatial filters with frequency-selective surfaces embedded in dielectric media, *IEEE Trans. Electromagn. Compat.* **44**(2):338–346 (May 2002).

Z. Li, P. Y. Papalambros, and J. L. Volakis, Frequency selective surface design by integrating optimization algorithms with fast full wave numerical methods, *IEE Proc. Microw. Anten. Propag.* **149**(3):175–180 (June 2002).

L. Lanuzza, A. Monorchio, and G. Manara, Synthesis of high-impedance FSSs using genetic algorithms, *Proc. IEEE Antennas and Propagation Society Int. Symp.*, June 2002, Vol. 4, pp. 364–367.

J. Yeo, R. Mittra, and S. Chakravarty, A GA-based design of electromagnetic bandgap (EBG) structures utilizing frequency selective surfaces for bandwidth enhancement of microstrip antennas, *Proc. IEEE Antennas and Propagation Society Int. Symp.*, June 2002, Vol. 2, pp. 400–403.

M. Bozzi, G. Manara, A. Monorchio, and L. Perregrini, Automatic design of inductive FSSs using the genetic algorithm and the MoM/BI-RME analysis, *Anten. Wireless Propag. Lett.* **1**(1):91–93 (2002).

J. Yeo, J. Ma, and R. Mittra, Design of artificial magnetic ground planes (AMGs) utilizing frequency selective surfaces embedded in multilayer structures with electric and magnetic losses, *Proc. IEEE Antennas and Propagations Society Int. Symp.*, June 2003, Vol. 3, pp. 343–346.

L. Lanuzza, A. Monorchio, D. J. Kern, and D. H. Werner, A robust GA-FSS technique for the synthesis of optimal multiband AMCs with angular stability, *Proc. IEEE Antennas and Propagation Society Int. Symp.*, June 2003, Vol. 2, pp. 419–422.

S. Chakravarty and R. Mittra, Design of a frequency selective surface (FSS) with very low cross-polarization discrimination via the parallel micro-genetic algorithm (PMGA), *IEEE Trans. Anten. Propag.* **51**(7):1664–1668 (July 2003).

D. J. Kern, D. H. Werner, M. J. Wilhelm, and K. H. Church, Genetically engineered multiband high-impedance frequency selective surfaces, *Microw. Opt. Technol. Lett.* **38**(5):400–403 (Sept. 2003).

Y. Yuan, C. H. Chan, K. F. Man, and K. M. Luk, Meta-material surface design using the hierarchical genetic algorithm, *Microw. Opt. Technol. Lett.* **39**(3):226–230 (Nov. 2003).

X. F. Luo, A. Qing, and C. K. Lee, The design of frequency selective surfaces (FSS) using real-coded genetic algorithm (RGA), *Proc. Int. Conf. Information, Communications and Signal Processing and Pacific Rim Conf. Multimedia*, Dec. 2003, Vol. 1, pp. 391–395.

M. A. Gingrich, D. H. Werner, and A. Monorchio, The synthesis of planar left-handed metamaterials from frequency selective surfaces using genetic algorithms, Abstract, *Proc. 2004 USNC/URSI Natl. Radio Science Meeting*, Monterey, CA, June 19–25, 2004, p. 166.

D. J. Kern, D. H. Werner, and P. L. Werner, Optimization of multi-band AMC surfaces with magnetic loading, *Proc. IEEE Antennas and Propagation Society Int. Symp.*, June 2004, Vol. 1, pp. 20–25.

J. A. Bossard, D. H. Werner, T. S. Mayer, and R. P. Drupp, Reconfigurable infrared frequency selective surfaces, *Proc. IEEE Antennas and Propagation Society Int. Symp.*, June 2004, Vol. 2, pp. 1911–1914.

M. Ohira, H. Deguchi, M. Tsuji, and H. Shigesawa, Multiband single-layer frequency selective surface designed by combination of genetic algorithm and geometry-refinement technique, *IEEE Trans. Anten. Propag.* **52**(11):2925–2931 (Nov. 2004).

D. J. Kern, D. H. Werner, A. Monorchio, and M. J. Wilhelm, The design synthesis of multiband artificial magnetic conductors using high impedance frequency selective surfaces, *IEEE Trans. Anten. Propag.*, **53**(1): 8–17 (Jan. 2005).

J. Yeo, J.-F. Ma, and R. Mittra, GA-based design of artificial magnetic ground planes (AMGs) utilizing frequency-selective surfaces for bandwidth enhancement of microstrip antennas, *Microw. Opt. Technol. Lett.* **44**(1):6–13 (Jan. 2005).

R. P. Drupp, J. A. Bossard, D. H. Werner, and T. S. Mayer, Single-layer multi-band infrared metallodielectric photonic crystals designed by genetic algorithm optimization, *Appl. Phys. Lett.* **86**:811021–811023 (Feb. 2005).

J. A. Bossard, D. H. Werner, T. S. Mayer, and R. P. Drupp, A novel design methodology for reconfigurable frequency selective surfaces using genetic algorithms, *IEEE Trans. Anten. Propag.* **53**(4):1390–1400 (April 2005).

A. Monorchio, G. Manara, U. Serra, G. Marola, and E. Pagana, Design of waveguide filters by using genetically optimized frequency selective surfaces, *IEEE Microw. Wireless Compon. Lett.* **15**(6):407–409 (June 2005).

J. A. Bossard, J. A. Smith, D. H. Werner, T. S. Mayer, and R. P. Drupp, Multiband planar infrared metallodielectric photonic crystals designed using genetic algorithms with fabrication constraints, *Proc. 2005 IEEE Antennas and Propagation Society Int. Symp. and USNC/URSI Nat. Radio Science Meeting*, Washington, DC, July 3–8, 2005, Vol. 1A, pp. 705–708.

M. A. Gingrich and D. H. Werner, Synthesis of zero index of refraction metamaterials via frequency-selective surfaces using genetic algorithms, *Proc. 2005 IEEE Antennas and Propagation Society Int. Symp. and USNC/URSI Natl. Radio Science Meeting*, Washington, DC, July 3–8, 2005, Vol. 1A, pp. 713–716.

D. J. Kern, T. G. Spence, and D. H. Werner, The design optimization of antennas in the presence of EBG AMC ground planes, *Proc. 2005 IEEE Antennas and Propagation Society Int. Symp. and USNC/URSI Natl. Radio Science Meeting*, Washington, DC, July 3–8, 2005, Vol. 3A, pp. 10–13.

L. Li and D. H. Werner, Design of all-dielectric frequency selective surfaces using genetic algorithms combined with the finite element-boundary integral method, *Proc. 2005 IEEE Antennas and Propagation Society Int. Symp. and USNC/URSI Natl. Radio Science Meeting*, Washington, DC, July 3–8, 2005, Vol. 4A, pp. 376–379.

G. Qiang, D.-B. Yan, Y.-Q. Fu, and N.-C. Yuan, A novel genetic-algorithm-based artificial magnetic conductor structure, *Microw. Opt. Technol. Lett.* **47**(1):20–22 (Oct. 2005).

M. A. Gingrich and D. H. Werner, Synthesis of low/zero index of refraction metamaterials from frequency selective surfaces using genetic algorithms, *Electron. Lett.* **41**(23):13–14 (Nov. 2005).

L. Li, D. H. Werner, J. A. Bossard, and T. S. Mayer, A model-based parameter estimation technique for wideband interpolation of periodic moment method impedance matrices with application to genetic algorithm optimization of frequency selective surfaces, *IEEE Trans. Anten. Propag.* **54**(3):908–924 (March 2006).

J. A. Bossard, D. H. Werner, T. S. Mayer, J. A. Smith, Y. U. Tang, R. Drupp, and L. Li, The design and fabrication of planar multiband metallodielectric frequency selective surfaces for infrared applications, *IEEE Trans. Anten. Propag.* **54**(4):1265–1276 (April 2006).

7. Electromagnetic Absorbers

E. Michielssen, J. Sajer, S. Ranjithan, and R. Mittra, Design of lightweight, broad-band microwave absorbers using genetic algorithms, *IEEE Trans. Microwave Theory Tech.* **41**(6/7):1024–1031 (June–July 1993).

E. Michielssen, J. Sajer, and R. Mittra, Pareto-optimal design of broadband microwave absorbers using genetic algorithms, *Proc. IEEE Antennas and Propagation Society Int. Symp.*, June 1993, Vol. 3, pp. 1167–1170.

B. Chambers and A. Tennant, Design of wideband Jaumann radar absorbers with optimum oblique incidence performance, *Electron. Lett.* **30**(18):1530–1532 (Sept. 1994).

R. L. Haupt, An introduction to genetic algorithms for electromagnetics, *IEEE Anten. Propag. Mag.* **37**:7–15 (April 1995).

R. L. Haupt, Backscattering synthesis from tapered resistive grids, *Proc. 1995 SBMO/IEEE MTT-S Int. Conf. Microwave and Optoelectronics*, July 24–27, 1995, Vol. 2, pp. 893–896.

B. Chambers, A. P. Anderson, and R. J. Mitchell, Application of genetic algorithms to the optimisation of adaptive antenna arrays and radar absorbers, *Proc. IEEE Conf. Genetic Algorithms in Engineering Systems: Innovations and Applications*, Sept. 1995, pp. 94–99.

B. Chambers and A. Tennant, Optimised design of Jaumann radar absorbing materials using a genetic algorithm, *IEE Proceedings of Radar Sonar Nav.* **143**(1):23–30 (Feb. 1996).

E. Michielssen, W. C. Chew, and D. S. Weile, Genetic algorithm optimized perfectly matched layers for finite difference frequency domain applications, *Proc. IEEE Antennas and Propagation Society Int. Symp.* July 1996, Vol. 3, pp. 2106–2109.

D. S. Weile, E. Michielssen, and D. E. Goldberg, Genetic algorithm design of Pareto optimal broadband microwave absorbers, *IEEE Trans. Electromagn. Compat.* **38**(3):518–525 (Aug. 1996).

A. Cheldavi and M. Kamarei, Practical optimum design for a single-layer electromagnetic wave absorber at C and X-band using genetic algorithm, *Proc. IEEE Antennas and Propagation Society Int. Symp.*, July 1997, Vol. 3, pp. 1708–1711.

A. Cheldavi and M. Kamarei, Optimum design of N sheet capactive Jaumann absorber using genetic algorithm, *Proc. IEEE Antennas and Propagation Society Int. Symp.*, July 1997, Vol. 4, pp. 2296–2299.

A. R. Foroozesh, A. Cheldavi, and F. Hodjat, Design of Jaumann absorbers using adaptive genetic algorithm, *ISAPE Int. Symp. Antennas, Propagation and EM Theory*, Aug. 2000, pp. 227–230.

Y. Lee and R. Mittra, Investigation of an artificially synthesized electromagnetic absorber, *Microw. Opt. Technol. Lett.* **27**(6):384–386 (Dec. 2000).

S. Chakravarty, R. Mittra, and N. R. Williams, On the application of the microgenetic algorithm to the design of broad-band microwave absorbers comprising frequency-selective surfaces embedded in multilayered dielectric media, *IEEE Trans. Microw. Theory Tech.* **49**(6)(Part 1):1050–1059 (June 2001).

A. Bajwa, T. Williams, and M. A. Stuchly, Design of broadband radar absorbers with genetic algorithm, *Proc. IEEE Antennas and Propagation Society Int. Symp.*, July 2001, Vol. 4, pp. 672–675.

H. Choo, H. Ling, and C. S. Liang, Design of corrugated absorbers for oblique incidence using genetic algorithm, *Proc. IEEE Antennas and Propagation Society Int. Symp.*, July 2001, Vol. 4, pp. 708–711.

S. Chakravarty, R. Mittra, and N. R. Williams, Application of a microgenetic algorithm (MGA) to the design of broad-band microwave absorbers using multiple frequency selective surface screens buried in dielectrics, *IEEE Trans. Anten. Propag.* **50**(3):284–296 (March 2002).

C. Erbas, T. Gunel, and S. Kent, Optimization on wedge type absorbers via genetic algorithm, *Proc. IEEE Int. Symp. Electromagnetic Compatibility*, May 2003, Vol. 2, pp. 1255–1258.

K. Matous and G. J. Dvorak, Optimization of electromagnetic absorption in laminated composite plates, *IEEE Trans. Magn.* **39**(3)(Part 2):1827–1835 (May 2003).

D. J. Kern and D. H. Werner, Ultra-thin electromagnetic bandgap absorbers synthesized via genetic algorithms, *Proc. IEEE AP-S/URSI Symp.* June 2003, Vol. 2, pp. 1119–1122.

D. J. Kern and D. H. Werner, A genetic algorithm approach to the design of ultra-thin electromagnetic bandgap absorbers, *Microw. Opt. Technol. Lett.* **38**:61–64 (July 2003).

R. Haupt and Y. C. Chung, Optimizing backscattering from arrays of perfectly conducting strips, *IEEE Anten. Propag. Mag.* **45**(5):26–33 (Oct. 2003).

S. Cui and D. S. Weile, Robust design of absorbers using genetic algorithms and the finite element-boundary integral method, *IEEE Trans. Anten. Propag.* **51**:3249–3258 (Dec. 2003).

S. Cui, A. Mohan, and D. S. Weile, Pareto optimal design of absorbers using a parallel elitist nondominated sorting genetic algorithm and the finite element-boundary integral method, *IEEE Trans. Anten. Propag.* **53**(6):2099–2107 (June 2005).

T. Liang, L. Li, J. A. Bossard, D. H. Werner, and T. S. Mayer, Reconfigurable ultra-thin EBG absorbers using conducting polymers, *Proc. IEEE Antennas and Propagation Society Int. Symp.*, July 2005, Vol. 2B, pp. 204–207.

S. Cui, D. S. Weile, and J. L. Volakis, Novel planar absorber designs using genetic algorithms, *Proc. IEEE Antennas and Propagation Society Int. Symp.*, July 2005, Vol. 2B, pp. 371–274.

Index

Genetic Algorithms in Electromagnetics, by Randy L. Haupt and Douglas H. Werner